Three-Particle Physics and Dispersion Relation Theory

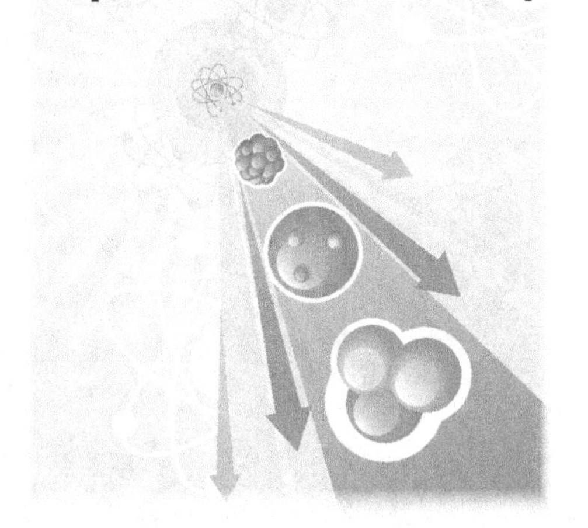

Three-Particle Physics and Dispersion Relation Theory

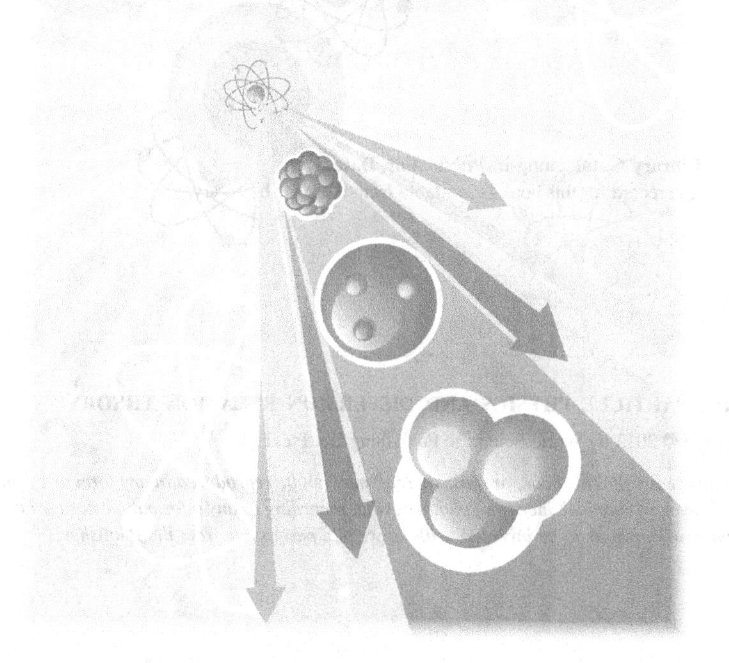

A V Anisovich • V V Anisovich • M A Matveev • V A Nikonov

Petersburg Nuclear Physics Institute, Russian Academy of Science, Russia

J Nyiri

Institute for Particle and Nuclear Physics, Wigner RCP,
Hungarian Academy of Sciences, Hungary

A V Sarantsev

Petersburg Nuclear Physics Institute, Russian Academy of Science, Russia

NEW JERSEY • LONDON • SINGAPORE • BEIJING • SHANGHAI • HONG KONG • TAIPEI • CHENNAI

Published by

World Scientific Publishing Co. Pte. Ltd.
5 Toh Tuck Link, Singapore 596224
USA office: 27 Warren Street, Suite 401-402, Hackensack, NJ 07601
UK office: 57 Shelton Street, Covent Garden, London WC2H 9HE

British Library Cataloguing-in-Publication Data
A catalogue record for this book is available from the British Library.

THREE-PARTICLE PHYSICS AND DISPERSION RELATION THEORY

ISBN 978-981-4478-80-9

Printed in Singapore by World Scientific Printers.

Preface

When we consider hadrons – their spectra and their interactions, the strong interactions – we are confronted with one of the most challenging problems in modern theoretical physics. Investigating the strong interactions of hadrons, we see an astonishing mixture of understandable and non-understandable phenomena, or observations which are not explained satisfactorily. But there are also problems which have to and can be solved, on which we have to concentrate, creating approaches and corresponding methods. The systematization of mesons and baryons [1] presents us many problems of this type in soft hadron interaction processes. One of them is the extension of the methods applied in two-particle systems to many-particle ones. Our book is considering mainly this topic.

We can easily describe a one-particle state: write its propagator even if the given particle exists in an external field (although a particle in an external field is not quite a one-particle state).

Two-particle systems, two-particle processes are the part of the strong interactions we are able to handle well; there are many well-understood problems and the methods for solving them are mostly well-founded also mathematically. A two-particle state can be successfully described. This is true not only for a non-relativistic system where the Schrödinger equation is used. In the relativistic case there are several possibilities, the most popular method is probably the dispersion relation technique. It was invented about half a century ago, but it became a frequently used, standard method only in the last decades.

In Chapter 1 of our book we give a short description of the appearing questions and the possible ways of solving them.

We concentrate our attention on the physics of three-hadron systems; this requires many, principally new steps compared to the description of a two-particle system. On the other hand, these steps are not only necessary but also sufficient for the possible generalization of the method to the investigation of many-particle systems.

Processes with the participation of hadrons are considered in the framework of dispersion integration technique (like in [1]), or of the similar method of spectral integration. The dispersion integration for hadrons was introduced by Gell-Mann, Goldberger and Thirring [2], the next principal step was made by Mandelstam [3]; a review of all the results can be found in [4]. The spectral integration was introduced by Lehman [5], Källen [6], and Gribov [7] (for four-point amplitudes).

One could think that it is not too important in the framework of which method we are working. However, in many-body problems the choice of variables is quite essential – if they are not appropriate, we may end up easily in the region of ambiguous solutions. The existence of ambiguities was already clear in the case of wave functions for two-particle systems when considering the Lippmann-Schwinger [8] and Bethe-Salpeter [9] equations. Still, for two-body systems this problem can be easily solved in a standard way – we consider it in Chapter 2, investigating spectral integral equations. We remind the reader how the dispersion method can be applied to two-particle systems. The chapter contains some additional information compared to that given in [1]; first of all, the interdependence of the confinement singularity and the σ-meson one and cusp physics in $\pi\pi$-systems.

Chapter 3 is devoted to three-particle systems. In the three-body problems the question of ambiguity of solutions occur again, and in the fifties and sixties it required a lot of efforts to answer it. The existence of ambiguities was obvious when the wave function of a three-nucleon system (short-range forces, non-relativistic approximation) was considered by Skornyakov and Ter-Martirosyan [10]. A different problem – the appearance of an infinite binding energy at point-like pair forces – was found even earlier [11]. Investigating short-range binding forces, these difficulties were removed by Danilov [12]: he suggested a solution based on the introduction of a new parameter (related to the binding energy of the basic state). A general solution of the Skornyakov–Ter-Martirosyan equation was given by [13]. A method to avoid the problem of ambiguous solutions in the case of a pair potential interaction was given by Faddeev (Faddeev equation [14], see also the monograph [15]). However, and this is essential from our point of view, these problems do not appear in methods used in the framework of the dispersion relation technique.

The dispersion relations are formulated in a way which makes it possible from the very beginning that both the problem of ambiguity and that of of the divergences of the amplitudes can be avoided. A serious advantage is that in the framework of dispersion technique the transition from the non-relativistic case to the relativistic one can be carried out in a simple and standard way. On the other hand, it requires a rather careful consideration of the reactions in the "neighbourhood", the virtual reactions and the corresponding singularities. If these peculiarities are not taken into account, one can easily transfer the considered amplitude to other sheets of complex energy and momenta – an example for that is presented by the Khuri-Treiman equation [16], in which the solution for the amplitude of a three-particle process occurs on the non-physical sheet.

The dispersion technique gives a possibility to write amplitudes for many-particle processes, satisfying the conditions of analyticity and unitarity, *i.e.* the causality condition. For the amplitudes of two-particle reactions the unitarity determines the proper relations between the real and imaginary parts, and in the general case of many-particle processes it provides the correct connection between the amplitudes of related reactions.

It is possible to formulate the dispersion relation approach as a selection of the singularities of the amplitudes which are subsequently taken into account. Hence, a special attention has to be paid to the selection of the leading and next-to-leading order singularities in three-particle amplitudes. In Chapter 4 we investigate this problem in the non-relativistic case.

The dispersion technique was applied to the expansion of the decay amplitude $K \to 3\pi$ over the relative momenta of the produced pions [17]. Though the threshold expansion in [17] was presented in the framework of the non-relativistic approximation, this technique is very convenient in general, since the transition from the non-relativistic case to the relativistic one is carried out in a standard and unambiguous way. The relativistic consideration of the leading and next-to-leading order singularities in three meson production amplitudes is carried out in Chapter 5. We investigate here the rather essential processes $p\bar{p} \to \pi\pi\pi, \pi\eta\eta, \pi K\bar{K}$ for which there exists already very high statistics, and the experimental research of which is expected also in the future.

The method of operating with two-particle systems can be generalized to many-particle ones on the basis of three processes:

(i) the decay of a massive particle into three particle states,

(ii) low-energy resonance production in the isobar models,

(iii) high energy production of massive hadron clusters.

These are closely related processes, and their connection is clearly demonstrated in the dispersion representation of the amplitude of the process (i) - the decay of a massive particle into a three-particle state. Indeed, the dispersion integral of the amplitude (i) includes not only the same amplitude at different energies but also those of the processes (ii) and (iii). This problem is described in detail in the introductory Chapter 1.

Considering processes of the type of (ii), let us mention those steps which were made in the fifties and sixties, developing the isobar model. A procedure for selecting the leading singularities (the pole ones) [18, 19] was created; the isobar model was proposed for $NN \to NN\pi$ [20] and for $\pi N \to N\pi\pi$ [21], and the next-to-leading singularities resulting from diagrams with resonances in intermediate states [22, 23] were singled out.

The isobar models turned out to be rather popular in the nineties, when experimental data on hadron spectra appeared with millions of events. For the analysis of the data multichannel K-matrix approaches were used, and not only leading (pole) singularities but also next-to-leading (logarithmic) ones [24]. The results of the K-matrix analyses were presented in PDG publications [25] and in review papers like [1, 26, 27]. By now, however, methods based on the dispersion technique are required for the isobar models; we present them for relatively low energies in Chapter 6, for high energies – in Chapter 7.

Obviously, if hadron physics can not be described by a new great idea, there remains only the way of building it up step by step, on the basis of the experimental data, and correcting ourselves by the appearing new results – *i.e.* by the phenomenological approach to construct an effective theory. This was the way how nuclear physics developed, arriving at the works of Aage Bohr and Mottelson [28]; the effective Hamiltonian suggested by Nilsson [29] is one of the most successful schemes for calculating the levels of excited nuclei.

Such a method has to be successful also for the investigation of the structure of baryons. Baryons as three-quark systems, qqq, have been investigated not only at times when the quark model appeared, e.g. [30, 31], but also later, trying to give predictions for the spectra of excited qqq states, see [32, 33] and references therein.

The systematization based on the solution of equations for three-quark states gave us a good description of the lowest baryons (S-wave, the 56-plet members of the SU(6)-symmetry). However, the equations predict a much larger number of baryons than experimentally observed. And, of course, it is the theory that should describe the experimental data. One of the possibilities for that is the transformation of three-quark states into quark-diquark ones, such a scheme is discussed in Chapter 8.

Though we describe in the present book only a relatively small part of strong interaction physics, we hope that it is just the investigation of relativistic three-particle states that may provide a breakthrough. A definite and standardized construction of the propagation functions for both stable (or bound) and free (or interacting) three-particle systems is a necessary step in the right direction. Of course, there are already serious results, but the aim to produce a standardized description of the three-body propagation in the relativistic approach is not achieved yet. This requires a preliminary overview of the present situation, which motivated us to write the book.

The names of those who contributed to the results during the more than fifty years of creating the dispersion approach to the three-particle problem, and with whom we had the good luck to work, should be mentioned with our special gratitude. They are, in a chronological order, V.N. Gribov, A.A. Anselm, I.J.R. Aitchison, L.G. Dakhno, M.N. Kobrinsky, A.K. Likhoded, E.M. Levin, A.N. Moskalev, P.E. Volkovitsky, D.V. Bugg, B.S. Zou.

We would like to thank our colleagues G.S. Danilov, V.R. Shaginyan, A. Frenkel, A. Lukács, N.Ya. Smorodinskaya for helpful discussions and G.V. Stepanova for her technical assistance.

References

[1] A.V. Anisovich, V.V. Anisovich, J. Nyiri, V.A. Nikonov, M.A. Matveev and A.V. Sarantsev, *Mesons and Baryons. Systematization and Methods of Analysis*, World Scientific, Singapore (2008).
[2] M. Gell-Mann, M.L. Goldberger, W.E. Thirring, Phys. Rev. **95**, 1612 (1954).
[3] S. Mandelstam, Phys. Rev. **112**, 1344 (1958).
[4] G.F. Chew *The analytic S-matrix*, N.-Y., Benjamin, 1966.
[5] H. Lehman, Nuovo Cim. **11**, 342 (1954).
[6] G. Källen, Helvetica Phys. Acta, **25**, 417 (1952).
[7] V.N. Gribov, ZhETF **34**, 1310 (1958) [Sov. Phys. JETP **34**, 903 (1958)]; ZhETF **35**, 416 (1959) [Sov. Phys. JETP **35**, 287 (1959)].
[8] B.A. Lippmann and J. Schwinger, Phys. Rev. **79** 469 (1950).
[9] E. Salpeter and H. Bethe, Phys. Rev. **84** 1232 (1951).
[10] V.G. Skornyakov, K.A. Ter-Martirosyan, ZhETF **31** 775 (1956).
[11] L.H. Thomas, Phys. Rev. **47**, 903 (1935).
[12] G.S. Danilov, ZhETF **40** 498 (1961).
[13] R.A. Minlos and L.D. Faddeev, ZhETF **41**, 1850 (1960).

[14] L.D. Faddeev, ZhETF **39** 1459 (1960).
[15] L.D. Faddeev and S.P. Merkuriev, *Scattering Theory for Several Particle Systems*, Springer (1993).
[16] N.N. Khuri, S.B. Treiman, Phys. Rev. **119** 1115 (1960) .
[17] V.V. Anisovich, A.A. Anselm and V.N. Gribov, Nucl. Phys. **38**, 132 (1962).
[18] K. Watson, Phys. Rev. **88**, 1163 (1952).
[19] A.B. Migdal, ZhETP **28**, 10 (1955).
[20] S. Mandelstam, Proc. Roy. Soc. A **244**, 491 (1958).
[21] V.V. Anisovich, ZhETP **39**, 97 (1960).
[22] V.V. Anisovich and L.G. Dakhno, Phys. Lett. **10**, 221 (1964); Nucl. Phys. 76, 657 (1966).
[23] V.V. Anisovich, Yad. Fiz. **6**, 146 (1967),
V.V. Anisovich and M.N. Kobrinsky, Yad. Fiz. **13**, 168 (1971), [Sov. J. Nucl. Phys. **13**, 169 (1971)].
[24] V.V. Anisovich, D.V. Bugg, A.V. Sarantsev, B.S. Zou, Phys. Rev. **D 50**, 1972 (1994).
[25] J. Beringer et al. (PDG), Phys. Rev.**D 86**, 010001 (2012).
[26] D.V. Bugg, Phys. Rep. 397, 257 (2004).
[27] E. Klempt and A.V. Zaitsev, Phys.Rept. **454**, 1 (2007).
[28] A. Bohr and B.R. Mottelson, Dan. Mat. Fys. Medd. **27**, No. 16 (1953).
[29] S.G. Nilsson, Dan. Mat. Fys. Medd. **29**, No. 16 (1955).
[30] N. Isgur and G. Karl, Phys. Lett. **B 72**, 109 (1977); **B74**, 353 (1978).
[31] C.P. Forsyth and R.E. Cutkosky, Phys. Rev. Lett. **49**, 576 (1981).
[32] S. Capstick and W. Roberts, Prog. Part. Nucl. Phys. **45**, 5241 (2000).
[33] U. Löring, B.C. Metsch, H.R. Petry, Eur.Phys.**A 10**, 395 (2001); **A 10**, 447 (2001).

Contents

Chapter 1

Introduction

The three-particle problem in hadron physics was formulated in the beginning as a problem of three non-relativistic nucleons: that of the H_3 and He_3 nuclei and their wave functions and the characteristics of nd amplitudes. A natural development of this activity was the invention of the quark model. But even before that, in the early sixties, the discussion of three-pion systems (decay processes $K \to \pi\pi\pi$ and $\eta \to \pi\pi\pi$) began, the problem of including relativistic effects appeared together with the consideration of the decay amplitudes in the framework of the dispersion technique. The relativistic dispersion description of amplitudes is always taking into account processes which are connected with the investigated reaction by the unitarity condition or by virtual transitions. In the case of three-particle processes they are, as a rule, those in which other many-particle states and resonances are produced. The description of these interconnected reactions and ways of handling them is the main subject of the book.

This introductory chapter presents in the described order the contents of the book, in a logical way from the two-particle cases to the more and more complicated ones.

1.1 Non-relativistic three-nucleon and three-quark systems

The necessity to describe three-nucleon and three-quark systems lead to a constant interest to the problem of three particles. In the beginning, in the fifties and sixties, the question of three nucleons was intensely discussed; in the end of the sixties the interest shifted to the quark models. By the nineties it became obvious that the realistic description of three-nucleon and three-quark systems requires the account of the relativistic effects.

1.1.1 *Description of three-nucleon systems*

For the description of light nuclei the insufficient knowledge of excited baryon states, as well as the many-nucleon ones, is the main source of the problems.

In two-nucleon systems the interaction between the nucleons is a strong one. This leads to the appearance of two levels in the S-wave NN systems: to that of the deuteron (a triplet state, with spin $S = 1$ and isospin $I = 0$) and to a virtual level (a singlet state $S = 0$, $I = 1$). In the sixties and seventies the experimental investigations of the scattering amplitudes of the NN states were carried out in the hope to decipher the potential structure of the low-energy NN interaction. The new experimental results lead to new, more sophisticated potentials, like the Paris [1] and Bonn [2] potentials, and finally the necessity to use energy-dependent interactions was realized. These are, in fact, already not potential interactions, but quasi-potential ones [3] or some modified effective interactions.

At that time different ideas appeared to solve the problems of the NN-interaction. Rather popular was the idea to use information given by many-nucleon systems, and, especially, three-body systems, for resolving the NN-interaction. The motivation was based on the observation that the characteristics of the three-particle systems are rather sensitive to the details of the two-particle potential.

The first steps in the three-nucleon problem were made in the framework of the variational methods – investigations were carried out making use of several, sufficiently realistic potentials. A qualitative agreement with the data on the binding energies for H_3/He_3 were obtained; for the *nd*-scattering length rough estimations were made only, see for example [4, 5] and references therein.

In the Skornyakov – Ter-Martirosyan equation [6] the hypothesis of the relatively short-range NN forces is applied, $r_{NN} \to 0$. In the quantum mechanical approach the zero radius of forces results in an infinite binding energy of the three particle system [7]. As a consequence, a problem appeared in [6] for the lightest states of the NNN-system. By using the subtraction procedure and introducing a new parameter (the binding energy for H_3), it is possible to resolve this problem [8]. The value of the scattering length *nd* in the $S = 1/2$ state, which was found when carrying out the procedure, coincided with the experimental data.

By dividing the wave functions for the three-particle system, in the Faddeev equation [9] one could obtain a set of integral equations with an unambiguous solution. The introduction of separable potentials allowed us to turn the problem of finding the amplitudes to that of solving a one-dimensional integral equation (see, *e.g.*, [10, 11, 12]). The generalization of the Faddeev integral equations to the case of the many-nucleon systems was carried out in [13]. A detailed description of this method for non-relativistic systems can be found in [14].

Several different ways of investigating non-relativistic systems were presented in [4]. Let us mention also the method given [15, 16, 17] which is based on the expansion of the three-particle wave function over six-dimensional angular harmonics. Using this approach, the problem leads to the solution of one-dimensional differential equations.

Later there appeared more sophisticated ways of considering the problem. Among other approaches, an effective account of the structure of constituent particles was carried out [18], which was, of course, necessary, having in mind the quark structure of hadrons. By that time, in the end of the seventies and the beginning of the eighties, it became clear that even systems consisting of a few nucleons are rather complicated, containing many-quark components like $6q$-bag, $9q$-bag and so on [19, 20].

In [21] the three-body problem was considered in the case when one of the amplitudes of the pair interaction contains a large inelasticity, see the reaction $\bar{p}d$. This in fact means that many-body states are included in the three-particle problem.

From the end of the seventies the three-quark systems, *i.e.* the baryons, became the main object of the investigation to which the three-body problem was applied.

1.1.2 *Three-quark systems*

The non-relativistic models of three-quark systems turned out to be rather successful for the description of the lowest baryon states with $I = 1/2, J = 1/2$ and $I = 3/2, J = 3/2$ (see, *e.g.*, [22, 23, 24]). However, for higher states there was only a temporary success: the equations predicted many more states than what was observed. As time passed, this situation did not change; the different possible confinement potentials (for example, [25]) and the account of the relativistic effects ([26, 27]) did not help either.

One of the possibilities to decrease essentially the number of predicted baryon states is the transformation of the three-quark states into quark-diquark ones. Such a scheme is suggested in [28] and we discuss it in Chapter 8.

The description of the experimental situation in the baryon sector is given in [29, 30].

1.2 Dispersion relation technique for three particle systems

It is, of course, impossible to go through, even superficially, all the problems concerning the three-body problem. Hence we begin with the main subject of the book, the dispersion relation technique for three particle systems.

We concentrate our attention on the triad of the processes:

$$\begin{aligned} &(i)\ \text{massive particle decay into three particle state,} \\ &(ii)\ \text{low}-\text{energy resonance production in the isobar model}\,, \\ &(iii)\ \text{high energy production of hadron massive cluster.} \end{aligned} \tag{1.1}$$

These are closely related processes, and the connection is clearly demonstrated in the dispersion representation of the amplitude of the process (i) - the particle decay into a three-particle state. Indeed, the dispersion integral of the amplitude (i) includes not only the same amplitude at different energies but also those of the processes (ii) and (iii). We discuss this topic in the next subsection.

The dispersion relations are based on the selection of the singularities of the amplitudes. Consequently, serious attention has to be paid to the selection of the leading and next-to-leading order singularities in three-particle and many-particle amplitudes. Considering processes of the type of (ii), let us accentuate those steps which were made in the fifties and sixties, developing the isobar model. Here the procedure of selecting the leading singularities (the pole ones) [31, 32] was given; the isobar model was proposed for $NN \to NN\pi$ [33] and for $\pi N \to N\pi\pi$ [34], and the next-to-leading singularities resulting from diagrams with resonances in intermediate states [35, 36] were singled out.

The relativistic consideration of the three-body problem started with the dispersion relation approach to the expansion of the amplitude of a many-particle process near the threshold. The dispersion technique was applied in the expansion of the decay amplitude $K \to 3\pi$ over the relative momenta of the produced pions [37]. The problem of the threshold expansion was first considered in [38, 39, 40]. The threshold expansion in [37] was presented in the framework of the dispersion relation approach though in the non-relativistic approximation. But in the dispersion relation technique the transition from the non-relativistic case to the relativistic one is carried out in a well-defined way.

1.2.1 *Elements of the dispersion relation technique for two-particle systems*

In Chapter 2 we present some elements of the dispersion relation method of handling two-particle systems which are used hereafter.

Considering the transition amplitude of the two-particle system $1+2 \to 3+4$ (the incoming and outgoing particles are denoted by numbers, see Fig. 1.1a), first, we have to underline the difference between the Feynman method of handling the amplitudes and that in the dispersion integration technique. In the framework of the Feynman technique (and the Feynman diagrams) the processes are not ordered in time, but in the sub-processes (*e.g.* when cutting the diagrams) the conservation laws of the energy and the momentum are satisfied. On the contrary, in the dispersion technique (and, consequently, in the dispersion diagrams) the energy conservation law is not satisfied, but the processes can be considered as being ordered in time. In this sense the dispersion method reminds quantum mechanics, where the energy is not conserved in the transitions to intermediate states, but the processes are time-ordered. In Chapter 2, presenting the dispersion relation (spectral integral) equations, this point is underlined. Also, we give a graphical interpretation of the amplitudes written in the framework of dispersion technique. In some cases the versions with the ansatz of separable vertices turn out to be very useful; this ansatz is also discussed in Chapter 2.

The dispersion relation technique is in its principal points the construction of an amplitude taking into account certain singularities and satisfying the unitarity condition. In Chapter 2 we discuss these requirements for the $1+2 \to 3+4$ amplitude.

Final states in the intensively studied decays $K \to \pi\pi\pi$, $\eta \to \pi\pi\pi$, $p\bar{p}\,(\mathrm{S}-\mathrm{wave}) \to \pi\pi\pi$ are determined by pion-pion interactions. In Chapter 2 we discuss the pion-pion interaction, methods of extraction of the amplitude from data, as well as isospin violation in the low-energy amplitude due to pion mass difference.

Let us mention that the problem of the ambiguity of the solutions for the three-particle systems which were intensely discussed in the sixties occur in fact also in the (non-relativistic and relativistic) two-particle equations. In the non-relativistic case the energy levels of a two-particle system can be written as

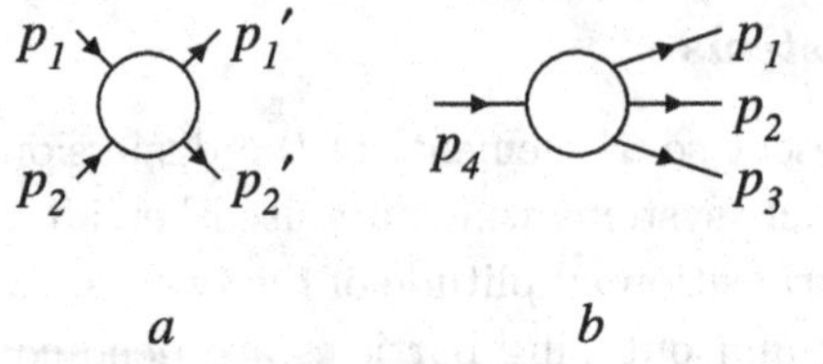

Fig. 1.1 Four-point amplitudes: a)transition process of a two-particle state into another two-particle state, $1+2 \to 3+4$; b) decay $4 \to 1+2+3$.

$$E_{tot} = \frac{P^2}{2(m_1+m_2)} + E_{12}, \tag{1.2}$$

where P is the total momentum of the system and E_{12} is the energy of the relative motion. The levels are multiple degenerate since to different E_{12} and P values at fixed E_{tot} correspond different states, *i.e.* the solution of the Lippmann-Schwinger equation [41] becomes ambiguous. An unambiguous solution can be obtained easily – it appears in the transition into the centre-of-mass system $P = 0$. This means the fixation of the invariant variable, the energy of the relative motion. In the dispersion technique the fixation of the invariant variables is a necessary element of the procedure not only in the two-particle amplitudes but also in the many-particle ones, leading to the way to the unambiguous solutions.

1.2.2 *Interconnection of three particle decay amplitudes and two-particle scattering ones in hadron physics*

After considering the transition amplitude of a two-particle state into a two-particle one, $1+2 \to 3+4$, in Chapter 3 we turn to the amplitude of the $4 \to 1+2+3$ decay in the case when the mass m_4 is large, see Fig. 1.1b. Both processes are determined by the same amplitude, though in different regions of the energy and the momentum. And, since we know how to write the spectral equation for $1+2 \to 3+4$ (it was discussed in Chapter 2), we are also able to write the equation for $4 \to 1+2+3$. There are, however, some problems which we could circumvent in [37]. Let us consider this in detail (even if it means to discuss here some technicalities).

In Fig. 1.2 the physical region of the process Figs. 1.1b is presented. When we write the dispersion representation for the amplitude of this pro-

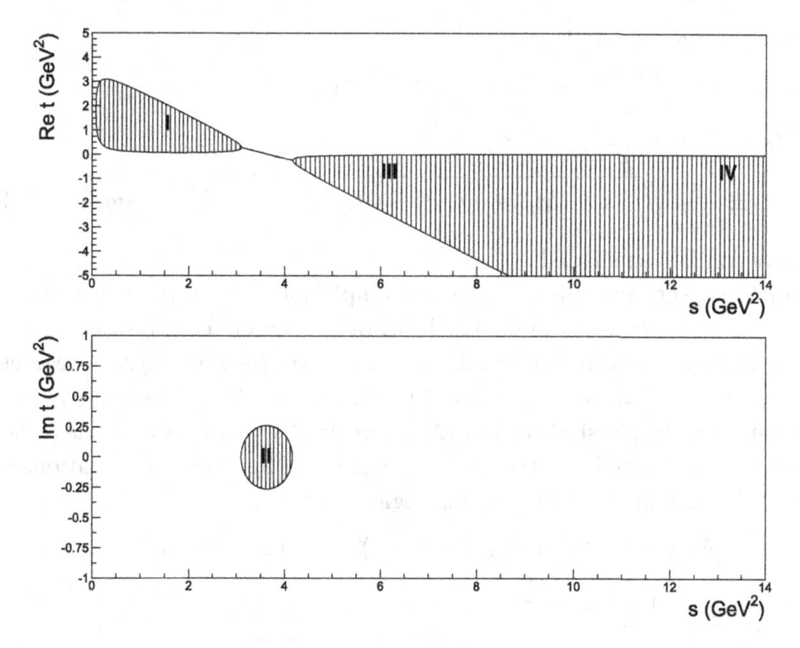

Fig. 1.2 (a) Mandelstam plane for the reactions $p\bar{p} \to \pi\pi\pi$ (the physical region is labeled as I) and $\pi\pi \to \pi + p\bar{p}$ with physical regions labeled as III for $\sqrt{s} \sim M_{p\bar{p}} + m_\pi$ and as IV for $\sqrt{s} >> M_{p\bar{p}} + m_\pi$. (b) Integration region in the spectral integral at $M_{p\bar{p}} - m_\pi < \sqrt{s} < M_{p\bar{p}} + m_\pi$.

cess, the dispersion integration has to be carried out necessarily over all the physical regions denoted in Fig. 1.2 by I, III and IV.

The physical regions III and IV (low and high energy reactions $\pi\pi \to \pi + p\bar{p}$) are constrained by the limits

$$
\begin{aligned}
s &\geq (M_{p\bar{p}} + m_\pi)^2 , \qquad t_- \leq t \leq t_+ \\
t_\pm &= \frac{3}{2}m_\pi^2 + \frac{1}{2}M_{p\bar{p}}^2 - \frac{1}{2}s \\
&\pm \frac{1}{2\sqrt{s}}\sqrt{[s - (M_{p\bar{p}} + m_\pi)^2][s - (M_{p\bar{p}} - m_\pi)^2][s - 4m_\pi^2]} . \qquad (1.3)
\end{aligned}
$$

The limits of the physical regions are marked by a solid line in Fig. 1.2a, the physical region is the shaded one.

Another physical region, which corresponds to the $p\bar{p} \to \pi\pi\pi$ decay, is situated inside the limits

$$\begin{aligned} & 4m_\pi^2 \le s \le (M_{p\bar{p}} - m_\pi)^2 , \qquad t_- \le t \le t_+ \\ & t_\pm = \frac{3}{2}m_\pi^2 + \frac{1}{2}M_{p\bar{p}}^2 - \frac{1}{2}s \\ & \quad \pm \frac{1}{2\sqrt{s}}\sqrt{[-s + (M_{p\bar{p}} + m_\pi)^2][-s + (M_{p\bar{p}} - m_\pi)^2][s - 4m_\pi^2]} . \end{aligned} \tag{1.4}$$

This region is also shaded.

In the dispersion integrals for the amplitudes of the processes shown in Fig. 1.1 the integrals are carried out over both of the physical regions. There is, however, also dispersion integration over the non-physical regions; this question is investigated in detail in Chapter 3. One of the integrations, that between the physical regions $(M_{p\bar{p}} - m_\pi)^2 < s < (M_{p\bar{p}} + m_\pi)^2$, is rather obvious: it is carried out inside the limits being analytic continuations of the boundaries in the neighbouring region:

$$\begin{aligned} & (M_{p\bar{p}} - m_\pi)^2 < s < (M_{p\bar{p}} + m_\pi)^2, \qquad t_- \le t \le t_+ , \\ & t_\pm = \frac{3}{2}m_\pi^2 + \frac{1}{2}M_{p\bar{p}}^2 - \frac{1}{2}s \\ & \quad \pm i\frac{1}{2\sqrt{s}}\sqrt{[-s + (M_{p\bar{p}} + m_\pi)^2][s - (M_{p\bar{p}} - m_\pi)^2][s - 4m_\pi^2]} . \end{aligned} \tag{1.5}$$

The integration is carried out over the complex values t, and its boundaries are shown in Fig. 1.2b.

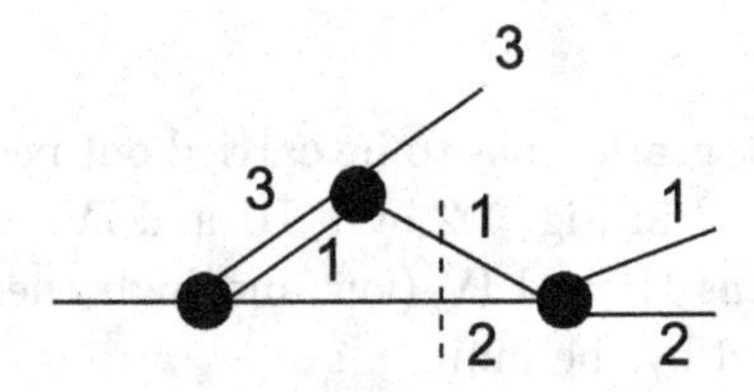

Fig. 1.3 Double rescattering diagram; the dashed line shows the cut in the dispersion integral.

Another integration, that in the region $\frac{1}{2}(M_{p\bar{p}}^2 - m_\pi^2) < s < (M_{p\bar{p}} - m_\pi)^2$, has to be identified carefully, because the subsequent rescatterings in the decay processes must be taken into account. A diagram with two subsequent rescatterings is presented in Fig. 1.3. In Chapter 4 we perform a consideration of this diagram in terms of the dispersion relation in the channel 12 (the cut in the dispersion integral is shown by the dashed line). The consideration of this diagram, as well as of analogous ones, demonstrates unambiguously that the singularities of the sub-processes in the beginning of the final state interaction (in the diagram Fig. 1.3 this is the sub-process on the left-hand side of the dotted line) influence seriously the integration region in the dispersion integral. Indeed, in the region $\frac{1}{2}(M_{p\bar{p}}^2 - m_\pi^2) < s < (M_{p\bar{p}} - m_\pi)^2$ there is, in addition to the integration over the physical region, another one over the interval

$$\begin{aligned}
&4m_\pi^2 \le s \le t_- , \\
&t_- = \frac{3}{2}m_\pi^2 + \frac{1}{2}M_{p\bar{p}}^2 - \frac{1}{2}s \\
&\quad - \frac{1}{2\sqrt{s}}\sqrt{[-s + (M_{p\bar{p}} + m_\pi)^2][-s + (M_{p\bar{p}} - m_\pi)^2][s - 4m_\pi^2]} .
\end{aligned} \tag{1.6}$$

This anomalous integral disappears at

$$4m_\pi^2 = t_- , \quad \text{or,} \quad s = \frac{1}{2}(M_{p\bar{p}}^2 - m_\pi^2) , \tag{1.7}$$

and after that (when there is a further decrease of s) the dispersion relation integral is carried out over the physical region only:

$$t_- \le s \le t_+ . \tag{1.8}$$

Therefore, considering examples like that of Fig. 1.3, we write the dispersion relations for decay amplitudes, *i.e.* for $M_{p\bar{p}} > 3m_\pi$. Decreasing $M_{p\bar{p}}$ into the region of the non-decaying mass, for example, to $M_{p\bar{p}} \sim m_\pi$, we face a standard diagram of a two particle transition $34 \to 12$ for which we can write a standard spectral integral equation. It is a way to connect a three particle spectral integral equation to two-particle standard spectral integral equation. In other words, it is a way from the consideration of two-particle systems (and corresponding relativistic equations) to three-particle systems and relativistic equations for them.

The possibility to go this way from the two-particle hadron physics to the three-particle one was noticed already in the sixties. An example for that is [42]. Still, the problem was not solved correctly - in the analytic continuation the anomalous integral (1.6) slipped on the un-physical plane. In Chapters 3 and 4 we consider analytical continuations of amplitudes into the decay mass region $M > 3m_\pi$ - it turned out to be a difficult problem that time. The correct way of the analytical continuation over the incident mass (over M) was given, as we mentioned above, in [37] and applied for the calculation of amplitudes of different processes [43, 44]. The analytic continuation of the spectral integral equation over M was carried out in [45] for the non-relativistic case (reaction $K \to \pi\pi\pi$) and in [46, 47] for relativistic particles, such as in decays $\eta \to 3\pi$ and $p\bar{p}(J^{PC} = 0^{-+}) \to 3$mesons.

1.2.3 *Quark-gluon language for processes in regions I, III and IV*

It is rather instructive to look at processes of low-energy three-meson production using the quark-gluon language and trace the connection of corresponding diagrams with that for the production of hadrons at low and intermediate energies (regions I and III on Fig. 1.2) and for diffractive production at high energies (region IV on Fig. 1.2).

The process $p\bar{p} \to \pi\pi\pi$ written in terms of quark-gluon states is shown in Fig. 1.4a. Here we demonstrate the planar diagram which is the leading one for the $1/N_c$ expansion [48] (N_c is the number of colours) while in Fig. 1.4b this diagram is shown schematically, using the constituent quark lines only. In this way the $q\bar{q}$ lines present the meson propagation. Meson propagators involve quark loops - they are shown in Fig. 1.4c. In terms of the $1/N$-expansion [49] ($N = N_c = N_f$ where N_f is the number of light flavours), the diagrams with quark loops are of the same order as that of Fig. 1.4a type. These loops form two-meson states and, if the energy of the meson propagator is sufficiently large, the loop diagrams have complex values. The propagator of the meson transforms into the propagator of a resonance, it may be the Breit–Wigner type pole. For a two-meson state one can write a Flatté type propagator [50]. In Chapters 2, 3 and others we discuss versions for the generalization of meson and baryon propagation amplitudes with the inclusion/exclusion of hadron states.

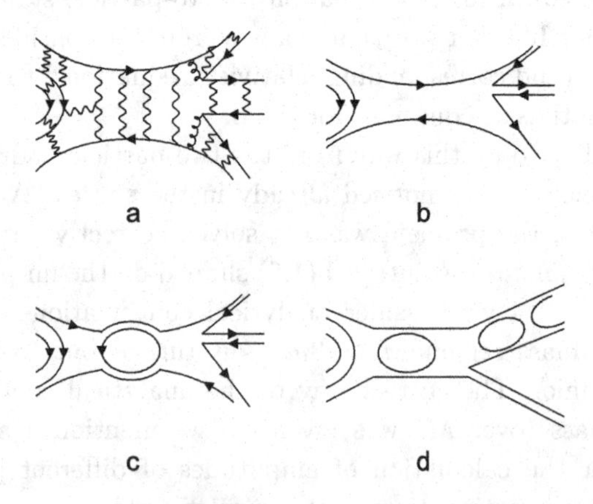

Fig. 1.4 Transition of a nucleon-antinucleon state into three mesons $N\bar{N} \to M_1M_2M_3$ written in terms of quark-gluon states: (a) Quark-gluon planar diagram, (b) the same diagram written for constituent quarks, (c) the same type of diagram with the inclusion of quark loops.

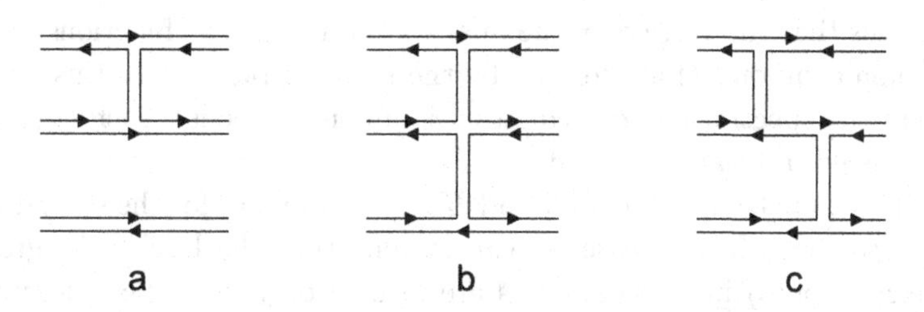

Fig. 1.5 Interactions in the transitions $M_1'M_2'M_3' \to M_1M_2M_3$ written in terms of constituent quarks: (a) Two-meson interaction, the third one is the spectator, (b) instantaneous interaction of all three mesons, (c) successive interactions of mesons.

Quark loops in gluonic nets can form diagrams with the rescattering of mesons - an example is shown in Fig. 1.4d. In this diagram the first (left-hand side) loop refers to the rescattering of mesons in a direct channel, the second one (right-hand side) describes the rescattering of the produced mesons. Rescatterings of the produced mesons are not suppressed in terms of the $1/N$ expansion, see Fig. 1.5, and there is no reason to neglect them. Rescatterings in the initial and final states form a singular structure of the unitary amplitudes. Very often it is rescatterings that create a genuine background for resonance production amplitudes.

1.2.4 *Spectral integral equation for three particles*

One of the main topics of this book is the relativistic spectral integral equation for three particle amplitude. The general consideration of it is presented in Chapter 3.

The standard formulation of the three-body problem is the following: three particles are produced in a certain state, after which the outgoing particles interact. The interaction shifts both the masses of the particles and the values of the vertices (or of the wave functions) of this state. Within this formulation we have to distinguish between two different problems.

1) An example of one type of problem is the description of decays like $K \to 3\pi$ or $\eta \to 3\pi$. Here the initial states, the K and η mesons, are quark-antiquark systems and, correspondingly, their masses are determined by the characteristics of the constituent quarks, their masses and quark-antiquark forces. The self-energy corrections corresponding to the transitions $K \to 3\pi \to K$ and $\eta \to 3\pi \to \eta$ are small and, hence, the shifts in the masses due to the three-pion interactions are also small. However, the pion spectra in the processes $K \to 3\pi$ and $\eta \to 3\pi$ are defined by pion interactions and

this means that the vertices of the final state (or the wave functions for the three-pion component) are formed by the pion interaction. In this type of problem the spectral integral equation is inhomogeneous - such equations are discussed in Chapters 3 and 4.

2) Three-nucleon systems H_3 and He_3 are examples for the second type of the problem. In this case we can assume that the levels (the ground and excited ones) in these systems are formed only by nucleon forces. If so, the homogeneous spectral integral equation provides both the location of these levels and the values of their wave functions. We discuss the H_3 and He_3 problem in Chapter 4. However, considering the nucleon systems, the question arises, whether the quark bags do not play a significant role in their formation. Arguments in favour of the presence of quark bags were given, *e.g.*, by Dakhno and Nikolaev [19] on the basis of diffraction scattering data of high energy pions and protons on light nuclei. Discussion of the role quark bags in other nuclei can be found, for example, in [20] and references therein.

The problem of cusps (threshold singularities connected with the electromagnetic differences between the masses of particles being in one isomultiplet – the mass differences of π^{+-} and π^0, p and n, etc.) were discussed already in the fifties and sixties. Nowadays we have statistics with millions of hadronic processes which allow us to fix these cusps. The problem of cusps is also discussed in Chapter 4.

The questions listed above present the classic content of the theory of three-body systems. In fact there are many more possible questions, first of all those raised by the isobar model at low, moderate and high energies.

1.2.5 *Isobar models*

The model, in its simplest form, was invented by Watson [31] and Migdal [32]. The essence of this model was the separation of the NN-pole singularities in $^1S_3$ and $^3S_1$ states from the multi-nucleon amplitude. Mandelstam has considered the isobar model for hadrons, for the $NN \to NN\pi$ reaction in the vicinity of the $N\Delta_{33}(1240)$ threshold [33]. The next steps were done for taking into account production of several resonances (K-matrix technique), for more correct representation of analytical amplitudes (N/D and D-matrix amplitudes).

One of the main tasks of multiparticle production studies is the fixation of non-stable hadrons (hadron resonances). The isobar models, which serve as reasonable tools for that, are discussed in Chapters 6 and 7. Isobar models are different for low-energy resonance production (Chapter 6) and for production at high energies (Chapter 7).

Hadron physics is still in the state when establishing the mass of a resonance (the mass of an unstable particle) and its quantum numbers is a great achievement which enables us to complete the systematization of hadrons; hadron systematization is one of the most important problems in recent hadron physics.

1.2.5.1 *Amplitude poles*

Searching for resonances means an extraction of amplitude pole singularities. The position of a pole gives mass and width of the corresponding state, the residue in the pole singularity tells us about couplings for this resonance production and decay. Masses and couplings are used for the identification of the nature of the resonance.

In an amplitude a single resonance is described by the Breit–Wigner pole which reads:

$$\sim \frac{1}{-s+M^2-iM\Gamma}. \tag{1.9}$$

This pole is formed by summing a set of diagrams of Fig. 1.6

$$\begin{aligned}&\frac{1}{-s+m^2}+\frac{1}{-s+m^2}B(s)\frac{1}{-s+m^2}+\\&+\frac{1}{-s+m^2}B(s)\frac{1}{-s+m^2}B(s)\frac{1}{-s+m^2}+...=\frac{1}{-s+m^2-B(s)},\end{aligned} \tag{1.10}$$

where $B(s)$ is the contribution of the loop diagram related to the resonance decay (here it is supposed to be a two-particle decay). If a resonance state decays into several channels, one should replace:

$$B(s)\to\sum_{a=1}^{n}B^{(a)}(s). \tag{1.11}$$

Here n is the number of open channels. In the standard Breit–Wigner approach we write:

$$M^2=m^2-\sum_{a=1}^{n}\mathrm{Re}B^{(a)}(M^2),\quad M\Gamma=\sum_{a=1}^{n}\mathrm{Im}B^{(a)}(M^2) \tag{1.12}$$

However, already at the beginning of the studies of hadron resonances it became clear that the shape of $\Delta_{33}(1240)$ is described rather well if an energy dependent width $M\Gamma\sim k^3$ is introduced where k is the relative

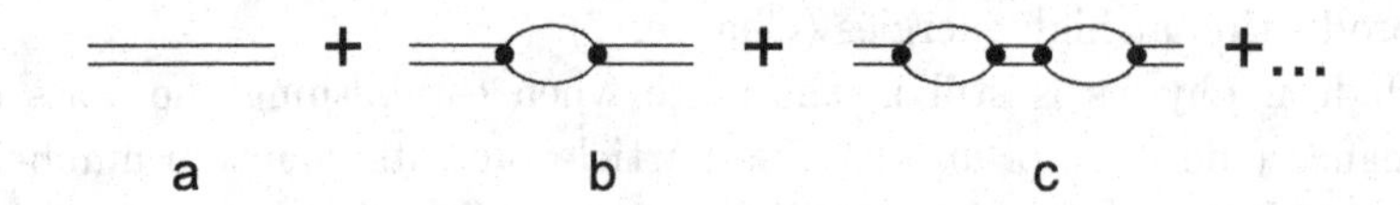

Fig. 1.6 Graphic representation of the Breit–Wigner pole term as an infinite set of transitions *particle → decay products → particle.*

momentum for the πN system. This means the introduction of the phase space factor k^{2L+1}. The imaginary part of the loop diagram reads:

$$\mathrm{Im}B(s) = \rho(s)g^2(s) \tag{1.13}$$

where $\rho(s)$ is the phase space for loop-diagram particles and $g(s)$ is the vertex for the transition *resonance → particles.*

1.2.5.2 *D-matrix propagator for an unstable particle and the K-matrix amplitude*

Equation (1.10) gives us the D-matrix propagator for a resonance in the simplest case, when we operate with a singled out state. For the scattering amplitude the transition to the K-matrix representation is realized as

$$\begin{aligned} A(s) &= \frac{g^2(s)}{-s+m^2 - ReB(s) - i\rho(s)g^2(s)} \simeq \\ &\simeq \frac{g^2(M^2)}{-s+m^2 - ReB(M^2) - i\rho(s)g^2(M^2)} \\ &= \frac{K(s)}{1-i\rho(s)K(s)}, \qquad K(s) = \frac{g^2}{M^2-s}. \end{aligned} \tag{1.14}$$

In the case of several resonances and the presence of a background we write:

$$K(s) = \sum_\alpha \frac{g_\alpha^2}{M_\alpha^2 - s} + f. \tag{1.15}$$

The multi-channel amplitude $A_{ab}(s)$, where a, b are channel indices, requires the introduction of the K-matrix

$$K_{ab}(s) = \sum_\alpha \frac{g_\alpha^{(a)} g_\alpha^{(b)}}{M_\alpha^2 - s} + f_{ab}. \tag{1.16}$$

Then we write for $A_{ab}(s) \to \hat{A}(s)$:

$$\begin{aligned} \hat{A}(s) &= \hat{K}(s)\frac{1}{1-i\hat{\rho}(s)\hat{K}(s)}, \\ \hat{\rho}(s) &= \mathrm{diag}\Big(\rho_1(s), \rho_2(s), ..., \rho_n(s)\Big) \end{aligned} \tag{1.17}$$

where the indices $(1, 2, ..., n)$ refer to channels.

1.2.5.3 *K-matrix and D-matrix masses and the amplitude pole*

We operate here with m and M, the D and K matrix masses, correspondingly. Both masses, generally speaking, differ from that given by the amplitude pole:

$$\Big[A_{ab}(s)\Big]_{\text{near pole }\alpha} \simeq \frac{G_a(\mu_\alpha^2)G_b(\mu_\alpha^2)}{\mu_\alpha^2 - s}, \qquad \mu_\alpha^2 \neq M_\alpha^2 \neq m_\alpha^2. \quad (1.18)$$

The amplitude poles do not always correspond to K-matrix and D-matrix poles, but let us assume now that it is the case.

Operation with different poles μ_α^2, M_α^2, m_α^2 has both prepotent and weak sides:

1) **Pole position**. The pole position, μ_α^2, is universal: it is the same for all reactions. Pole residues are factorized, $Res A_{ab}(s) = G_a(\mu_\alpha^2)G_b(\mu_\alpha^2)$, and complex vertices depend only on the type of the channel. In the fitting procedure the universality of μ_α and $G_a(\mu_\alpha^2)$ should be considered as a test that we deal with a particle.

2) **K-matrix pole.** In the K-matrix amplitude the removal of the phase space, $\rho_a(s) \to 0$, means the removal of a cloud of real particles which attend the resonance. In this limit amplitude poles approach the real s-axis and we have a set of single poles corresponding to stable particles:

$$\Big[A_{ab}(s)\Big]_{\hat{\rho}\to 0} \to K_{ab}(s)\,. \quad (1.19)$$

In [51, 52] the K-matrix poles are referred conventionally as poles of bare states. Actually, in the K-matrix bare states the coat of real hadrons is removed only while the coat of virtual hadrons is preserved.

A more detailed discussion of the K-matrix amplitude can be found in [51, 53, 54] and references therein.

3) **D-matrix bare masses**. In the D-matrix the bare state with mass m_α, the virtual coat related to particles which attend the decay is removed:

$$\Big[D_{\alpha\beta}(s)\Big]_{\hat{B}\to 0} = \delta_{\alpha\beta}\frac{g_\alpha^2}{m_\alpha^2 - s} \quad (1.20)$$

However, the choice of the loop diagrams which should be taken into account can create problems. The symmetry properties can be easily lost for a D-matrix bare state if into the loop diagram set $\hat{B}$ we do not include all particles of the multiplets but only some of them.

The K-matrix bare states do not suffer an obvious loss of the symmetry properties. Their couplings can obey constraints imposed by the quark-gluon structure of hadrons, these constraints for mesons are discussed in Chapter 6.

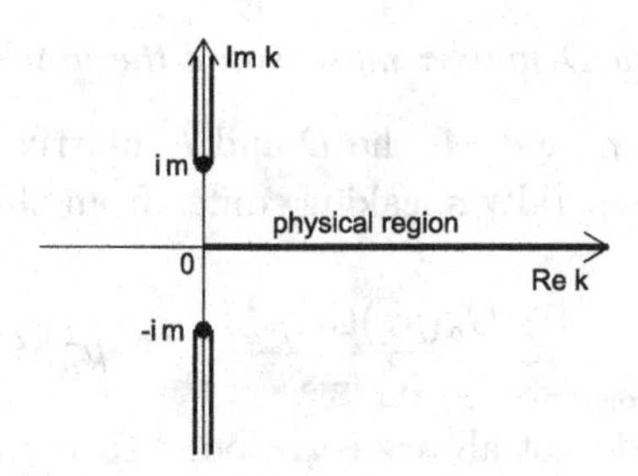

Fig. 1.7 Complex-k plane ($k = \sqrt{s/4 - m^2}$). Doubled lines correspond to left-hand cuts on the first sheet ($s = 0$ or $k = im$) and on the second one ($k = -im$). Location of cuts tells us about their peer influence on the physical region.

The K-matrix method, however, lets some analytical properties of amplitudes slip. An example is given by the K-matrix analysis of the 00^{++}-wave in states $\pi\pi$, $K\bar{K}$, $\eta\eta$, $\eta\eta'$ at $\sqrt{s} \sim 300 - 2000$ MeV [55]. The fit at low energies is problematic without the proper consideration of the left-hand side singularities (caused by forces in scattering amplitudes). In [55] the left-hand side singularities were taken into account effectively only by introducing poles at $s < 0$ into the background terms of the K matrix elements.

The analyticity of amplitudes at $s \sim 0$ is broken if in the K-matrix amplitude the invariant phase space is used. For equal mass states it is, up to a numerical factor, $\sim \sqrt{(s - 4m^2)/s}$ and the term $\sqrt{1/s}$ creates a singularity which is absent on the first sheet of the complex-s plane. One can eliminate the $\sqrt{1/s}$-singularity replacing $\sqrt{(s - 4m^2)/s} \to \sqrt{s - 4m^2}$ and thus restore analyticity on the first sheet but not on the second because this procedure eliminates singularity on the second sheet as well. But the $\sqrt{1/s}$-singularity exists on the second sheet - this can be seen by direct calculation of loop diagrams, see Chapter 3. To see the two-sheet structure of the scattering amplitude, we demonstrate the complex-k plane: its upper half-plane gives the first complex-s sheet and the bottom one is the second-s sheet. Fig. 1.7 shows directly that the second sheet singularities are to be considered together with those on the first sheet.

1.2.5.4 *Accumulation of widths of overlapping resonances*

The accumulation of widths of overlapping resonances by one of them is a well-known effect in nuclear physics [56, 57, 58]. In meson physics this phenomenon can play a rather important role, in particular, for exotic states which are beyond the $q\bar{q}$ systematics. Indeed, being among $q\bar{q}$ resonances, the exotic state creates a group of overlapping resonances. The exotic state, which is not orthogonal to its neighbours, after accumulating the "excess" of widths, turns into a broad one. This broad resonance should be accompanied by narrow states which are the descendants of states from which the widths have been taken off. In this way, the existence of a broad resonance accompanied by narrow ones may be a signature of exotics. This possibility, in context of searching for exotic states, was discussed in [59, 60].

The existence of this effect, an accumulation of widths of overlapping resonances by one of them, emphasizes a necessity to perform K-matrix analysis because that gives a most reasonable information about the quark-gluon content of the studied states. The broad states can easily escape the attention in the fitting procedure, to return them into consideration one simultaneously need to perform the systematization of these states.

The K-matrix fit of meson states accompanied by $q\bar{q}$ systematics is discussed in Chapter 6, see also [51] and references therein.

1.2.5.5 *Loop diagrams with resonances in the intermediate states*

Particles reveal themselves as poles, but not all pole singularities correspond to particles. An example is given by processes of Fig. 1.8. Loop diagrams with resonances in the intermediate states have singularities which can be located near the physical region.

The process Fig. 1.8a gives a logarithmic singularity $ln(s_{12}-s_{t0})$ [35, 61, 62], the one in Fig. 1.8b gives a square-root singularity $1/\sqrt{s_{12}-s_{b0}}$ [36, 63] - in the cross section this is a pole singularity. The five-point loop diagram Fig. 1.8c at fixed momenta k_3, k_4, k_5 has a pole singularity $1/(s_{12}-s_{f0})$.

We have to emphasize that these processes set up rather strong singularities and definitely are to be considered deeply. Triangle diagram singularities are considered here in Chapter 4, in the study of wave functions $H_3 \to NNN$ and $He_3 \to NNN$.

The isobar model, including the consideration of the pole and triangle diagrams (the Fig. 1.8a type), was applied to the $p\bar{p}$ annihilation at rest $p\bar{p}(^1S_0) \to \pi^0\pi^0\pi^0$, $\pi^0\eta\eta$ in [64].

In Chapter 5 we discuss the five-point amplitudes $p\bar{p} \to \pi\pi\pi, \pi^0\eta\eta, \pi^0K\bar{K}$. Let us underline that such processes were studied at LEAR (CERN), and their investigation is planned on the future hadron colliders. In the case when the $p\bar{p}$ system is at rest or in the low energy

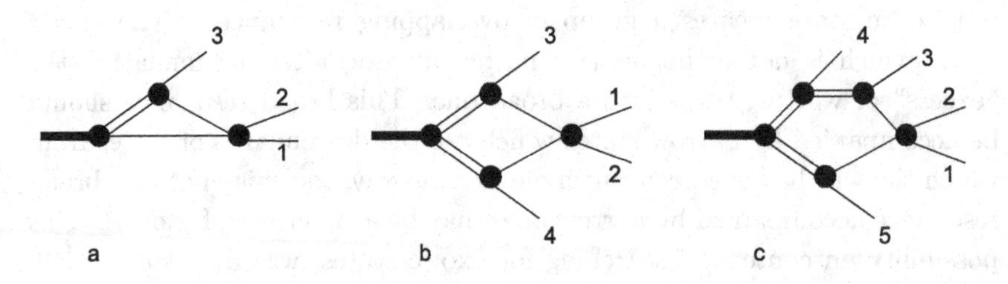

Fig. 1.8 Loop diagrams with singularities which can be near physical regions: (a) Triangle diagram singularity with the behaviour $\sim \ln(s - s_0)$, (b) Box diagram singularity, $(s - s_0)^{-1/2}$, (c) Five-point diagram with pole singularity, $(s - s_0)^{-1}$.

region, this reaction is investigated by the classical approach to the three-body problem. One has to take into account:
(i) the pion re-scatterings in several waves, for example in the S, P and D-waves,

(ii) the production of resonances in these waves (usually there is a number of resonances in each wave)
(iii) the transitions into other hadron states, *i.e.* transitions like $\pi\pi \to \eta\eta$, $\pi\pi \to K\bar{K}$, $\pi\pi \to \eta\eta'$ etc.
The resonances which have to be included in our considerations in the $\pi\pi$, $\eta\eta$, $K\bar{K}$ and $\eta\eta'$ channels are instable particles produced by quark-antiquark systems ($q\bar{q}$); that's why they are described in such considerations as input-poles of the amplitudes. The amplitudes corresponding to the transitions $\pi\pi \to \eta\eta$, $\pi\pi \to K\bar{K}$ etc. are also appearing on the quark level and hence, they also have to be considered as input-amplitudes. Specific expressions for the reactions $p\bar{p}$ (at rest) $\to \pi\pi\pi$ (and reactions connected to it by the transitions $\pi\pi \to \eta\eta, \pi\pi \to K\bar{K}$) are given in Chapter 5.

1.2.5.6 *Isobar model for high energy peripheral production processes*

The isobar model generalized to the case of high energy processes gives a rather advanced method for the search of resonances.

Diffraction processes provide a convenient way of investigating meson and baryon bound states at high energies. An example may be the peripheral $\pi N \to \pi\pi N$ reaction (Fig. 1.9) which is a process with a small momentum transferred to the nucleon. In the beginning it was considered as a pion exchange process. Further investigations show that there are t-channel exchanges by pion and a_1 reggeons, R_π and the R_{a_1}. Since the

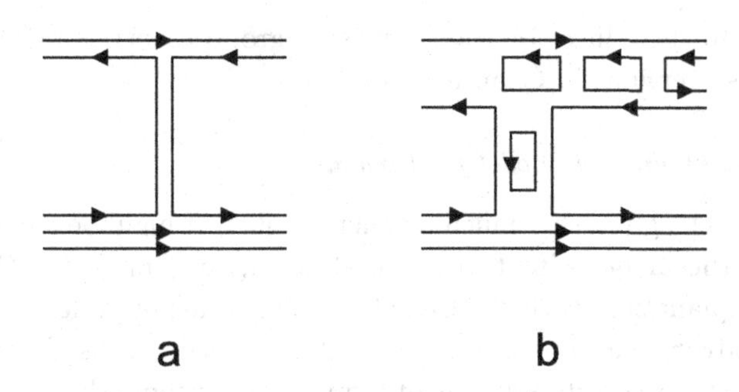

Fig. 1.9 Examples of constituent quark diagrams for isobar model processes at moderately high energies: peripheral production of hadrons with reggeon t-channel exchange.

reggeon is formed by all particles lying on the Regge trajectory, in this way all possible t-channel meson exchanges are taken into account. In the channel of the $\pi\pi$ production we have another sum of meson states in the form of resonances $\pi R \to \sum resonances \to \pi\pi$. The amplitude of this process can be described in the framework of dispersion technique.

As we see, three-particle states appear in different forms, and, in order to be able to make the next step, all they have to be collected. The aim is to construct a convenient method for the description of three-particle states and open the way for describing many-body processes with a large number of particles.

Considering three particle states, we in fact consider the problem of finding the distribution or propagation functions, or writing an equations for these functions. From a technical point of view it is rather convenient to write the distribution function in the momentum representation. It is in fact a generalization of a one-particle problem: for one-particle states we have Lehman-Källen spectral integrals [65, 66]. For two-particle states the spectral representation was written by Gribov [67].

1.2.6 *Quark-diquark model for baryons and group-theory approach*

The problem the physics of baryons is actually facing is to decrease the number of the predicted excited states *i.e.* to decrease the number of degrees of freedom in higher states.

There are several ways to do that. For example, we can introduce a diquark with the condition that it can not have excited states. Also, it is possible to introduce for (qqq) systems group-theory constraints. We discuss these versions in Chapter 8.

1.2.6.1 *Quark-diquark model for baryons*

The experiment gives us a much smaller number of highly excited baryons than what the model with three constituent quarks predicts. One of the plausible explanations is that the excited baryons do not prefer to be formed as three-body systems of spatially separated colored quarks. Instead, effectively they are two-body systems of a quark and a diquark:

$$q_\alpha D^\alpha = q_\alpha \left[\varepsilon^{\alpha\beta\gamma} q_\beta q_\gamma \right]. \tag{1.21}$$

Here $\varepsilon^{\alpha\beta\gamma}$ is a three-dimensional totally antisymmetrical tensor which acts in the color space.

It is an old idea that a qq-system inside the baryon can be regarded as a specific object – a diquark. Thus, interactions with a baryon may be considered as interactions with quark, q, and two-quark system, (qq): such a hypothesis was used in [68] for the description of the high-energy hadron–hadron collisions. In [69, 70, 71], baryons were described as quark-diquark systems. In hard processes on nucleons (or nuclei), the coherent qq state (composite diquark) can be responsible for interactions in the region of large Bjorken-x values, at $x \sim 2/3$; deep inelastic scatterings were considered in the framework of such an approach in [72, 73, 74, 75, 76]. More detailed considerations of the diquark and the applications to different processes may be found in [77, 78, 79].

Here we suppose that excited baryons are quark–diquark systems. It means that in the space of three colors ($\mathbf{c_3}$) the excited baryons, similarly to excited mesons, are $\left(\bar{\mathbf{c}}_3(D^0_{\bar{0}})\mathbf{c_3}(q)\right)$ or $\left(\bar{\mathbf{c}}_3(D^1_{\bar{1}})\mathbf{c_3}(q)\right)$ systems.

The two-particle system has considerably less degrees of freedom than three-particle one and, consequently, much less excited states. At the same time, the comparison of experimental data with model calculations [80, 81, 82] demonstrates that the number of predicted three-quark states is much larger than the number of observed ones.

The quark-diquark scheme discussed in Chapter 8 can be applied to relativistic systems *i.e.* to highly excited states. The different possibilities for such a scheme lead to a huge number of resonances with roughly the same masses – a typical tower of states. At large masses there can be found resonances with equal IJ^{PC} values which means that there is a possibility of mixing and thus a problem of identification may appear.

1.2.6.2 *Group-theory approach*

The three-body problem in quantum mechanics in general, and in nuclear physics in particular, provides an extremely interesting field for investigations, and the developed technics were useful for the solution of various, sometimes quite unexpected problems (see, for example, [3-6]).

In classical mechanics or, more precisely, in celestial mechanics the three-body problem was the subject of several rather successful theories. But, naturally, only systems with newtonian interactions were considered in detail. The case of other forces was practically not investigated.

In quantum mechanics the three-body problem was almost not taken into consideration at that time, or if, then only for practical purposes – calculations of energy levels in He^3, neutron-deuteron scattering – without formulating the problem in general. Generally speaking, a consequent way to handle the problem is to construct a basis first for non-interacting particles and, after that, to introduce interactions which give us the required baryon scheme.

From a group theoretical point of view the most interesting questions are related to the fifth quantum number Ω. This has to be introduced because the quantum numbers describing rotations and permutations are not sufficient to characterize the states in the three-body system.

In Chapter 8 a complete set of basis functions in the form of hyperspherical functions is discussed. It is characterized by quantum numbers corresponding to the chain $O(6) \supset SU(3) \supset O(3)$. Equations are derived to obtain the basis functions in an explicit form.

The problem of constructing a basis for a system of three free particles, realizing representations of the three-dimensional rotation group and of the permutation group was quite simple in principle. To solve the problem turned out to be, however, rather hard: it was very difficult to obtain a general solution for the calculated set of equations to determine the eigenfunctions. However, the eigenvalue equations could be simplified considerably, the solution was derived in a closed form, the coefficients were calculated in different ways, numerical results were obtained.

References

[1] M. Lacombe *et al.*, Phys. Rev. C **21**, 861 (1980).
[2] S. Machleidt, K. Hollande, C. Elsla, Phys. Rep. **149**, 1 (1987).
[3] A.A. Logunov and A.N. Tavkhelidze, Nuovo Cim. **29**, 380 (1963).

[4] A.G. Sitenko and V.F. Kharchenko, UFN **103**, 469 (1971).
[5] L.M. Delves, J.M. Blatt, I. Pask, B. Davies, Phys. Lett. B **28**, 472 (1969).
[6] V.G. Skornyakov, K.A. Ter-Martirosyan, JETP **31**, 775 (1956).
[7] L.H. Thomas, Phys. Rev. **47**, 903 (1935).
[8] G.S. Danilov, JETP, **40**, 498 (1961);
G.S. Danilov, V.I. Lebedev, JETP **44**, 1509 (1963).
[9] L.D. Faddeev, JETP **39**, 1459 (1960).
[10] H.P. Noyes, Phys. Rev. Lett. **15**, 538 (1965).
[11] K.L. Kowalski, Phys. Rev. Lett. **15**, 795 (965).
[12] J. Gillespie, Phys. Rev. **160**, 1432 (1967).
[13] O.A. Jakubovsky, Yad. Fiz. **5**, 1312 (1967).
[14] L.D. Faddeev and S.P. Merkuriev, *Scattering Theory for Several Particle Systems*, Springer (1993).
[15] A.M. Badalyan, Y.A. Simonov, Yad. Fiz. **3** 1032 (1966) [Sov. J. Nucl. Phys. **3** 1032 (1966)];
Yad. Fiz. **9** 69 (1969) [Sov. J. Nucl. Phys. **9** 69 (1969)].
[16] J. Nyiri, Ya.A. Smorodinsky, Yad. Fiz. **9** 882 (1969) [Sov. J. Nucl. Phys. **9** 515 (1969)];
Yad. Fiz. **12** 202 (1970) [Sov. J. Nucl. Phys. **12** 109 (1971)].
[17] V.V. Pustovalov, Ya.A. Smorodinsky, Yad. Fiz. **10** 1287 (1969) [Sov. J. Nucl. Phys. **10** 729 (1969)].
[18] Yu.A. Kuperin, K.A. Makarov, S.P. Merkuriev, *et al.*, Theor. and Math. Phys. **76**, 242 (1988).
[19] L.G. Dakhno and N.N. Nikolaev, Nucl. Phys. **A436**, 653 (1985).
[20] L.A. Sliv, M.I. Strickman and L.L. Frankfurt, UFN **128**, 281 (1985)
[21] Yu.A. Kuperin, S.B. Levin, Theor. and Math. Phys. **118**, 60 (1999).
[22] N. Isgur and G. Karl, Phys. Lett. **B72**, 109 (1977); **B74**, 353 (1978).
[23] C.P. Forsyth and R.E. Cutkosky, Phys. Rev. Lett. **49**, 576 (1981).
[24] R. Sator and F. Stancu, Phys. Rev. D**31**, 128 (1985).
[25] S.M. Gerasyuta, Yu.A. Kuperin, A.V. Sarantsev and V.A. Yarevsky, Yad. Fiz. **53** 1397 (1991) [Sov. J. Nucl. Phys. **53** 864 (1991)].
[26] S. Capstick and W. Roberts, Prog. Part. Nucl. Phys. **45**, 5241 (2000).
[27] U. Löring, B.C. Metsch, H.R. Petry, Eur.Phys. A**10**, 395 (2001); A**10**, 447 (2001).
[28] A.V. Anisovich, V.V. Anisovich, M.A. Matveev, V.A. Nikonov, A.V. Sarantsev and T.O. Vulfs, Int. J. Mod. Phys. **A25**, 2965 (2010);
Jad. Fiz. **74**, 438 (2011), [Part. Atomic Nucl. **74**, 418 (2011)].
[29] J. Beringer et al. (PDG), Phys. Rev.**D 86**, 010001 (2012).
[30] A.V. Anisovich, R. Beck, E. Klempt, V.A. Nikonov, A.V. Sarantsev, U. Thoma, Eur. Phys. J. **A48**, 15 (2012).
[31] K. Watson, Phys. Rev. **88**, 1163 (1952).
[32] A.B. Migdal, ZhETP **28**, 10 (1955).
[33] S. Mandelstam, Proc. Roy. Soc. A **244**, 491 (1958).
[34] V.V. Anisovich, ZhETP **39**, 97 (1960).
[35] V.V. Anisovich and L.G. Dakhno, Phys. Lett. **10**, 221 (1964); Nucl. Phys. 76, 657 (1966).

[36] V.V. Anisovich, Yad. Fiz. **6**, 146 (1967),
V.V. Anisovich and M.N. Kobrinsky, Yad. Fiz. **13**, 168 (1971), [Sov. J. Nucl. Phys. **13**, 169 (1971)].
[37] V.V. Anisovich, A.A. Anselm and V.N. Gribov, Nucl. Phys. **38**, 132 (1962).
[38] V.N. Gribov, Nucl. Phys. **5**, 653 (1958).
[39] A.A. Anselm and V.N. Gribov, ZhETP **36**, 1890 (1959); **37**, 501 (1959).
[40] I.T. Dyatlov, ZhETP **37**, 1330 (1959).
[41] B. A. Lippmann, J. Schwinger, Phys. Rev. **79**, 469 (1950).
[42] N.N. Khuri and S.B. Treiman, Phys. Rev. **119**, 1115 (1960).
[43] J. Nyiri, ZhETP **46**, 671 (1964).
[44] V.V. Anisovich and A.A. Anselm, UFN **88**, 287 (1966); [Sov. Phys. Usp. **88**, 117 (1966)].
[45] V.V. Anisovich, ZhETP **44**, 1593 (1963).
[46] A.V. Anisovich, Yad. Fiz. **58**, 1467 (1995) [Phys. Atom. Nucl. **58**, 1383 (1995)].
[47] A.V. Anisovich, Yad. Fiz. **66**, 175 (2003) [Phys. Atom. Nucl. **66**, 172 (2003)].
[48] G. 't Hooft, Nucl. Phys. B**72**, 461 (1972).
[49] G. Veneziano, Nucl. Phys. B**117**, 519 (1976).
[50] S.M. Flatté, Phys. Lett. B**63**, 224 (1976).
[51] A.V. Anisovich, V.V. Anisovich, J. Nyiri, V.A. Nikonov, M.A. Matveev and A.V. Sarantsev, *Mesons and Baryons. Systematization and Methods of Analysis*, World Scientific, Singapore (2008).
[52] A.V. Anisovich, V.V. Anisovich, A.V. Sarantsev, Nucl. Phys. B**A359**, 173 (1997).
[53] A. Švarc, arXiv:[nucl-thd]1202.30445v1 (2012).
[54] I.J.R. Aitchison, Nucl. Phys. B**A189**, 417 (1972).
[55] V.V. Anisovich, Yu.D. Prokoshkin, and A.V. Sarantsev, Phys. Lett. B **389** 388 (1996);
V.V. Anisovich, A.A. Kondashov, Yu.D. Prokoshkin, S.A. Sadovsky, and A.V. Sarantsev, Yad. Fiz. **60** 1489 (2000) [Phys. Atom. Nuclei **60** 1410 (2000)].
[56] I.S. Shapiro, Nucl. Phys. A **122**, 645 (1968).
[57] I.Yu. Kobzarev, N.N. Nikolaev, and L.B. Okun, Sov. J. Nucl. Phys. **10**, 499 (1970).
[58] L. Stodolsky, Phys. Rev. D **1**, 2683 (1970).
[59] V.V.Anisovich, D.V.Bugg, and A.V.Sarantsev, Phys. Rev. D **58**:111503 (1998).
[60] V.V.Anisovich, D.V.Bugg, and A.V.Sarantsev, Sov. J. Nucl. Phys. **62**, 1322 (1999) [Phys. Atom. Nuclei **62**, 1247 (1999).
[61] I.J.R. Aitchison, Phys. Rev. **133**, 1257 (1964).
[62] B.N. Valuev, ZhETP **47**, 649 (1964).
[63] P. Collas, R.E. Norton, Phys. Rev. **160**, 1346 (1967).
[64] V.V. Anisovich, D.V. Bugg, A.V. Sarantsev, B.S. Zou Phys. Rev. **D50**, 1972 (1994).
[65] H. Lehman, Nuovo Cim. **11**, 342 (1954).

[66] G. Källen, Helvetica Phys. Acta, **25**, 417 (1952).
[67] V.N. Gribov, Sov. Phys. ZhETP 36:384 (1959).
[68] V.V. Anisovich, Pis'ma ZhETF **2**, 439 (1965) [JETP Lett. **2**, 272 (1965)].
[69] M. Ida and R. Kobayashi, Progr. Theor. Phys. **36**, 846 (1966).
[70] D.B Lichtenberg and L.J. Tassie, Phys. Rev. **155**, 1601 (1967).
[71] S. Ono, Progr. Theor. Phys. **48** 964 (1972).
[72] V.V. Anisovich, Pis'ma ZhETF **21** 382 (1975) [JETP Lett. **21**, 174 (1975)]; V.V. Anisovich, P.E. Volkovitski, and V.I. Povzun, ZhETF **70**, 1613 (1976) [Sov. Phys. JETP **43**, 841 (1976)].
[73] A. Schmidt and R. Blankenbeckler, Phys. Rev. **D16**, 1318 (1977).
[74] F.E Close and R.G. Roberts, Z. Phys. C **8**, 57 (1981).
[75] T. Kawabe, Phys. Lett. B **114**, 263 (1982).
[76] S. Fredriksson, M. Jandel, and T. Larsen, Z. Phys. C **14**, 35 (1982).
[77] M. Anselmino and E. Predazzi, eds., *Proceedings of the Workshop on Diquarks*, World Scientific, Singapore (1989).
[78] K. Goeke, P.Kroll, and H.R. Petry, eds., *Proceedings of the Workshop on Quark Cluster Dynamics* (1992).
[79] M. Anselmino and E. Predazzi, eds., *Proceedings of the Workshop on Diquarks II*, World Scientific, Singapore (1992).
[80] N. Isgur and G. Karl, Phys. Rev. D**18**, 4187 (1978); D**19**, 2653 (1979); S. Capstick, N. Isgur, Phys. Rev. D**34**, 2809 (1986).
[81] L.Y. Glozman et al., Phys. Rev. D**58**:094030 (1998).
[82] U. Löring, B.C. Metsch, H.R. Petry, Eur.Phys. A**10**, 395 (2001); A**10**, 447 (2001).

Chapter 2

Elements of Dispersion Relation Technique for Two-Body Scattering Reactions

In this chapter we present methods for the investigation of four-point processes, concentrating our attention mainly on the scattering process $1+2 \to 1'+2'$, see Fig. 2.1a. We consider the analytic properties of the scattering amplitude. To write it, we make use of the dispersion integrals, give graphical representations of the dispersion relation diagrams, present spectral integral equations for vertex functions "composite system $\to$ constituents". These elements of the dispersion relation technique are used in the next chapters when we consider three-body states - in this sense this chapter, being introductory, is a necessary step in our presentation.

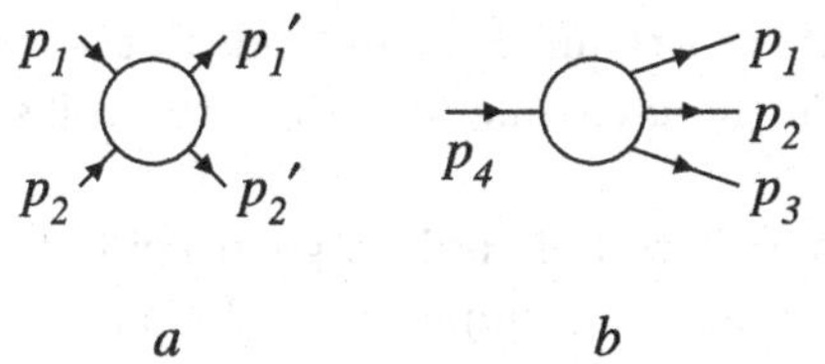

p_1 p_1' p_2 p_2' p_4 p_1 p_2 p_3

a *b*

Fig. 2.1 Four-point amplitudes: (a) scattering process $1+2 \to 1'+2'$; (b) decay $4 \to 1+2+3$.

2.1 Analytical properties of four-point amplitudes

Analytical properties of amplitudes are provided by Feynman diagrams (see [1, 2] for detail). Four-point reactions are shown in Fig. 2.1: as it is seen, we face in these reactions both the problems of two-body and three-body interactions, Figs. 2.1a and 2.1b correspondingly.

2.1.1 *Mandelstam planes for four-point amplitudes*

The analytical properties of the scattering amplitude $1 + 2 \to 1' + 2'$ (Fig. 2.1a) can be considered conveniently if we use the Mandelstam plane [1, 3]. For the sake of simplicity, let us take the masses of the spinless scattered particles in the process of Fig. 2.1a to be equal:

$$p_1^2 = p_2^2 = p_1'^2 = p_2'^2 = m^2. \tag{2.1}$$

The scattering amplitude of spinless particles depends on two independent variables. However, there are three variables for the description of the scattering amplitude on the Mandelstam plane:

$$\begin{aligned} s &= (p_1 + p_2)^2 = (p_1' + p_2')^2 \ , \\ t &= (p_1 - p_1')^2 = (p_2 - p_2')^2 \ , \\ u &= (p_1 - p_2')^2 = (p_2 - p_1')^2 . \end{aligned} \tag{2.2}$$

These variables obey the condition

$$s + t + u = 4m^2. \tag{2.3}$$

The Mandelstam plane of the variables s, t and u is shown in Fig. 2.2. The physical region of the s-channel corresponds to the case shown in Fig. 2.1a: the incoming particles are 1 and 2, while particles $1'$ and $2'$ are the outgoing ones. s is the energy squared, t and u are the momentum transfers squared. The physical region of the t-channel corresponds to the case when particles 1 and $1'$ collide, while the u-channel describes the collision of particles 1 and $2'$.

The Feynman diagram technique is a good guide for finding the analytical properties of scattering amplitudes. Below, we consider typical singularities as examples.

(i) One-particle exchange diagrams are shown in Fig. 2.3a,b,c: they provide pole singularities of the scattering amplitude, which are written as

$$\frac{g^2}{\mu^2 - t}, \qquad \frac{g^2}{\mu^2 - s}, \qquad \frac{g^2}{\mu^2 - u}, \tag{2.4}$$

where μ is the mass of a particle in the intermediate state, while g is its coupling constant with external particles.

(ii) The two-particle exchange diagram is shown in Fig. 2.4a. It has square-root singularities in the s-channel (the corresponding cut is shown in Fig. 2.4b) and in the t-channel (the cutting marked by crosses is shown

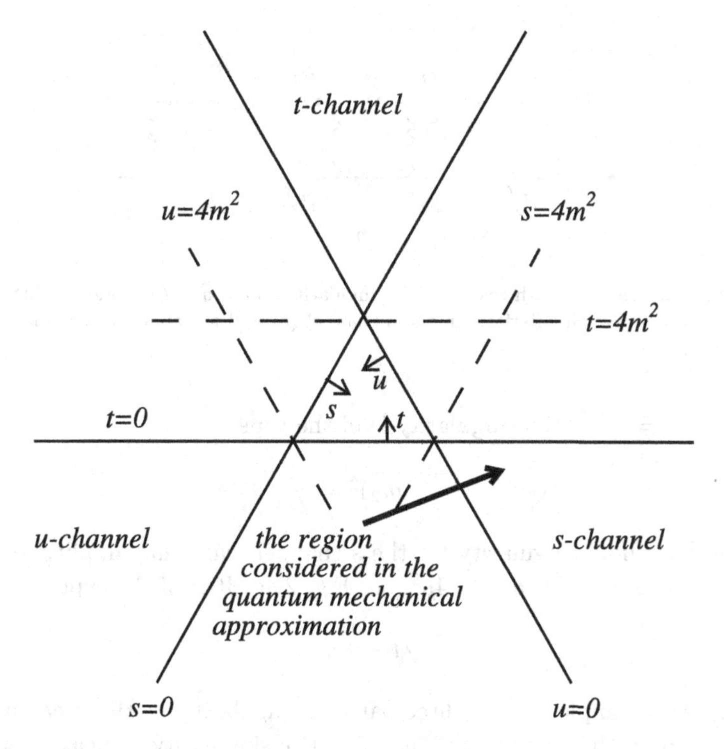

Fig. 2.2 The Mandelstam plane for the scattering processes: $1+2 \rightarrow 1'+2'$ (s-channel), $1+1' \rightarrow 2+2'$ (t-channel), $1+2' \rightarrow 2+1'$ (u-channel) .

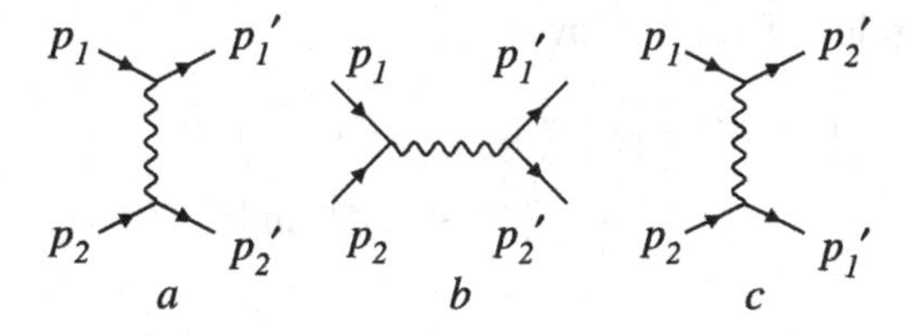

Fig. 2.3 One-particle exchange diagrams, with pole singularities in: (a) t-channel, (b) s-channel, and (c) u-channel.

in Fig. 2.4c). The s-channel cutting corresponds to the replacement of the Feynman propagators in the following way:

$$(k^2 - m^2)^{-1} \rightarrow \delta(k^2 - m^2) \,, \tag{2.5}$$

thus providing us with the imaginary part of the diagram Fig. 2.4a in the s-channel. The s-channel two-particle singularity is located at $s =$

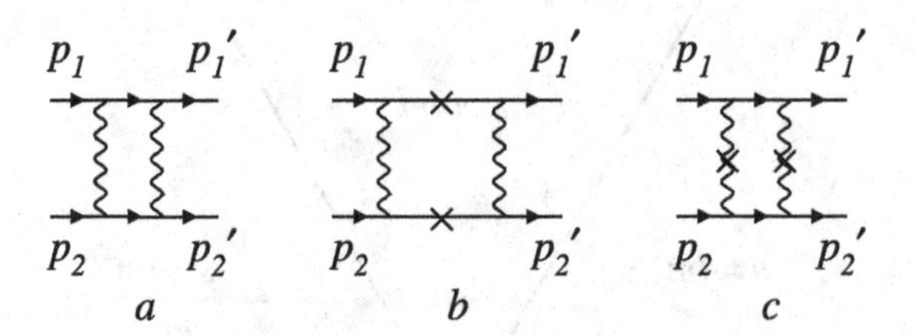

Fig. 2.4 Box diagrams with two-particle singularities in s- and t-channels. Cuttings of diagrams which give singularities in the s-channel (b) and t-channel (c) are marked by crosses.

$(m_1+m_2)^2 = 4m^2$; the singularity is of the type

$$\sqrt{s-(m_1+m_2)^2} = \sqrt{s-4m^2}; \tag{2.6}$$

it is the threshold singularity for the s-channel scattering amplitude. The t-channel singularity is at $t = 4\mu^2$, see Fig. 2.4c. It is of the type

$$\sqrt{t-4\mu^2}. \tag{2.7}$$

(iii) An example of the three-particle singularity in the s-channel is represented by the diagram of Fig. 2.5a. The singularity is located at

$$s = (m_1+m_2+\mu)^2 = (2m+\mu)^2. \tag{2.8}$$

The type of singularity is as follows:

$$\left[s-(m_1+m_2+\mu)^2\right]^2 \ln\left[s-(m_1+m_2+\mu)^2\right] = \\ = \left[s-(2m+\mu)^2\right]^2 \ln\left[s-(2m+\mu)^2\right]. \tag{2.9}$$

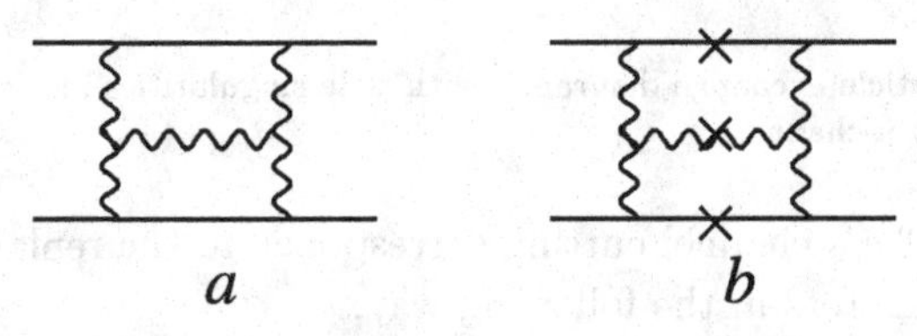

Fig. 2.5 Examples of the diagram with three-particle intermediate state in the s-channel; the crosses mark the state which is responsible for the appearance of the three-particle threshold singularity.

2.1.1.1 *Dalitz plot for the* $4 \to 1+2+3$ *decay*

The four-point amplitude has an additional physical region when particle masses (m_1, m_2, m_3) and m_4 are different and one of them is larger than the sum of all others:

$$m_4 > m_1 + m_2 + m_3. \tag{2.10}$$

It leads to a possibility of the decay process, see Fig. 2.1b (as before, we put $m_1 = m_2 = m_3 = m$):

$$4 \to 1+2+3. \tag{2.11}$$

Schematically the Mandelstam plane is shown for this case in Fig. 2.6. The physical region of the decay process is located in the centre of the plane (shadowed area). One can see also the physical regions for two-particle transitions: $4+1 \to 2+3$, $4+2 \to 3+1$, $4+3 \to 1+2$.

For a realistic decay process $4 \to 1+2+3$ with $m_4 = 1.9$ GeV and $m_1 = m_2 = m_3 = 0.14$ GeV we show the Dalitz plot in Fig. 2.7.

The energies squared of the outgoing particles, $s_{ij} = (p_i + p_j)^2$, obey the constraint

$$s_{12} + s_{13} + s_{23} - (m_1^2 + m_2^2 + m_3^2) == s_{12} + s_{13} + s_{23} - 3m^2 = m_4^2. \tag{2.12}$$

The threshold singularities at

$$s_{ij} = (m_i + m_j)^2 = 4m^2 \tag{2.13}$$

are touching the physical region of the decay.

2.1.2 *Bethe–Salpeter equations in the momentum representation*

We discuss here the Bethe–Salpeter (BS) equation [4], which is widely used for scattering processes and bound systems, and compare it with a treatment of the same amplitudes based on dispersion relations.

The non-homogeneous BS-equation in the momentum representation reads:

$$A(p_1', p_2'; p_1, p_2) = V(p_1', p_2'; p_1, p_2) + \int \frac{d^4k_1\, d^4k_2}{i(2\pi)^4}\, A(p_1', p_2'; k_1 k_2)$$
$$\times \frac{\delta^4(k_1 + k_2 - P)}{(m^2 - k_1^2 - i0)(m^2 - k_2^2 - i0)} V(k_1, k_2; p_1, p_2), \tag{2.14}$$

or in the graphical form:

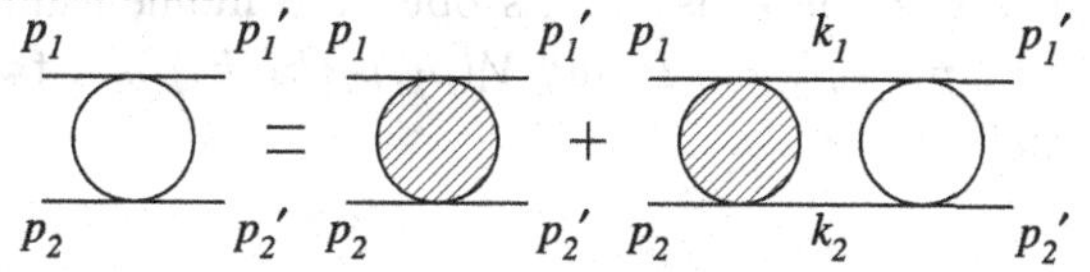

p_1 p_1' p_1 p_1' p_1 k_1 p_1'

= +

p_2 p_2' p_2 p_2' p_2 k_2 p_2' (2.15)

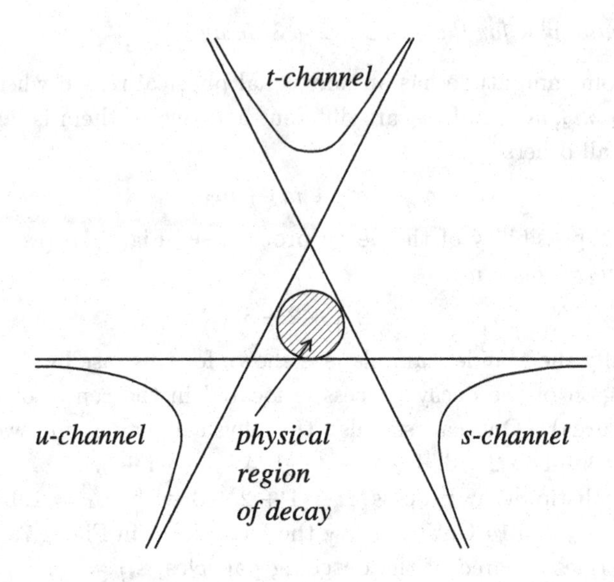

Fig. 2.6 Mandelstam plane and physical regions at $m_4 > m_1 + m_2 + m_3$ for processes $4 \to 1+2+3$ (shadowed area) and $4+1 \to 2+3$, $4+2 \to 3+1$, $4+3 \to 1+2$.

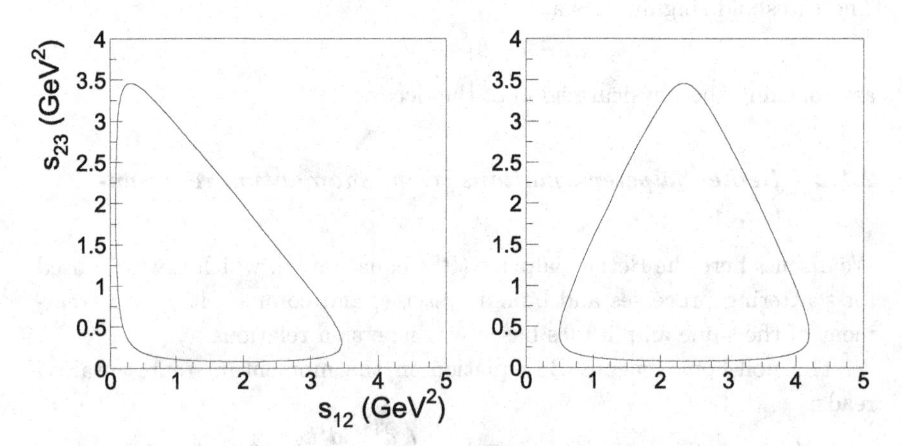

Fig. 2.7 Different representations for the Dalitz plot of the decay $4 \to 1+2+3$, $m_4 = 1.9$ GeV, $m_1 = m_2 = m_3 = m_\pi$

Here the momenta of the constituents obey the momentum conservation law $p_1 + p_2 = p_1' + p_2' = P$ and $V(p_1, p_2; k_1, k_2)$ is a two-constituent irreducible kernel:

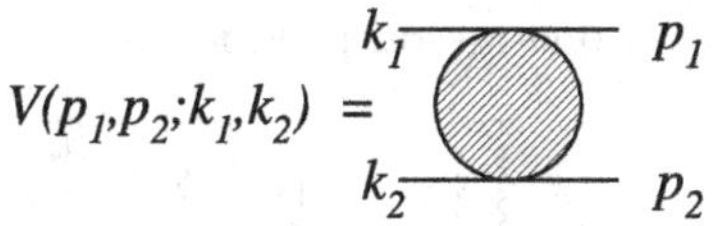

$$V(p_1,p_2;k_1,k_2) = \quad \text{(diagram with } k_1, k_2, p_1, p_2\text{)} \tag{2.16}$$

For example, it can be a kernel induced by the meson-exchange interaction

$$\frac{g^2}{\mu^2 - (k_1 - p_1)^2} . \tag{2.17}$$

Generally, $V(p_1, p_2; k_1, k_2)$ is an infinite sum of irreducible two-particle graphs.

We would like to emphasize that the amplitude A determined by the BS-equation is an off-mass-shell amplitude. Even if we put $p_1^2 = p_1'^2 = p_2^2 = p_2'^2 = m^2$ in the left-hand side of (2.14), the right-hand side contains the amplitude $A(k_1', k_2; p_1', p_2')$ for $k_1^2 \neq m^2$, $k_2^2 \neq m^2$.

Let us restrict ourselves to the one-meson exchange in the irreducible kernel V. By iterating Eq. (2.14), we come to infinite series of ladder diagrams:

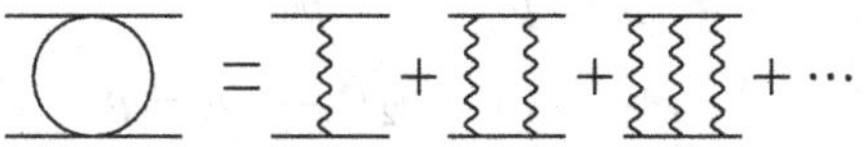

(2.18)

So we investigate the intermediate states in these ladder diagrams. Note that these diagrams have two-particle intermediate states which can appear as real states at c.m. energies $\sqrt{s} > 2m$. This corresponds to the cutting of the ladder diagrams across constituent lines:

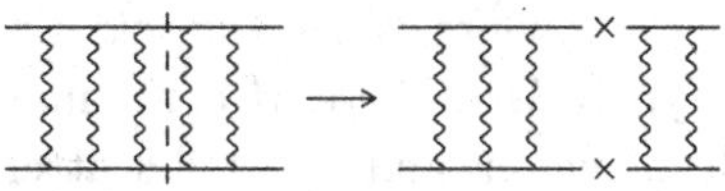

(2.19)

Such a two-particle state manifests itself as a singularity of the scattering amplitude at $s = 4m^2$.

2.1.2.1 *Multiparticle intermediate states*

However, the amplitude A being a function of s has not only this singularity but also an infinite set of singularities which correspond to the ladder diagram cuts across meson lines of the type:

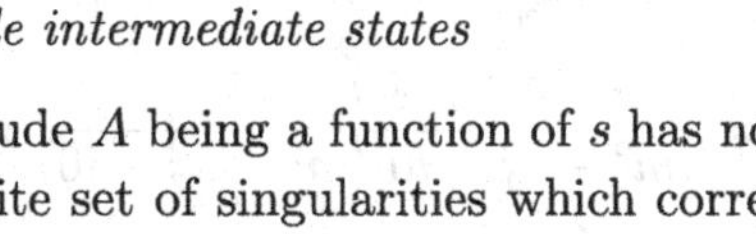

(2.20)

The diagrams, which appear after this cutting procedure, are meson production diagrams, *e.g.*, one-meson production diagrams:

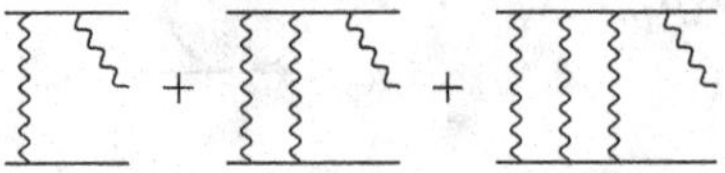

$$+ \quad + \tag{2.21}$$

Hence, the amplitude $A(p_1', p_2'; p_1, p_2)$ has the following cut singularity in the complex-s plane:

$$s = 4m^2 , \tag{2.22}$$

which is related to the rescattering process. Other singularities are related to the meson production processes with the cuts starting at

$$s = (2m + n\mu)^2; \qquad n = 1, 2, 3, \ldots \tag{2.23}$$

The four-point amplitude, which is the subject of the BS-equation, depends on six variables:

$$\begin{aligned} & p_1^2,\ p_2^2,\ p_1'^2,\ p_2'^2, \\ s = (p_1 + p_2)^2 \ &= \ (p_1' + p_2')^2, \\ t = (p_1 - p_1')^2 \ &= \ (p_2 - p_2')^2, \end{aligned} \tag{2.24}$$

while the seventh variable, $u = (p_1 - p_2')^2 = (p_1' - p_2)^2$, is not independent because of the relation

$$s + t + u \ = \ p_1^2 + p_2^2 + p_1'^2 + p_2'^2. \tag{2.25}$$

2.1.2.2 *Composite systems*

If a bound state of the constituents exists, the scattering partial amplitude has a pole at $s = (p_1 + p_2)^2 = M^2$, where M is the mass of the bound state. This pole appears both in the on- and off-shell scattering amplitudes.

The vertex for the transition "bound state $\to$ constituents", $\chi(p_1, p_2; P)$, satisfies the homogeneous BS-equation

$$\begin{aligned} \chi(p_1, p_2; P) = & \int \frac{d^4k_1\, d^4k_2}{i(2\pi)^4}\, V(p_1, p_2; k_1, k_2) \\ & \times \frac{\delta^4(k_1 + k_2 - P)}{(m^2 - k_1^2 - i0)(m^2 - k_2^2 - i0)}\, \chi(k_1, k_2; P), \end{aligned} \tag{2.26}$$

the graphical form of which is

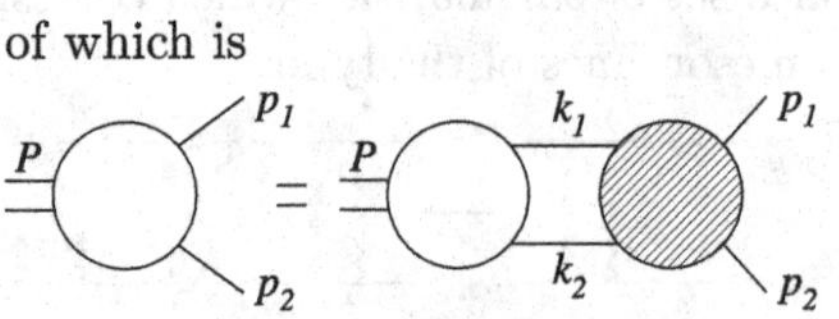

(2.27)

The n iterations of (2.27) with a meson-exchange kernel give

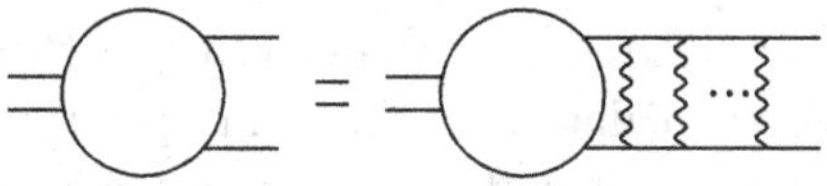

$$(2.28)$$

The cutting procedure of the interaction block in the right-hand side of (2.28) shows us that the amplitude $\chi(p_1, p_2; P)$ contains all the singularities of the amplitude A given by Eqs. (2.19), (2.20).

The three-point amplitude $\chi(p_1, p_2; P)$ depends on three variables

$$P^2 \ (\text{or } s)\ ,\ p_1^2\ ,\ p_2^2\ , \tag{2.29}$$

and again, as in the case of the scattering amplitude A, the BS-equation contains the off-mass-shell amplitude $\chi(p_1, p_2; P)$. Let us remark that $\chi(p_1, p_2; P)$ is a solution of the homogeneous equation, hence the normalization condition for it should be imposed independently. For the normalization, one can use the connection between χ and A at $P^2 \to M^2$:

$$A(p_1, p_2; p_1', p_2') = \frac{\chi(p_1, p_2; P^2 = M^2)\chi(P^2 = M^2; p_1', p_2')}{P^2 - M^2} + \text{regular terms}. \tag{2.30}$$

In the formulation of scattering theory, we start from a set of asymptotic states, containing constituent particles (with mass m) and mesons (with mass μ) only. We do not include in such a formulation of the scattering theory the composite particles as asymptotic states: we simply cannot know beforehand whether such bound states exist or not. But if we consider the production or decay of particles which are bound states, they should be included into the set of asymptotic states.

2.1.2.3 *Zoo-diagrams in the BS-equation*

The Bethe–Salpeter equation for the composite system

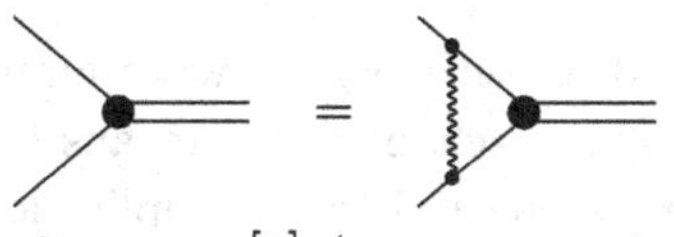

$$(2.31)$$

includes so-called zoo-diagrams [5] (we mean tad-pole, penguin-type and other configurations).

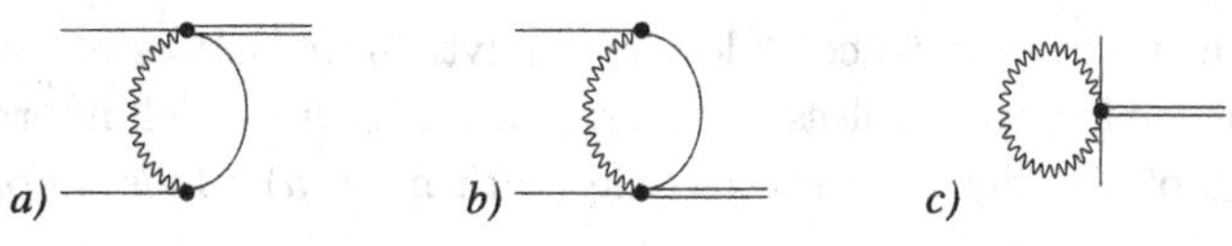

a) *b)* *c)* (2.32)

The matter is that the BS-equation includes in the intermediate state the factor $V(s;p_1'^2,p_2'^2,k_1^2,k_2^2)$ which may contain in the denominator k_1^2, k_2^2 or even products $k_1^2k_2^2$. Expanding these terms in the numerator: $k_a^2 = k_a^2-m^2+m^2$, we have a conciliation of the term $k_a^2-m^2$ with the propagator in the BS-equation and, as a result, the appearance of the zoo-diagrams of the (2.32) type.

2.1.2.4 *Miniconclusion: two-particle composite systems in the BS-equation*

We see that it is impossible to describe a pure two-particle composite system in terms of the Bethe–Salpeter equation. Ladder diagrams give us always additional multiparticle states. Additional multiparticle states may be eliminated if we consider a t-channel instantaneous interaction; however, this procedure does not eliminate zoo-diagrams – it is easily seen in the consideration of constituents with spin, for example fermion-antifermion systems.

2.2 Dispersion relation N/D-method and ansatz of separable interactions

In this section the basic features of the dispersion integration method are considered for the scattering amplitude $1+2 \to 1'+2'$. We concentrate our attention on a version which uses the ansatz of separable interactions — separable interactions allow us to build easily a bridge between the dispersion relation N/D-method and the spectral integral technique.

2.2.1 *N/D-method for the one-channel scattering amplitude of spinless particles*

In the N/D-method we deal with partial wave amplitudes. We consider the partial amplitudes in the s-channel – it means they depend on s only. In the complex-s plane the partial wave amplitudes have the s-channel right-hand side singularities and the left-hand side ones. The left-hand side singularities appear due to singularities in t and u channels of the amplitude $A(s,t)$.

For an illustration let us consider the analytical properties of the scattering amplitude for two spinless particles (with mass m) which interact via the exchange of another spinless particle (with mass μ). This amplitude,

$A(s,t)$, has singularities in s, t and u channels. In the t channel there are singularities at $t = \mu^2, 4\mu^2, 9\mu^2$, etc., which correspond to one- or many-particle exchanges; in the s-plane the amplitude has a singularity at $s = 4m^2$ (elastic rescattering) and singularities at $s = (2m + n\mu)^2$, with $n = 1, 2, \ldots$, corresponding to the production of n particles with mass μ in the s-channel intermediate state. If a bound state with mass M exists, the pole singularity is at $s = M^2$. If the mass of this bound state $M > 2m$, this is a resonance and the corresponding pole is located on the second sheet of the complex s-plane.

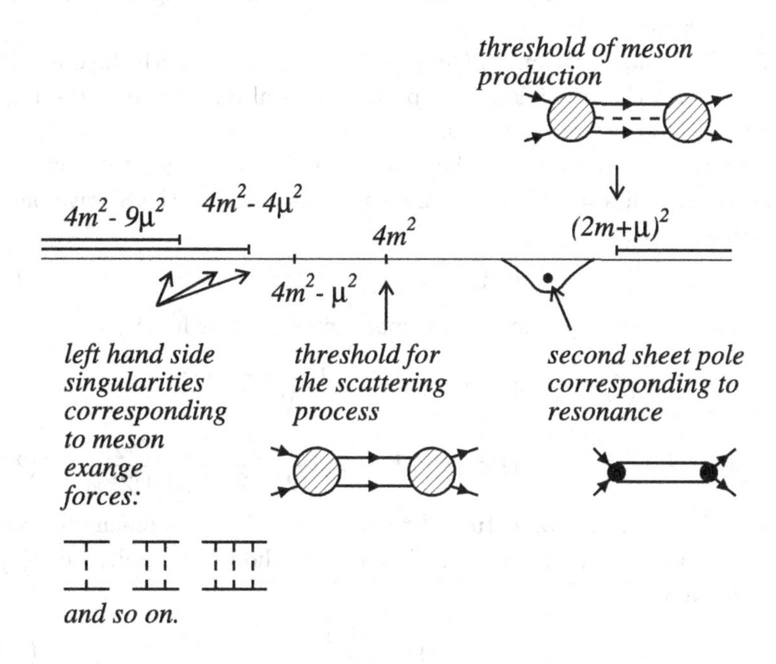

Fig. 2.8 Singularities of partial wave amplitudes in the s-plane.

The partial wave s-channel amplitude, depending on s only, has all the s-channel right-hand side singularities of $A(s,t)$: at $s = M^2$, $s = 4m^2$, $s = (2m + \mu)^2, \ldots$ shown in Fig. 2.8. Left-hand side singularities of the partial amplitudes are related to the t and u channel exchanges contributing to $A(s,t)$. The S-wave partial amplitude is equal to

$$A(s) = \int_{-1}^{1} \frac{dz}{2} A(s, t(z)), \tag{2.33}$$

where $t(z) = -2(s/4 - m^2)(1-z)$ and $z = \cos\theta$. Left-hand side singularities related to the t-channel exchanges correspond to

$$t(z=-1) = s(t - \text{singularity}) \qquad (2.34)$$

$$\text{Ladder diagrams}: \quad s(t - \text{singularity}) = 4m^2 - (n^2\mu^2) \qquad n = 1, 2, 3, ...$$

The u-channel singularities produce on the s-plane the left-hand side singularities at

$$t(z=1) = s(u - \text{singularity}). \qquad (2.35)$$

For the scattering of identical particles the left-hand singularities from u and t channels coincide.

The dispersion relation N/D-method [6] provides us with the possibility to reconstruct the relativistic two-particle partial amplitude in the region of low and intermediate energies.

Let us restrict ourselves to the region of elastic scattering (or, the region in the vicinity of $s = 4m^2$). The unitarity condition for the S-wave partial amplitude reads:

$$\text{Im}A(s) = \rho(s) \mid A(s) \mid^2 . \qquad (2.36)$$

Here $\rho(s)$ is the two-particle phase space integrated at fixed s:

$$\rho(s) = \int d\Phi_2(P; k_1, k_2) = \frac{1}{16\pi}\sqrt{\frac{s-4m^2}{s}},$$

$$d\Phi_2(P; k_1, k_2) = \frac{1}{2}(2\pi)^4\delta^4(P - k_1 - k_2)\frac{d^3k_1}{(2\pi)^3 2k_{10}}\frac{d^3k_2}{(2\pi)^3 2k_{20}}, \qquad (2.37)$$

where P is the total momentum, $P^2 = s$; k_1 and k_2 are momenta of particles in the intermediate state. In the N/D-method the amplitude $A(s)$ is represented as

$$A(s) = \frac{N(s)}{D(s)}. \qquad (2.38)$$

Here $N(s)$ has only left-hand side singularities, whereas $D(s)$ has only right-hand side ones. So, the N-function is real in the physical region $s > 4m^2$. The unitarity condition can be rewritten as:

$$\text{Im}\, D(s) = -\rho(s)N(s). \qquad (2.39)$$

The solution of this equation is

$$D(s) = 1 - \int\limits_{4m^2}^{\infty} \frac{d\tilde{s}}{\pi}\frac{\rho(\tilde{s})N(\tilde{s})}{\tilde{s} - s} \equiv 1 - B(s). \qquad (2.40)$$

In Eq. (2.40) we neglect the so-called CDD-poles [7] and normalize $N(s)$ by the condition $D(s) \to 1$ as $s \to \infty$.

Let us introduce the vertex function

$$G(s) = \sqrt{N(s)}. \tag{2.41}$$

We assume here that $N(s)$ is positive (the cases with negative $N(s)$ or if $N(s)$ changes sign need a special and more cumbersome treatment). Then the partial wave amplitude $A(s)$ can be expanded into a series

$$A(s) = G(s)[1 + B(s) + B^2(s) + B^3(s) + \cdots]G(s)\ , \tag{2.42}$$

where $B(s)$ is a loop-diagram

$$B(s) = \tag{2.43}$$

The graphical interpretation of Eq. (2.42) is as follows:

$$= \quad + \quad + \quad + \cdots \tag{2.44}$$

so the amplitude $A(s)$ is a set of terms with different numbers of rescatterings.

2.2.2 *Scattering amplitude and energy non-conservation in the spectral integral representation*

The off-energy-shell amplitude emerges when the cutting procedure of the series (2.44) is performed:

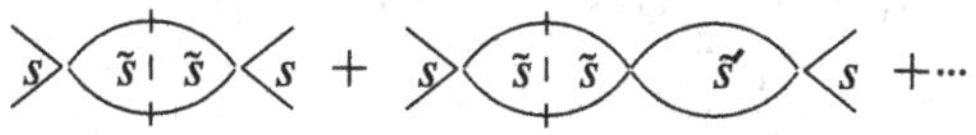

$$s \quad \tilde{s} \mid \tilde{s} \quad s \quad + \quad s \quad \tilde{s} \mid \tilde{s} \quad \tilde{s}' \quad s \quad + \cdots \tag{2.45}$$

The right-hand side of the cut amplitude is also represented as an infinite sum of loop diagrams, where, however, the initial and final values $\tilde{s}$ and s are different:

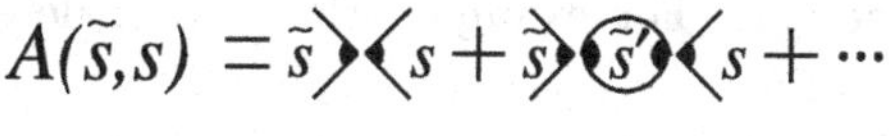

$$A(\tilde{s}, s) = \tilde{s} \quad s + \tilde{s} \quad \tilde{s}' \quad s + \cdots \tag{2.46}$$

It is the off-energy-shell amplitude which has to be considered in the general case. This amplitude satisfies the equation

$$A(\tilde{s}, s) = G(\tilde{s})G(s) + G(\tilde{s}) \int_{4m^2}^{\infty} \frac{d\tilde{s}'}{\pi} \frac{G(\tilde{s}')\rho(\tilde{s}')A(\tilde{s}', s)}{\tilde{s}' - s}. \tag{2.47}$$

Let us emphasize that in the dispersion approach we deal with the on-mass-shell amplitudes, *i.e.* amplitudes for real constituents, whereas in the BS-equation (2.14) the amplitudes are off-mass-shell. The appearance of the off-energy-shell amplitude in the dispersion method, Eq. (2.47), is the price we have to pay for keeping all the constituents on the mass shell.

The solution of Eq. (2.47) reads:

$$A(\tilde{s}, s) = G(\tilde{s}) \frac{G(s)}{1 - B(s)}. \tag{2.48}$$

For the physical processes $\tilde{s} = s$, so the partial wave amplitude $A(s)$ is determined as $A(s) = A(s, s)$.

Consider the partial amplitude near the pole corresponding to the bound state. The pole appears when

$$B(M^2) = 1\,, \tag{2.49}$$

and in the vicinity of this pole we have:

$$A(s) = G(s)\frac{1}{1 - B(s)}G(s) \simeq \frac{G(s)}{\sqrt{B'(M^2)}} \cdot \frac{1}{M^2 - s} \cdot \frac{G(s)}{\sqrt{B'(M^2)}} + \ldots \tag{2.50}$$

Here we take into account that $1 - B(\tilde{s}) \simeq 1 - B(M^2) - B'(M^2)(s - M^2)$.

The homogeneous equation for the bound state vertex $G_{vertex}(s, M^2)$ reads:

$$\begin{aligned} G_{vertex}(s, M^2) &= G(s) \int_{4m^2}^{\infty} \frac{d\tilde{s}}{\pi} G(\tilde{s}) \frac{\rho(\tilde{s})}{\tilde{s} - M^2} G_{vertex}(\tilde{s}, M^2) \\ &= G(s) \int_{4m^2}^{\infty} \frac{d\tilde{s}}{\pi} G(\tilde{s}) \frac{\rho(\tilde{s})}{\tilde{s} - M^2} \frac{G(\tilde{s}, M^2)}{\sqrt{B'(M^2)}} = \frac{G(s, M^2)}{\sqrt{B'(M^2)}}. \end{aligned} \tag{2.51}$$

because $B(M^2) = 1$.

The vertex function $G_{vertex}(s, M^2)$ enters all processes containing the bound state interaction. For example, this vertex determines the form factor of a bound state.

2.2.3 *Composite system wave function and its form factors*

The vertex function represented by (2.51) gives way to a subsequent description of composite systems in terms of dispersion relations with separable interactions. To see this, one should consider not only the two-particle interaction (what we have dealt with before) but to go off the frame of this problem: we have to study the interaction of the two-particle composite system with the electromagnetic field. In principle, this is not a difficult task when interactions are separable.

Consider the dispersion representation of the triangle diagram shown in Fig. 2.9a. It can be written in a way similar to the one-fold representation for the loop diagram with a certain necessary complication (as before, we consider a simple case of equal masses $m_1 = m_2 = m$).

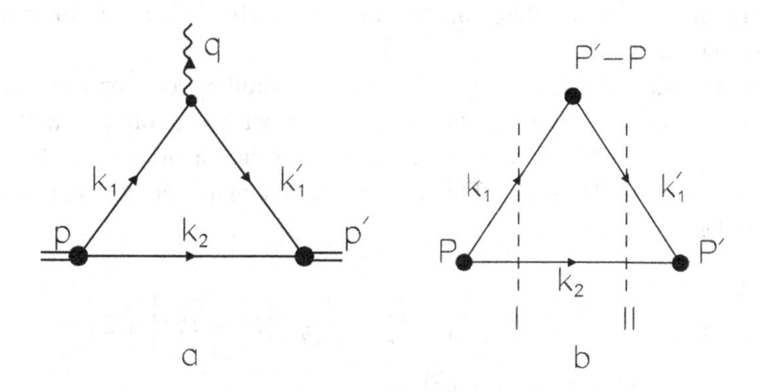

Fig. 2.9 (a) Additive quark model diagram for a composite system: one of the constituents interacts with electromagnetic field; (b) cut triangle diagram in the double spectral representation: $P^2 = s$, $P'^2 = s'$ and $(P' - P)^2 = q^2$.

First, a double dispersion integral should be written in terms of the masses of the incoming and outgoing particles:

$$\int\limits_{4m^2}^{\infty} \frac{ds}{\pi} \frac{1}{s - p^2 - i0} \int\limits_{4m^2}^{\infty} \frac{ds'}{\pi} \frac{1}{s' - p'^2 - i0} \times \ldots \tag{2.52}$$

The double spectral representation is inevitable when the interaction of the photon, though with one constituent only, divides the loop diagram into two pieces. Dots in (2.52) stand for the double discontinuity of the triangle diagram, with cutting lines I and II (see Fig. 2.9b); let us denote it as $disc_s\, disc_{s'}\, F(s, s', q^2)$. This double discontinuity is written analogously to the discontinuity of the loop diagram. Namely,

$$\begin{aligned}
&disc_s\, disc_{s'} F(s,s',q^2) \sim \\
&\sim G_{vertex}(s,M^2) d\Phi_{tr}(P,P';k_1,k'_1,k_2) G_{vertex}(s',M^2), \\
&d\Phi_{tr}(P,P';k_1,k'_1,k_2) = \\
&= d\Phi_2(P;k_1,k_2) d\Phi_2(P';k'_1,k'_2)(2\pi)^3 2k'_{20}\delta^3(\mathbf{k}_2 - \mathbf{k}'_2)
\end{aligned} \tag{2.53}$$

Here the vertex G_{vertex} is defined according to (2.51), the two-particle phase volume is written following (2.37) and the factor $2(2\pi)^3 k'_{20}\delta^3(\mathbf{k}_2 - \mathbf{k}'_2)$ reflects the fact that the constituent spectator line was cut twice (that is, of course, impossible and requires to eliminate in (2.53) the extra phase space integration). Let us stress that in (2.53) the constituents are on the mass shell: $k_1^2 = k_2^2 = k_1'^2 = m^2$, the momentum transfer squared is fixed $(k'_1 - k_1)^2 = (P' - P)^2 = q^2$ but $P' - P \neq q$.

We did not write in (2.53) an equality sign, since there is one more factor in Fig. 2.9b.

In the diagram of Fig. 2.9b, the gauge invariant vertex for the interaction of a scalar (or pseudoscalar) constituent with a photon is written as $(k_{1\mu} + k'_{1\mu})$, from which one should separate a factor orthogonal to the momentum transfer $P'_\mu - P_\mu$. This is not difficult using the kinematics of real particles:

$$\begin{aligned}
&k_{1\mu} + k'_{1\mu} = \alpha(s,s',q^2)\left[P_\mu + P'_\mu - \frac{s'-s}{q^2}(P'_\mu - P_\mu)\right] + k_{\perp\mu}, \\
&\alpha(s,s',q^2) = -\frac{q^2(s+s'-q^2)}{\lambda(s,s',q^2)}, \\
&\lambda(s,s',q^2) = -2q^2(s+s') + q^4 + (s'-s)^2,
\end{aligned} \tag{2.54}$$

where $k_{\perp\mu}$ is orthogonal to both $(P_\mu + P'_\mu)$ and $(P_\mu - P'_\mu)$. Hence,

$$\begin{aligned}
&disc_s\, disc_{s'} F(s,s',q^2) = \\
&= G_{vertex}(s,M^2) G_{vertex}(s',M^2) d\Phi_{tr}(P,P';k_1,k'_1,k_2)\alpha(s,s',q^2),
\end{aligned} \tag{2.55}$$

and the form factor of the composite system reads:

$$F(q^2) = \int\limits_{4m^2}^{\infty} \frac{ds}{\pi} \int\limits_{4m^2}^{\infty} \frac{ds'}{\pi} \frac{disc_s\, disc_{s'} F(s,s',q^2)}{(s-M^2)(s'-M^2)}, \tag{2.56}$$

where we took into account that $p^2 = p'^2 = M^2$ and the term $k_{\perp\mu}$ equals zero after integrating over the phase space.

Let us underline that the full amplitude of the interaction of the photon with a composite system, when the charge of the composite system equals one, is:

$$A_\mu(q^2) = (p_\mu + p'_\mu)F(q^2) \ , \tag{2.57}$$

that is, the form factor of the composite system is an invariant coefficient before the transverse part of the amplitude A_μ:

$$(p + p') \perp q \ . \tag{2.58}$$

Likewise, the invariant coefficient $\alpha(s, s'q^2)$ defines the transverse part of the diagram shown in Fig. 2.9b:

$$\left[P + P' - \frac{s' - s}{q^2}(P' - P)\right] \perp (P' - P) \ . \tag{2.59}$$

The formula (2.56) has a remarkable property: for the vertex $G_{vertex}(s)$ (2.51) it gives a correct normalization of the charge form factor,

$$F(0) = 1 \ . \tag{2.60}$$

It is easy to carry out the derivation of this normalization condition, we shall do that below. For $F(q^2)$, after integrating in (2.56) over the momenta k_1, k'_1 and k_2 at fixed s and s', we obtain the following expression:

$$\begin{aligned} F(q^2) \ &= \int\limits_{4m^2}^{\infty} \frac{ds\, ds'}{\pi^2} \frac{G_{vertex}(s, M^2)}{s - M^2} \frac{G_{vertex}(s', M^2)}{s' - M^2} \\ &\times \frac{\Theta\left(-ss'q^2 - m^2\lambda(s, s', q^2)\right)}{16\sqrt{\lambda(s, s', q^2)}} \alpha(s, s', q^2) \ . \end{aligned} \tag{2.61}$$

Here the Θ-function is defined as follows: $\Theta(X) = 1$ at $X \geq 0$ and $\Theta(X) = 0$ at $X < 0$.

To calculate (2.61) in the limit $q^2 \to 0$, let us introduce new variables:

$$\sigma = \frac{1}{2}(\tilde{s} + \tilde{s}') \ ; \quad \Delta = \tilde{s} - \tilde{s}', \quad Q^2 = -q^2 \ , \tag{2.62}$$

and then consider the case of interest, $Q^2 \to 0$. The form factor formula reads:

$$F(-Q^2 \to 0) = \int\limits_{4m^2}^{\infty} \frac{d\sigma}{\pi} \frac{G^2_{vertex}(\sigma, M^2)}{(\sigma - M^2)(\sigma - M^2)} \int\limits_{-b}^{b} d\Delta \frac{\alpha(\sigma, \Delta, Q^2)}{16\pi\sqrt{\Delta^2 + 4\sigma Q^2}}, \tag{2.63}$$

where

$$b = \frac{Q}{m}\sqrt{\sigma(\sigma - 4m^2)} \ , \quad \alpha(\sigma, \Delta, Q^2) \ = \ \frac{2\sigma\, Q^2}{\Delta^2 + 4\sigma Q^2} \ . \tag{2.64}$$

As a result we have:

$$F(0) = 1 = \int_{4m^2}^{\infty} \frac{ds}{\pi} \Psi^2(s)\rho(s),$$

$$\rho(s) = \frac{1}{16\pi}\sqrt{1 - 4m^2/s}\,, \qquad \Psi(s) = \frac{G_{vertex}(s, M^2)}{s - M^2}. \tag{2.65}$$

We see that the condition $F(0) = 1$ means actually the normalization condition for the wave function of the composite system $\Psi(s)$.

Formulae (2.61) and (2.65) show that we have for composite systems a unified triad:
(i) the hypothesis of separable interaction,
(ii) the method of spectral integration,
(iii) the technique for calculation of radiative transitions in the additive model.

This triad opens future prospects for the calculation of both wave functions (or vertices) of the composite systems and radiative processes with this composite systems.

Of course, the use of separable interactions imposes a model restriction on the treatment of physical processes (for example, within the above triad we do not account for the interaction of photons with exchange currents). But for composite systems the most important are additive processes, and the discussed model opens a possibility to carry out subsequent calculations of interaction processes with the electromagnetic field taking into account the gauge invariance.

The procedure of construction of gauge invariant amplitudes within the framework of the spectral integration method has been realized for the deuteron in [8, 9], and, correspondingly, for the elastic scattering and the photodisintegration process. A generalization of the method for the composite quark systems has been performed in [10, 11, 12].

2.2.4 *Scattering amplitude with multivertex representation of separable interaction*

A separable representation of the interaction block can be generalized:
(i) instead of one vertex we can introduce a sum of vertices,
(ii) the left-hand side and right-hand side vertices can be chosen differently (without violating the T-invariance for the physical amplitude).

In this way we write for the interaction block:

$$G(\tilde{s}')\,G(s) \to \sum_j G_{Lj}(\tilde{s}')\,G^{Rj}(s) \equiv G_j(\tilde{s}')\,G^j(s) \tag{2.66}$$

Here, for simplicity, we omit indices L and R allowing the left G_j and right G^j vertex functions to be different.

For the solution of the equation $A(\tilde{s}', s)$ it is convenient to use the amplitude with a removed vertex of the outgoing particles. We denote these amplitudes as $a^j(s)$:

$$A(\tilde{s}', s) = \sum_j G_j(\tilde{s}')a^j(s)\ . \tag{2.67}$$

The amplitude $a^j(s)$ satisfies the following equation:

$$a^j(s) = \sum_{j'} a^{j'}(s)B^j_{j'}(s) + G^j(s)\ ,$$

$$B^j_{j'}(s) = \int\limits_{4m^2}^{\infty} \frac{ds'}{\pi}\, \frac{G_{j'}(s')\rho(s')G^j(s')}{s'-s}\ . \tag{2.68}$$

The equation (2.68) can be rewritten in the matrix form:

$$\hat{a}(s) = \ \hat{B}(s)\hat{a}(s) + \hat{g}^R(s), \tag{2.69}$$

where

$$\hat{a}(s) = \begin{vmatrix} a^1(s) \\ a^2(s) \\ \cdot \\ \cdot \\ \cdot \end{vmatrix},\ \hat{g}_L(s) \ = \begin{vmatrix} G_1(s) \\ G_2(s) \\ \cdot \\ \cdot \\ \cdot \end{vmatrix},\ \hat{g}^R(s) \ = \begin{vmatrix} G^1(s) \\ G^2(s) \\ \cdot \\ \cdot \\ \cdot \end{vmatrix},$$

$$\hat{B}(s) \ = \begin{vmatrix} B^1_1(s) & B^2_1(s) & \cdot \\ B^1_2(s) & B^2_2(s) & \cdot \\ \cdot & \cdot & \cdot \\ \cdot & \cdot & \cdot \\ \cdot & \cdot & \cdot \end{vmatrix}. \tag{2.70}$$

Thus, we have the following expression for the amplitude $\hat{A}(\tilde{s}, s)$:

$$\hat{A}(\tilde{s}, s) \ = \ \hat{g}^T_L(\tilde{s})\frac{1}{\hat{I} - \hat{B}(s)}\hat{g}^R(s)\ , \tag{2.71}$$

where $\hat{I}$ is a unit matrix in the vertex space and the symbol T means the transposition

$$\hat{g}^T_L(\tilde{s}) \ = \ \begin{vmatrix} G_1(\tilde{s}), G_2(\tilde{s}), \ldots \end{vmatrix}. \tag{2.72}$$

The physical amplitude is equal to:

$$\hat{A}(s) \equiv \hat{A}(s,s) \ = \ \hat{g}^T_L(s)\frac{1}{I - \hat{B}(s)}\hat{g}^R(s)\ . \tag{2.73}$$

2.2.4.1 *Generalization for an arbitrary angular momentum state, $L = J$*

The partial wave amplitude for arbitrary $L = J$ can be written quite similarly to the considered case $J = 0$. Instead of (2.66), we have the following interaction block:

$$\sum_J X^J_{\mu_1\ldots\mu_J}(\tilde{k})\hat{g}^T_{J,L}(\tilde{s})\hat{g}^R_J(s)X^J_{\mu_1\ldots\mu_J}(k) \ . \tag{2.74}$$

Let us remind that k and $\tilde{k}$ are relative momenta in initial and final states $k_\mu = g^{\perp P}_{\mu\nu}p_{1\nu} = -g^{\perp P}_{\mu\nu}p_{2\nu}$ and $\tilde{k}_\mu = g^{\perp \tilde{P}}_{\mu\nu}\tilde{p}_{1\nu} = -g^{\perp \tilde{P}}_{\mu\nu}\tilde{p}_{2\nu}$. The angular operator $X^J_{\mu_1\ldots\mu_J}(\tilde{k})$ is determined in Appendix A. The scattering amplitude, expanded over partial waves using operators $X^{(J)}_{\mu_1\ldots\mu_J}$, reads:

$$\sum_J X^J_{\mu_1\ldots\mu_J}(\tilde{k})\hat{A}_J(\tilde{s},s)X^J_{\mu_1\ldots\mu_J}(k) \ , \tag{2.75}$$

As before, we can represent $A_J(\tilde{s},s)$ as follows:

$$A_J(\tilde{s},s)\sum_j G^J_j(\tilde{s})a^j_J(s) \ . \tag{2.76}$$

The amplitudes a^j_J satisfy the following equations:

$$a^j_J(s) = \sum_{j'} a^{j'}_J(s)B^j_{j'}(IJ;s) + G^j_J(s) \ ,$$

$$B^j_{j'}(J;s) = \int\limits_{4m^2}^{\infty} \frac{ds'}{\pi}\,\frac{G^J_{j'}(s')\rho^{(J)}(s')G^j_J(s')}{s'-s} \ ,$$

$$\rho^{(J)}(s') = \int d\Phi_2(P' = k'_1 + k'_2; k'_1, k'_2)X^J_{\mu_1\ldots\mu_J}(k')X^J_{\mu_1\ldots\mu_J}(k'). \tag{2.77}$$

The equation (2.77) can be rewritten in the matrix form:

$$\hat{a}_J(s) = \ \hat{B}_J(s)\hat{a}_J(s) + \hat{g}^R_J(s), \tag{2.78}$$

with $\hat{a}^J(s)$, $\hat{g}_J(s)$, $\hat{B}(J;s)$ determined according Eq. (2.70) but with an additional label J. Thus we have the following expression for the partial wave amplitude:

$$\hat{A}_J(s) \equiv \hat{A}_J(s,s) \ = \ \hat{g}^T_{LJ}(s)\frac{1}{\hat{I} - \hat{B}(J;s)}\hat{g}^R_J(s) \ , \tag{2.79}$$

2.3 Instantaneous interaction and spectral integral equation for two-body systems

If we consider a strictly fixed number of constituents in composite systems, we should use instantaneous interactions, otherwise the carriers of interaction would be constituents as well (see the discussion in Section 1).

2.3.1 *Instantaneous interaction*

In the instantaneous approximation the interaction block depends on $t_\perp$:

$$\widehat{V} \longrightarrow \widehat{V}(t_\perp), \qquad t_\perp = (k_1^{\perp P} - k_1'^{\perp P'})_\mu(-k_2^{\perp P} + k_2'^{\perp P'})_\mu \, . \quad (2.80)$$

The absence of retardation effects in the interaction (2.80) is easily seen in the c.m. system where the time components of the constituent momenta disappear in $t_\perp$ (see also the discussion in [13, 14, 15, 16, 17] for more details).

Fitting to quark–antiquark states [13, 18, 19, 20], we expand the interaction blocks using the following $t_\perp$-dependent terms:

$$I_{-1} = \frac{4\pi}{\mu^2 - t_\perp}, \qquad I_0 = \frac{8\pi\mu}{(\mu^2 - t_\perp)^2},$$

$$I_1 = 8\pi\left(\frac{4\mu^2}{(\mu^2 - t_\perp)^3} - \frac{1}{(\mu^2 - t_\perp)^2}\right),$$

$$I_2 = 96\pi\mu\left(\frac{2\mu^2}{(\mu^2 - t_\perp)^4} - \frac{1}{(\mu^2 - t_\perp)^3}\right),$$

$$I_3 = 96\pi\left(\frac{16\mu^4}{(\mu^2 - t_\perp)^5} - \frac{12\mu^2}{(\mu^2 - t_\perp)^4} + \frac{1}{(\mu^2 - t_\perp)^3}\right), \quad (2.81)$$

or, in the general case,

$$I_N = \frac{4\pi(N+1)!}{(\mu^2 - t_\perp)^{N+2}} \sum_{n=0}^{N+1} (\mu + \sqrt{t_\perp})^{N+1-n}(\mu - \sqrt{t_\perp})^n \, . \quad (2.82)$$

2.3.1.1 *Coordinate representation*

Traditionally, the interaction of quarks in the instantaneous approximation is represented in terms of the potential $V(r)$. The form of the potential can be obtained with the help of the Fourier transform of (2.81) in the centre-of-mass system. Thus, we have

$$t_\perp = -(\mathbf{k} - \mathbf{k}')^2 = -\mathbf{q}^2 \, ,$$

$$I_N^{(\mathrm{coord})}(r, \mu) = \int \frac{d^3 q}{(2\pi)^3} \, e^{-i\mathbf{q}\cdot\mathbf{r}} \, I_N(t_\perp) \, , \quad (2.83)$$

that gives

$$I_N^{(\text{coord})}(r,\mu) = r^N\, e^{-\mu r}\ . \tag{2.84}$$

In the fitting of the quark-antiquark states [18] the following types of $V(r)$ were used:

$$V(r) = a + b\,r + c\,e^{-\mu_c\,r} + d\frac{e^{-\mu_d\,r}}{r}\ , \tag{2.85}$$

where the constant and linear (confinement) terms read:

$$\begin{aligned} a &\to a\, I_0^{(\text{coord})}(r,\mu_{\text{constant}} \to 0)\ , \\ br &\to b\, I_1^{(\text{coord})}(r,\mu_{\text{linear}} \to 0)\ . \end{aligned} \tag{2.86}$$

The limits μ_{constant}, $\mu_{\text{linear}} \to 0$ mean that in the fitting procedure the parameters μ_{constant} and μ_{linear} are chosen to be small enough, of the order of 1–10 MeV. It was checked that the solutions for the states with radial excitation $n \le 6$ are practically stable, when μ_{constant} and μ_{linear} change within this interval.

2.3.1.2 *Instantaneous interaction – transformation into a set of separable vertices*

Sometimes it is convenient to transform the instantaneous interaction into that described by separable vertices – this procedure is especially helpful in the consideration of three-body systems [21, 22].

The expansion of $V(t_\perp)$ into a set of separable vertices is a two-step procedure

$$\begin{aligned} V(t_\perp) &= \sum_J X^J_{\mu_1\ldots\mu_J}(k')V_J(s',s)X^J_{\mu_1\ldots\mu_J}(k) \\ &= \sum_J X^J_{\mu_1\ldots\mu_J}(k')\Big(\sum_{a,b} G^J_{L,a}(s')G^{R,b}_J(s)\Big)X^J_{\mu_1\ldots\mu_J}(k). \end{aligned} \tag{2.87}$$

The expansion $V_J(s',s)$ over $G^J_{L,a}(s')G^{R,b}_J(s)$ can be performed using a full set of orthogonal functions. However, it is possible to work with non-orthogonal functions as well – the pivotal requirement is the close description of $V_J(\tilde{s},s)$ in the region of interest using a possibly minimal number of the terms $G^J_{L,a}(s')G^{R,b}_J(s)$.

One can re-denote

$$[a,b] \to j\ , \tag{2.88}$$

transforming the consideration exactly to the case considered in Section 2.3.1.

2.3.1.3 *An example: expressions for $q_\perp$ and $t_\perp$ in the centre-of-mass system*

To make calculations more apparent, let us present here the expressions for $q_\perp$ and $t_\perp$ in the centre-of-mass system. The momenta of interacting constituents have the following form in the centre-of-mass frame:

$$k_1 = \left(\frac{\sqrt{s}}{2}, \vec{n}\sqrt{\frac{s}{4} - m^2}\right), \qquad k'_1 = \left(\frac{\sqrt{s'}}{2}, \vec{n}'\sqrt{\frac{s'}{4} - m^2}\right),$$
$$k_2 = \left(\frac{\sqrt{s}}{2}, -\vec{n}\sqrt{\frac{s}{4} - m^2}\right), \qquad k'_2 = \left(\frac{\sqrt{s'}}{2}, -\vec{n}'\sqrt{\frac{s'}{4} - m^2}\right), \qquad (2.89)$$

In accordance with this,

$$k_1^\perp = \left(0, \vec{n}\sqrt{\frac{s}{4} - m^2}\right), \qquad k_1'^\perp = \left(0, \vec{n}'\sqrt{\frac{s'}{4} - m^2}\right),$$
$$k_2^\perp = \left(0, -\vec{n}\sqrt{\frac{s}{4} - m^2}\right), \qquad k_2'^\perp = \left(0, -\vec{n}'\sqrt{\frac{s'}{4} - m^2}\right), \qquad (2.90)$$

and in the c.m. system we write

$$q^\perp = k_1^\perp - k_1'^\perp = -k_2^\perp + k_2'^\perp = \left(0, \vec{n}\sqrt{\frac{s}{4} - m^2} - \vec{n}'\sqrt{\frac{s'}{4} - m^2}\right). \qquad (2.91)$$

The absence of retardation effects in the case of the use of $V(t_\perp)$ is seen from (2.91) directly.

2.3.2 ***Spectral integral equation for a composite system***

First, we consider the case of $L = 0$ for scalar constituents with equal masses; these constituents are, however, not identical. The bound system is treated as a composite system of them. Henceforth the case $L \neq 0$ is considered in detail.

2.3.2.1 *Spectral integral equation for vertex function with $L = 0$*

The equation for the vertex *composite system* $\to$ *constituents*, shown graphically in Fig. 2.31, reads:

$$G(s) = \int\limits_{4m^2}^{\infty} \frac{ds'}{\pi} \int d\Phi_2(P'; k'_1, k'_2) V(k_1, k_2; k'_1, k'_2) \frac{G(s')}{s' - M^2 - i0}. \qquad (2.92)$$

Let us remind that the phase space is determined by Eq. (2.37). Scalar constituents are supposed not to be identical, so we do not write an additional identity factor 1/2 in the phase space.

The equation (2.92) written in the spectral-representation form deals with the off-energy-shell states $s' = (k'_1 + k'_2)^2 \neq M^2$, $s = (k_1 + k_2)^2 \neq M^2$ and $s \neq s'$; the constituents are on-mass-shell, $k'^2_1 = m^2$ and $k'^2_2 = m^2$. We can use an alternative expression for the phase space:

$$d\Phi_2(P'; k'_1, k'_2) = \rho(s')\frac{dz}{2} \equiv d\Phi(k') , \qquad z = \frac{(kk')}{\sqrt{k^2}\sqrt{k'^2}} , \tag{2.93}$$

where $k = (k_1 - k_2)/2$ and $k' = (k'_1 - k'_2)/2$. Then

$$G(s) = \int\limits_{4m^2}^{\infty} \frac{ds'}{\pi} \int d\Phi(k')\, V\left(s, s', (kk')\right) \frac{G(s')}{s' - M^2 - i0} . \tag{2.94}$$

In the c.m. system $(kk') = -(\mathbf{k}\mathbf{k}')$ and $\sqrt{k^2} = \sqrt{-\mathbf{k}^2} i|\mathbf{k}|$ and $\sqrt{k'^2} = \sqrt{-\mathbf{k}'^2} = i|\mathbf{k}'|$ so $z = (\mathbf{k}\mathbf{k}')/(|\mathbf{k}||\mathbf{k}'|)$. The phase space and the spectral integrations can be written as follows:

$$\int\limits_{4m^2}^{\infty} \frac{ds'}{\pi} \int d\Phi_2(P'; k'_1, k'_2) = \int \frac{d^3\mathbf{k}'}{(2\pi)^3 k'_0} , \tag{2.95}$$

where $k'_0 = \sqrt{m^2 + \mathbf{k}'^2}$. In the c.m. system Eq. (2.92) reads

$$G(s) = \int \frac{d^3\mathbf{k}'}{(2\pi)^3 k'_0} V\left(s, s', -(\mathbf{k}\mathbf{k}')\right) \frac{G(s')}{s' - M^2 - i0} . \tag{2.96}$$

2.3.2.2 *Spectral integral equation for the $(L = 0)$-wave function*

Consider now the wave function of a composite system,

$$\psi(s) = \frac{G(s)}{s - M^2} . \tag{2.97}$$

To this aim, the identity transformation upon the equation (2.92) should be carried out as follows:

$$(s - M^2)\frac{G(s)}{s - M^2} = \int\limits_{4m^2}^{\infty} \frac{ds'}{\pi} \int d\Phi(k') V\left(s, s', (kk')\right) \frac{G(s')}{s' - M^2 - i0} . \tag{2.98}$$

Using the wave functions, the equation (2.98) can be written as:

$$(s - M^2)\psi(s) = \int\limits_{4m^2}^{\infty} \frac{ds'}{\pi} \int d\Phi(k') V\left(s, s', (kk')\right) \psi(s') . \tag{2.99}$$

Finally, using $\mathbf{k}'^2$ and $\mathbf{k}^2$ instead of s' and s,

$$\psi(s) \to \psi(\mathbf{k}^2),$$

we have:

$$(4\mathbf{k}^2 + 4m^2 - M^2)\psi(\mathbf{k}^2) = \int \frac{d^3\mathbf{k}'_1}{(2\pi)^3 k'_0} V\left(s, s', -(\mathbf{k}\mathbf{k}')\right) \psi(\mathbf{k}'^2). \quad (2.100)$$

This is a basic equation for the set of states with $L = 0$. The set is formed by the levels with different radial excitations $n = 1, 2, 3, ...$, and the relevant wave functions are as follows:

$$\psi_1(\mathbf{k}^2),\ \psi_2(\mathbf{k}^2),\ \psi_3(\mathbf{k}^2), ...$$

The wave functions are normalized and orthogonal to each other. The normalization/orthogonality condition reads:

$$\int \frac{d^3\mathbf{k}}{(2\pi)^3 k_0} \psi_n(\mathbf{k}^2)\psi_{n'}(\mathbf{k}^2) = \delta_{nn'} \ . \quad (2.101)$$

Here $\delta_{nn'}$ is the Kronecker symbol. The equation (2.101) is due to the consideration of the charge form factors of composite systems with the gauge-invariance requirement imposed. This normalization–orthogonality condition looks as in quantum mechanics.

Therefore, the spectral integral equation for the S-wave mesons can be written as

$$4(\mathbf{k}^2 + m^2)\psi_n(\mathbf{k}^2) = \int_0^\infty \frac{d\mathbf{k}'^2}{\pi} V_0(\mathbf{k}^2, \mathbf{k}'^2)\phi(\mathbf{k}'^2)\psi_n(\mathbf{k}'^2) M^2 \psi_n(k^2), \quad (2.102)$$

where

$$\phi(\mathbf{k}'^2) = \frac{1}{4\pi}\frac{|\mathbf{k}'|}{k'_0} \ . \quad (2.103)$$

Note that the new phase space, $\phi(\mathbf{k}'^2)$, differs from the standard $\rho(s')$ by a numerical coefficient.

The $\psi_n(\mathbf{k}^2)$ presents a full set of wave functions which are orthogonal and normalized:

$$\int_0^\infty \frac{d\mathbf{k}^2}{\pi} \psi_a(\mathbf{k}^2)\phi(k)\psi_b(\mathbf{k}^2) = \delta_{ab} \ . \quad (2.104)$$

The function $V_0(\mathbf{k}^2, \mathbf{k}'^2)$ is the projection of potential $V(s, s', (kk'))$ on the S-wave:

$$V_0(k, k') = \int \frac{d\Omega_{\mathbf{k}}}{4\pi} \int \frac{d\Omega_{\mathbf{k}'}}{4\pi} V\left(s, s', -(\mathbf{k}\mathbf{k}')\right) \ . \quad (2.105)$$

Let us expand $V_0(\mathbf{k}^2,\mathbf{k}'^2)$ with respect to the full set of wave functions:

$$V_0(\mathbf{k}^2,\mathbf{k}'^2)=\sum_{a,b}\psi_a(\mathbf{k}^2)v^{(0)}_{ab}\psi_b(\mathbf{k}'^2)\,, \tag{2.106}$$

where the numerical coefficients $v^{(0)}_{ab}$ are defined by the inverse transformation as follows:

$$v^{(0)}_{ab}=\int\limits_0^\infty\frac{d\mathbf{k}^2}{\pi}\frac{d\mathbf{k}'^2}{\pi}\psi_a(\mathbf{k}^2)\phi(\mathbf{k}^2)V_0(\mathbf{k}^2,\mathbf{k}'^2)\phi(\mathbf{k}'^2)\psi_b(\mathbf{k}'^2)\,. \tag{2.107}$$

Taking into account the series (2.106), the equation (2.102) is re-written as follows:

$$4(\mathbf{k}^2+m^2)\psi_n(\mathbf{k}^2)-\sum_a\psi_a(\mathbf{k}^2)v^{(0)}_{an}=M^2\psi_n(k^2)\,. \tag{2.108}$$

Such a transformation should be carried out upon the kinetic-energy term, it is also expanded in a series with respect to a full set of wave functions:

$$4(\mathbf{k}^2+m^2)\psi_n(\mathbf{k}^2)=\sum_a K_{na}\psi_a(\mathbf{k}^2)\,, \tag{2.109}$$

where

$$K_{na}=\int\limits_0^\infty\frac{d\mathbf{k}^2}{\pi}\psi_a(\mathbf{k}^2)\phi(\mathbf{k}^2)\,4(\mathbf{k}^2+m^2)\psi_n(\mathbf{k}^2)\,. \tag{2.110}$$

Finally, the spectral integral equation takes the form:

$$\sum_a K_{na}\psi_a(\mathbf{k}^2)-\sum_a v^{(0)}_{na}\psi_a(\mathbf{k}^2)=M_n^2\psi_n(\mathbf{k}^2)\,. \tag{2.111}$$

We take into account that $v^{(0)}_{na}=v^{(0)}_{an}$.

The equation (2.111) is a standard homogeneous equation:

$$\sum_a s_{na}\psi_a(\mathbf{k}^2)=M_n^2\psi_n(\mathbf{k}^2)\,, \tag{2.112}$$

with $s_{na}=K_{na}-v^{(0)T}_{na}$. The values M^2 are defined as zeros of the determinant

$$det|\hat{s}-M^2I|=0\,, \tag{2.113}$$

where I is the unit matrix.

2.3.2.3 *The spectral integral equation for the states with angular momentum L*

For the wave with arbitrary angular momentum L, the wave function reads as follows:

$$\psi^{(L)}_{(n)\mu_1,\dots,\mu_L}(s) = X^{(L)}_{\mu_1,\dots,\mu_L}(k)\psi^{(L)}_n(s)\ . \tag{2.114}$$

The momentum operator $X^{(L)}_{\mu_1,\dots,\mu_L}(k)$ features will be shown in Appendix A.

The spectral integral equation for the (L,n)-state, presented in the form similar to (2.102), reads:

$$4(\mathbf{k}^2+m^2)X^{(L)}_{\mu_1,\dots,\mu_L}(k)\psi^{(L)}_n(\mathbf{k}^2) = X^{(L)}_{\mu_1,\dots,\mu_L}(k)$$
$$\times\int\limits_0^\infty \frac{d\mathbf{k}'^2}{\pi}V_L(s,s')\times X^2_L(k'^2)\phi(\mathbf{k}'^2)\psi^{(L)}_n(\mathbf{k}'^2)M^2X^{(L)}_{\mu_1,\dots,\mu_L}(k)\psi^{(L)}_n(\mathbf{k}^2)\ . \tag{2.115}$$

The potential is expanded into a series with respect to the operators $X^{(L)}_{\mu_1,\dots,\mu_L}(k)X^{(L)}_{\mu_1,\dots,\mu_L}(k')$, that is,

$$V(s,s',(kk')) = \sum_{L,\mu_1\dots\mu_L} X^{(L)}_{\mu_1,\dots,\mu_L}(k)V_L(s,s')X^{(L)}_{\mu_1,\dots,\mu_L}(k')\ ,$$
$$X^2_L(k^2)V_L(s,s')X^2_L(k'^2) = \int\frac{d\Omega_{\mathbf{k}}}{4\pi}\frac{d\Omega_{\mathbf{k}'}}{4\pi}X^{(L)}_{\nu_1,\dots,\nu_L}(k)$$
$$\times V(s,s',(kk'))\,X^{(L)}_{\nu_1,\dots,\nu_L}(k')\ . \tag{2.116}$$

Hence, the formula (2.115) reads as follows:

$$4(\mathbf{k}^2+m^2)\psi^{(L)}_n(\mathbf{k}^2) - \int\limits_0^\infty\frac{d\mathbf{k}'^2}{\pi}V_L(s,s')\alpha(L)(-\mathbf{k}'^2)^L\phi(\mathbf{k}'^2)\psi^{(L)}_n(\mathbf{k}'^2) =$$
$$= M^2_n\psi^{(L)}_n(\mathbf{k}^2). \tag{2.117}$$

As compared to (2.102), this equation contains the additional factor $X^2_L(k'^2)$; still, the same factor is in the normalization condition, so it would be reasonable to insert it into the phase space. Finally, we have:

$$4(\mathbf{k}^2+m^2)\psi^{(L)}_n(\mathbf{k}^2) - \int\limits_0^\infty\frac{d\mathbf{k}'^2}{\pi}\widetilde{V}_L(s,s')\phi_L(\mathbf{k}'^2)\psi^{(L)}_n(\mathbf{k}'^2) = M^2_n\psi^{(L)}_n(\mathbf{k}^2)\ , \tag{2.118}$$

where

$$\phi_L(\mathbf{k}'^2) = \alpha(L)(\mathbf{k}'^2)^L\phi(\mathbf{k}'^2), \qquad \widetilde{V}_L(s,s') = (-1)^L V_L(s,s'). \quad (2.119)$$

The normalization condition for a set of wave functions with orbital momentum L is:

$$\int_0^\infty \frac{d\mathbf{k}^2}{\pi}\psi_a^{(L)}(\mathbf{k}^2)\phi_L(\mathbf{k}^2)\psi_b^{(L)}(\mathbf{k}^2) = \delta_{ab}\,. \quad (2.120)$$

One can see that it is similar to the case of $L = 0$, the only difference consists in the redefinition of the phase space $\phi \to \phi_L$. The spectral integral equation is:

$$\sum_a s_{na}^{(L)}\psi_a^{(L)}(\mathbf{k}^2) = M_{n,L}^2\psi_n^{(L)}(\mathbf{k}^2)\,, \quad (2.121)$$

with

$$s_{na}^{(L)} = K_{na}^{(L)} - v_{na}^{(L)T}\,,$$

$$v_{ab}^{(L)} = \int_0^\infty \frac{d\mathbf{k}^2}{\pi}\frac{d\mathbf{k}'^2}{\pi}\psi_a^{(L)}(\mathbf{k}^2)\phi_L(\mathbf{k}^2)\widetilde{V}_L(s,s')\phi_L(\mathbf{k}'^2)\psi_b^{(L)}(\mathbf{k}'^2)\,,$$

$$K_{na}^{(L)} = \int_0^\infty \frac{d\mathbf{k}^2}{\pi}\psi_a^{(L)}(\mathbf{k}^2)\phi_L(\mathbf{k})4(\mathbf{k}^2+m^2)\psi_n^{(L)}(\mathbf{k}^2)\,. \quad (2.122)$$

Using radial excitation levels one can reconstruct the potential in the L-wave and then re-write, with the help of (2.116), the t-dependent potential.

2.4 Appendix A. Angular momentum operators

The angular-dependent part of the two-body wave function is described by operators constructed for the relative momenta of particles and the metric tensor [5, 23]. Such operators (we denote them as $X^{(L)}_{\mu_1\ldots\mu_L}$, where L is the angular momentum) are called angular momentum operators; they correspond to irreducible representations of the Lorentz group. They satisfy the following properties:

(i) Symmetry with respect to the permutation of any two indices:

$$X^{(L)}_{\mu_1\ldots\mu_i\ldots\mu_j\ldots\mu_L} = X^{(L)}_{\mu_1\ldots\mu_j\ldots\mu_i\ldots\mu_L}. \quad (2.123)$$

(ii) Orthogonality to the total momentum of the system, $P = k_1 + k_2$:

$$P_{\mu_i}X^{(L)}_{\mu_1\ldots\mu_i\ldots\mu_L} = 0. \quad (2.124)$$

(iii) Tracelessness with respect to the summation over any two indices:

$$g_{\mu_i\mu_j} X^{(L)}_{\mu_1\ldots\mu_i\ldots\mu_j\ldots\mu_L} = 0. \tag{2.125}$$

Let us consider a one-loop diagram describing the decay of a composite system into two spinless particles, which propagate and then form again a composite system. The decay and formation processes are described by angular momentum operators. Owing to the quantum number conservation, this amplitude must vanish for initial and final states with different spins. The S-wave operator is a scalar and can be taken as a unit operator. The P-wave operator is a vector. In the dispersion relation approach it is sufficient that the imaginary part of the loop diagram, with S- and P-wave operators as vertices, equals 0. In the case of spinless particles, this requirement entails

$$\int \frac{d\Omega}{4\pi} X^{(1)}_\mu = 0 \,, \tag{2.126}$$

where the integral is taken over the solid angle of the relative momentum. In general, the result of such an integration is proportional to the total momentum P_μ (the only external vector):

$$\int \frac{d\Omega}{4\pi} X^{(1)}_\mu = \lambda P_\mu \,. \tag{2.127}$$

Convoluting this expression with P_μ and demanding $\lambda = 0$, we obtain the orthogonality condition (2.124). The orthogonality between the D- and S-waves is provided by the tracelessness condition (2.125); equations (2.124), (2.125) provide the orthogonality for all operators with different angular momenta.

The orthogonality condition (2.124) is automatically fulfilled if the operators are constructed from the relative momenta $k^\perp_\mu$ and tensor $g^\perp_{\mu\nu}$. Both of them are orthogonal to the total momentum of the system:

$$k^\perp_\mu = \frac{1}{2} g^\perp_{\mu\nu}(k_1 - k_2)_\nu \,, \qquad g^\perp_{\mu\nu} = g_{\mu\nu} - \frac{P_\mu P_\nu}{s} \,. \tag{2.128}$$

In the c.m. system, where $P = (P_0, \vec{P}) = (\sqrt{s}, 0)$, the vector $k^\perp$ is space-like: $k^\perp = (0, \vec{k})$.

The operator for $L = 0$ is a scalar (for example, a unit operator), and the operator for $L = 1$ is a vector, which can be constructed from $k^\perp_\mu$ only.

The orbital angular momentum operators for $L = 0$ to 3 are:

$$
\begin{aligned}
&X^{(0)}(k^\perp) = 1, \\
&X^{(1)}_\mu(k^\perp) = k^\perp_\mu, \\
&X^{(2)}_{\mu_1\mu_2}(k^\perp) = \frac{3}{2}\left(k^\perp_{\mu_1}k^\perp_{\mu_2} - \frac{1}{3}k^2_\perp g^\perp_{\mu_1\mu_2}\right), \\
&X^{(3)}_{\mu_1\mu_2\mu_3}(k^\perp) = \frac{5}{2}\left[k^\perp_{\mu_1}k^\perp_{\mu_2}k^\perp_{\mu_3} - \frac{k^2_\perp}{5}\left(g^\perp_{\mu_1\mu_2}k^\perp_{\mu_3} + g^\perp_{\mu_1\mu_3}k^\perp_{\mu_2} + g^\perp_{\mu_2\mu_3}k^\perp_{\mu_1}\right)\right].
\end{aligned}
\tag{2.129}
$$

The operators $X^{(L)}_{\mu_1\ldots\mu_L}$ for $L \geq 1$ can be written in the form of a recurrency relation:

$$
\begin{aligned}
X^{(L)}_{\mu_1\ldots\mu_L}(k^\perp) &= k^\perp_\alpha Z^\alpha_{\mu_1\ldots\mu_L}(k^\perp), \\
Z^\alpha_{\mu_1\ldots\mu_L}(k^\perp) &= \frac{2L-1}{L^2}\Big(\sum_{i=1}^{L} X^{(L-1)}_{\mu_1\ldots\mu_{i-1}\mu_{i+1}\ldots\mu_L}(k^\perp)g^\perp_{\mu_i\alpha} \\
&\quad - \frac{2}{2L-1}\sum_{\substack{i,j=1\\ i<j}}^{L} g^\perp_{\mu_i\mu_j}X^{(L-1)}_{\mu_1\ldots\mu_{i-1}\mu_{i+1}\ldots\mu_{j-1}\mu_{j+1}\ldots\mu_L\alpha}(k^\perp)\Big).
\end{aligned}
\tag{2.130}
$$

The convolution equality reads

$$
X^{(L)}_{\mu_1\ldots\mu_L}(k^\perp)k^\perp_{\mu_L} = k^2_\perp X^{(L-1)}_{\mu_1\ldots\mu_{L-1}}(k^\perp). \tag{2.131}
$$

On the basis of Eq.(2.131) and taking into account the tracelessness property of $X^{(L)}_{\mu_1\ldots\mu_L}$, one can write down the orthogonality–normalization condition for orbital angular operators

$$
\begin{aligned}
\int\frac{d\Omega}{4\pi}X^{(L)}_{\mu_1\ldots\mu_L}(k^\perp)X^{(L')}_{\mu_1\ldots\mu'_L}(k^\perp) &= \delta_{LL'}\alpha_L k^{2L}_\perp, \\
\alpha_L &= \prod_{l=1}^{L}\frac{2l-1}{l}.
\end{aligned}
\tag{2.132}
$$

Iterating equation (2.130), one obtains the following expression for the operator $X^{(L)}_{\mu_1\ldots\mu_L}$:

$$
\begin{aligned}
X^{(L)}_{\mu_1\ldots\mu_L}(k^\perp) = \alpha_L\Big[&k^\perp_{\mu_1}k^\perp_{\mu_2}k^\perp_{\mu_3}k^\perp_{\mu_4}\ldots k^\perp_{\mu_L} \\
&- \frac{k^2_\perp}{2L-1}\left(g^\perp_{\mu_1\mu_2}k^\perp_{\mu_3}k^\perp_{\mu_4}\ldots k^\perp_{\mu_L} + g^\perp_{\mu_1\mu_3}k^\perp_{\mu_2}k^\perp_{\mu_4}\ldots k^\perp_{\mu_L} + \ldots\right) \\
&+ \frac{k^4_\perp}{(2L-1)(2L-3)}\Big(g^\perp_{\mu_1\mu_2}g^\perp_{\mu_3\mu_4}k^\perp_{\mu_5}k^\perp_{\mu_6}\ldots k^\perp_{\mu_L} \\
&+ g^\perp_{\mu_1\mu_2}g^\perp_{\mu_3\mu_5}k^\perp_{\mu_4}k^\perp_{\mu_6}\ldots k^\perp_{\mu_L} + \ldots\Big) + \ldots\Big].
\end{aligned}
\tag{2.133}
$$

2.4.1 *Projection operators and denominators of the boson propagators*

The projection operator $O^{\mu_1\ldots\mu_L}_{\nu_1\ldots\nu_L}$ is constructed of the metric tensors $g^{\perp}_{\mu\nu}$. It has the properties as follows:

$$
\begin{aligned}
X^{(L)}_{\mu_1\ldots\mu_L} O^{\mu_1\ldots\mu_L}_{\nu_1\ldots\nu_L} &= X^{(L)}_{\nu_1\ldots\nu_L} \,, \\
O^{\mu_1\ldots\mu_L}_{\alpha_1\ldots\alpha_L} O^{\alpha_1\ldots\alpha_L}_{\nu_1\ldots\nu_L} &= O^{\mu_1\ldots\mu_L}_{\nu_1\ldots\nu_L} \,.
\end{aligned}
\tag{2.134}
$$

Taking into account the definition of projection operators (2.134) and the properties of the X-operators (2.133), we obtain

$$
k_{\mu_1}\ldots k_{\mu_L} O^{\mu_1\ldots\mu_L}_{\nu_1\ldots\nu_L} = \frac{1}{\alpha_L} X^{(L)}_{\nu_1\ldots\nu_L}(k^{\perp}). \tag{2.135}
$$

This equation is the basic property of the projection operator: it projects any operator with L indices onto the partial wave operator with angular momentum L.

For the lowest states

$$
\begin{aligned}
O &= 1 \,, \qquad O^{\mu}_{\nu} = g^{\perp}_{\mu\nu} \,, \\
O^{\mu_1\mu_2}_{\nu_1\nu_2} &= \frac{1}{2}\left(g^{\perp}_{\mu_1\nu_1} g^{\perp}_{\mu_2\nu_2} + g^{\perp}_{\mu_1\nu_2} g^{\perp}_{\mu_2\nu_1} - \frac{2}{3} g^{\perp}_{\mu_1\mu_2} g^{\perp}_{\nu_1\nu_2} \right).
\end{aligned}
\tag{2.136}
$$

For higher states the operator can be calculated using the recurrent expression:

$$
\begin{aligned}
O^{\mu_1\ldots\mu_L}_{\nu_1\ldots\nu_L} = \frac{1}{L^2}\Bigg(& \sum_{i,j=1}^{L} g^{\perp}_{\mu_i\nu_j} O^{\mu_1\ldots\mu_{i-1}\mu_{i+1}\ldots\mu_L}_{\nu_1\ldots\nu_{j-1}\nu_{j+1}\ldots\nu_L} \\
& - \frac{4}{(2L-1)(2L-3)} \sum_{\substack{i<j \\ k<m}}^{L} g^{\perp}_{\mu_i\mu_j} g^{\perp}_{\nu_k\nu_m} O^{\mu_1\ldots\mu_{i-1}\mu_{i+1}\ldots\mu_{j-1}\mu_{j+1}\ldots\mu_L}_{\nu_1\ldots\nu_{k-1}\nu_{k+1}\ldots\nu_{m-1}\nu_{m+1}\ldots\nu_L} \Bigg).
\end{aligned}
\tag{2.137}
$$

The product of two X-operators integrated over a solid angle (that is equivalent to the integration over internal momenta) depends only on the external momenta and the metric tensor. Therefore, it must be proportional to the projection operator. After straightforward calculations we obtain

$$
\int \frac{d\Omega}{4\pi} X^{(L)}_{\mu_1\ldots\mu_L}(k^{\perp}) X^{(L)}_{\nu_1\ldots\nu_L}(k^{\perp}) = \frac{\alpha_L\, k_{\perp}^{2L}}{2L+1} O^{\mu_1\ldots\mu_L}_{\nu_1\ldots\nu_L} \,. \tag{2.138}
$$

Let us introduce the positive valued $|\vec{k}|^2$:

$$
|\vec{k}|^2 = -k_{\perp}^2 = \frac{[s-(m_1+m_2)^2][s-(m_1-m_2)^2]}{4s} \,. \tag{2.139}
$$

In the c.m.s. of the reaction, $\vec{k}$ is the momentum of a particle. In other systems we use this definition only in the sense of $|\vec{k}| \equiv \sqrt{-k_\perp^2}$; clearly, $|\vec{k}|^2$ is a relativistically invariant positive value. If so, equation (2.138) can be written as

$$\int \frac{d\Omega}{4\pi} X^{(L)}_{\mu_1\dots\mu_L}(k^\perp) X^{(L)}_{\nu_1\dots\nu_L}(k^\perp) = \frac{\alpha_L\, |\vec{k}|^{2L}}{2L+1} (-1)^L O^{\mu_1\dots\mu_L}_{\nu_1\dots\nu_L}. \tag{2.140}$$

The tensor part of the numerator of the boson propagator is defined by the projection operator. Let us write it as follows:

$$F^{\mu_1\dots\mu_L}_{\nu_1\dots\nu_L} = (-1)^L\, O^{\mu_1\dots\mu_L}_{\nu_1\dots\nu_L}\,, \tag{2.141}$$

with the definition of the propagator

$$\frac{F^{\mu_1\dots\mu_L}_{\nu_1\dots\nu_L}}{M^2 - s}. \tag{2.142}$$

This definition guarantees that the width of a resonance (calculated using the decay vertices) is positive.

2.4.2 *Useful relations for $Z^\alpha_{\mu_1\dots\mu_n}$ and $X^{(n-1)}_{\nu_2\dots\nu_n}$*

Here we list a few useful expressions:

$$\begin{aligned}
Z^\alpha_{\mu_1\dots\mu_n} &= X^{(n-1)}_{\nu_2\dots\nu_n} O^{\alpha\nu_2\dots\nu_n}_{\mu_1\dots\mu_n} \frac{2n-1}{n}\,, \\
Z^\alpha_{\mu_1\dots\mu_n}(q) &= (-1)^n O^{\mu_1\dots\mu_n}_{\nu_1\dots\nu_n} Z^\beta_{\nu_1\dots\nu_n}(k) \frac{\alpha_n}{n^2}(-1)^n \\
&\times \left(\sqrt{k_\perp^2}\sqrt{q_\perp^2}\right)^{n-1} \Bigg[g^\perp_{\alpha\beta} P'_n - \left(\frac{q^\perp_\alpha q^\perp_\beta}{q^2_\perp} + \frac{k^\perp_\alpha k^\perp_\beta}{k^2_\perp} \right) P''_{n-1} \\
&+ \frac{q^\perp_\alpha k^\perp_\beta}{\sqrt{k^2_\perp}\sqrt{q^2_\perp}} (P''_{n-2} - 2P'_{n-1}) + \frac{k^\perp_\alpha q^\perp_\beta}{\sqrt{k^2_\perp}\sqrt{q^2_\perp}} P''_n \Bigg]\,, \\
X_{\alpha\mu_1\dots\mu_n}(q) &= (-1)^n O^{\mu_1\dots\mu_n}_{\nu_1\dots\nu_n} X_{\beta\nu_1\dots\nu_n}(k) \frac{\alpha_n}{(n+1)^2}(-1)^n \\
&\times \left(\sqrt{k_\perp^2}\sqrt{q_\perp^2}\right)^{n+1} \Bigg[g^\perp_{\alpha\beta} P'_{n+1} - \left(\frac{q^\perp_\alpha q^\perp_\beta}{q^2_\perp} + \frac{k^\perp_\alpha k^\perp_\beta}{k^2_\perp} \right) P''_{n+1} \\
&+ \frac{q^\perp_\alpha k^\perp_\beta}{\sqrt{k^2_\perp}\sqrt{q^2_\perp}} (P''_{n+2} - 2P'_{n+1}) + \frac{k^\perp_\alpha q^\perp_\beta}{\sqrt{k^2_\perp}\sqrt{q^2_\perp}} P''_n \Bigg]\,,
\end{aligned}$$

$$Z^{\alpha}_{\mu_1\ldots\mu_n}(q^\perp) = (-1)^n O^{\mu_1\ldots\mu_n}_{\nu_1\ldots\nu_n} X_{\beta\nu_1\ldots\nu_n}(k) = \frac{\alpha_{n-1}}{n(n+1)}(-1)^n$$
$$\times(-k_\perp^2)\left(\sqrt{k_\perp^2}\sqrt{q_\perp^2}\right)^{n+1}\left[g^\perp_{\alpha\beta}P'_n - \frac{q^\perp_\alpha q^\perp_\beta}{q^2_\perp}P''_{n-1}\right.$$
$$\left. - \frac{k^\perp_\alpha k^\perp_\beta}{k^2_\perp}P''_{n+1} + \frac{q^\perp_\alpha k^\perp_\beta}{\sqrt{k^2_\perp}\sqrt{q^2_\perp}}P''_n + \frac{k^\perp_\alpha q^\perp_\beta}{\sqrt{k^2_\perp}\sqrt{q^2_\perp}}P''_n\right]. \quad (2.143)$$

Consider now a few expressions used in the one-loop diagram calculations. In our case, the operators are constructed of $X^{(n+1)}_{\alpha\mu_1\ldots\mu_n}$ and $Z^{\beta}_{\mu_1\ldots\mu_n}$, where α and β indices are to be convoluted with tensors. Let us start with the loop diagram with the Z-operator:

$$\int\frac{d\Omega}{4\pi}Z^{\alpha}_{\mu_1\ldots\mu_n}(k^\perp)T_{\alpha\beta}Z^{\beta}_{\nu_1\ldots\nu_n}(k^\perp)=\Lambda O^{\mu_1\ldots\mu_n}_{\nu_1\ldots\nu_n}(-1)^n. \quad (2.144)$$

For different tensors $T_{\alpha\beta}$, one has the following Λ:

$$T_{\alpha\beta} = g_{\alpha\beta}, \qquad \Lambda = -\frac{\alpha_n}{n}|\vec{k}|^{2n-2}, \quad (2.145)$$

$$T_{\alpha\beta} = k^\perp_\alpha k^\perp_\beta, \qquad \Lambda = \frac{\alpha_n}{2n+1}|\vec{k}|^{2n}. \quad (2.146)$$

The equation (2.145) can be easily obtained using (2.143) and (2.138), while equation(2.146) can be written using (2.130) and (2.138). For the X operators, one has

$$\int\frac{d\Omega}{4\pi}X^{(n+1)}_{\alpha\mu_1\ldots\mu_n}(k^\perp)T_{\alpha\beta}X^{(n+1)}_{\beta\nu_1\ldots\nu_n}(k^\perp)=\Lambda O^{\mu_1\ldots\mu_n}_{\nu_1\ldots\nu_n}(-1)^n, \quad (2.147)$$

where

$$T_{\alpha\beta} = g_{\alpha\beta}, \qquad \Lambda = -\frac{\alpha_n}{n+1}|\vec{k}|^{2n+2},$$
$$T_{\alpha\beta} = k^\perp_\alpha k^\perp_\beta, \qquad \Lambda = \frac{\alpha_n}{2n+1}|\vec{k}|^{2n+4}. \quad (2.148)$$

To derive (2.148), the properties of the projection operator

$$O^{\alpha\mu_1\ldots\mu_n}_{\alpha\nu_1\ldots\nu_n} = \frac{2n+3}{2n+1}O^{\mu_1\ldots\mu_n}_{\nu_1\ldots\nu_n} \quad (2.149)$$

and Eq. (2.131) are used. The interference term between X and Z operators is given by

$$\int\frac{d\Omega}{4\pi}X^{(n+1)}_{\alpha\mu_1\ldots\mu_n}(k^\perp)T_{\alpha\beta}Z^{\beta}_{\nu_1\ldots\nu_n}(k^\perp)=\Lambda O^{\mu_1\ldots\mu_n}_{\nu_1\ldots\nu_n}(-1)^n, \quad (2.150)$$

with

$$T_{\alpha\beta} = g_{\alpha\beta}, \qquad \Lambda = 0\,,$$
$$T_{\alpha\beta} = k^\perp_\alpha k^\perp_\beta, \qquad \Lambda = -\frac{\alpha_n}{2n+1}|\vec{k}|^{2n+2}\,.+ \quad (2.151)$$

Equation (2.151) is derived using (2.143) and the orthogonality (2.132) of the X operators.

2.5 Appendix B: The $\pi\pi$ scattering amplitude near the two-pion thresholds, $\pi^+\pi^-$ and $\pi^0\pi^0$

Here we consider the $\pi\pi$ scattering amplitude near the two-pion thresholds taking into account the mass difference of charged and neutral pion systems, $\pi^+\pi^-$ and $\pi^0\pi^0$. The mass difference of charged and neutral pions results in violation of the isotopic relations for pion scattering amplitudes. Here we consider a version suggested in [24] when the main contribution into the isotopic violation comes from the phase space factors.

The following $\pi\pi$-amplitudes describe scattering reactions near the thresholds:

$$\begin{aligned}
&\pi^+\pi^- \to \pi^+\pi^- \ : \\
&A^{++}_{--} = \frac{a^{++}_{--} + ik^0_0[(a^{+0}_{-0})^2 - a^{++}_{--}a^{00}_{00}]}{1 - ik^+_- a^{++}_{--} - ik^0_0 a^{00}_{00} + k^0_0 k^+_-[-a^{00}_{00}a^{++}_{--} + (a^{0+}_{-0})^2]}, \\
&\pi^0\pi^0 \to \pi^+\pi^- \ : \\
&A^{0+}_{0-} = \frac{a^{0+}_{0-}}{1 - ik^+_- a^{++}_{--} - ik^0_0 a^{00}_{00} + k^0_0 k^+_-[-a^{00}_{00}a^{++}_{--} + (a^{0+}_{0-})^2]}, \\
&\pi^0\pi^0 \to \pi^0\pi^0 \ : \\
&A^{00}_{00} = \frac{a^{00}_{00} + ik^+_-[(a^{+0}_{-0})^2 - a^{++}_{--}a^{00}_{00}]}{1 - ik^+_- a^{++}_{--} - ik^0_0 a^{00}_{00} + k^0_0 k^+_-[-a^{00}_{00}a^{++}_{--} + (a^{0+}_{-0})^2]},
\end{aligned} \tag{2.152}$$

with

$$k^+_- = \sqrt{\frac{s}{4} - \mu^2_{\pi^+}} \equiv k, \quad k^0_0 = \frac{1}{2}\sqrt{\frac{s}{4} - \mu^2_{\pi^0}} = \frac{1}{2}\sqrt{k^2 + \Delta^2}. \tag{2.153}$$

Here $\Delta^2 = \mu^2_{\pi^+} - \mu^2_{\pi^0} \simeq 0.07\mu^2_{\pi^+}$. The factor 1/2 in k^0_0 arises due to the identity of pions in the $\pi^0\pi^0$ state.

We impose on the scattering length values the standard isotopic relations:

$$\begin{aligned}
a^{++}_{--} &= \frac{2}{3}a_0(s) + \frac{1}{3}a_2(s), \\
a^{+0}_{-0} &= -\frac{2}{3}a_0(s) + \frac{2}{3}a_2(s), \\
a^{00}_{00} &= 2a^{++}_{--} + a^{+0}_{-0} = \frac{2}{3}a_0(s) + \frac{4}{3}a_2(s).
\end{aligned} \tag{2.154}$$

Then at large k^2, when $k^2 >> \Delta^2$, the unitary amplitudes of Eq. (2.152) obey the isotopic relations:

$$A^{++}_{--} = \frac{\frac{2}{3}a_0(s)}{1-ika_0(s)} + \frac{\frac{1}{3}a_2(s)}{1-ika_2(s)},$$

$$A^{+0}_{-0} = \frac{-\frac{2}{3}a_0(s)}{1-ika_0(s)} + \frac{\frac{2}{3}a_2(s)}{1-ika_2(s)},$$

$$A^{00}_{00} = \frac{\frac{2}{3}a_0(s)}{1-ika_0(s)} + \frac{\frac{4}{3}a_2(s)}{1-ika_2(s)}. \tag{2.155}$$

The $(I=0)$-amplitude and the corresponding S-matrix read:

$$\frac{a_0(s)}{1-ika_0(s)} = 2A^{++}_{--} - \frac{1}{2}A^{00}_{00} = A^{++}_{--} - \frac{1}{2}A^{+0}_{-0},$$

$$\exp[2i\delta^0_0(s)] = \frac{A^{++}_{--} - \frac{1}{2}A^{+0}_{-0}}{(A^{++}_{--} - \frac{1}{2}A^{+0}_{-0})^*} = \frac{A^{++}_{--} - \frac{1}{2}A^{+0}_{-0}}{(2A^{++}_{--} - \frac{1}{2}A^{00}_{00})^*} = \frac{2A^{++}_{--} - \frac{1}{2}A^{00}_{00}}{(2A^{++}_{--} - \frac{1}{2}A^{00}_{00})^*}. \tag{2.156}$$

In the $K^+ \to e^+\nu(\pi^+\pi^-)$ decay the S-wave pions are $I=0$ states, and the amplitude can be written as follows:

$$A\left(K^+ \to e^+\nu(\pi^+\pi^-)_{I=0,S-wave}\right) = \lambda[1 - ik^0_0 A^{0+}_{0-} + ik^+_- A^{++}_{--}] =$$

$$= \lambda\left[\frac{1 - ik^0_0 a^{00}_{00} - ik^0_0 a^{0+}_{0-}}{1 - ik^+_- a^{++}_{--} - ik^0_0 a^{00}_{00} + k^0_0 k^+_- [-a^{00}_{00} a^{++}_{--} + (a^{0+}_{0-})^2]}\right]. \tag{2.157}$$

Here the first term, λ, is a direct production amplitude while the second and third terms take into account pion rescatterings.

At large pion relative momentum, when $k^2 >> \Delta^2$, we have:

$$A\left(K^+ \to e^+\nu(\pi^+\pi^-)_{I=0,S-wave}\right)_{k^2>>\Delta^2} = \lambda\frac{1}{1-ika_0(s)}. \tag{2.158}$$

Recall that the factor $(1-ika_0(s))^{-1}$ is due to rescatterings of pions in the $I=0$ state.

2.6 Appendix C: Four-pole fit of the $\pi\pi(00^{++})$ wave in the region $M_{\pi\pi} < 900$ MeV

Here we give an example of a very simple, and formally correct, consideration of the 00^{++} wave in the low-energy region. We write an analytical and unitary amplitude as follows:

$$A^I_{thr}(s) = \frac{g^2}{s^2 - \left(a_I + b_I s + ig^2\sqrt{s-4\mu^2_\pi}\right)}. \tag{2.159}$$

Hence,

$$e^{2i\delta_0^0(s)} = \frac{D^I(s)}{D^{I*}(s)} \frac{[k-(a+ib)][k-(-a+ib)][k-(c+id)][k-(-c+id)]}{[k-(a-ib)][k-(-a-ib)][k-(c-id)][k-(-c-id)]}$$
$$\text{with} \quad a>0, \quad b>0\,. \tag{2.160}$$

Considering (a,b,c,b) as parameters, we fit the data for $\delta_0^0(s)$ in the energy interval $280 \le \sqrt{s} \le 950$ MeV, see Fig. 2.10. We obtain the following parameters and amplitude pole positions, M_I and M_{II}:

$$\begin{aligned}
\text{Fig.2.10a}: \quad & a=3.1\mu_\pi, \quad b=1.0\mu_\pi, \quad c=7.7\mu_\pi, \quad d=9.0\mu_\pi, \\
& M_I=(896.3-i274.3)\text{MeV}, \quad M_{II}=(2163.7-i2511.0)\text{MeV} \\
\text{Fig.2.10b}: \quad & a=2.5\mu_\pi, \quad b=1.3\mu_\pi, \quad c=50.0\mu_\pi, \quad d=2.0\mu_\pi, \\
& M_I=(742.1-i352.2)\text{MeV}, \quad M_{II}=(14002.8-i559.9)\text{MeV}\,.
\end{aligned} \tag{2.161}$$

In all solutions the scalar-isoscalar scattering length is not small: $a_0^0 \sim (0.3-0.4)\mu_\pi^{-1}$.

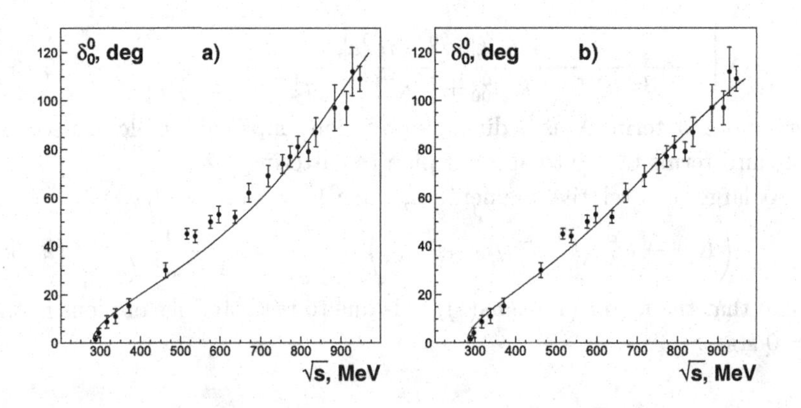

Fig. 2.10 Examples of the fit of low energy $\pi\pi$ scattering data [25, 26].

In the fit of Fig. 2.10 we use the values for δ_0^0 found in [25] in order to perform a more visual comparison of the obtained here results, Eq. (2.161), with those in [27]. Let us recall that we fit in [27] the amplitude 00^{++} in the region $280 \le \sqrt{s} \le 900$ MeV in the framework of the dispersion relation approach sewing the N/D-solution with the K-matrix one at $450 \le \sqrt{s} \le 1950$ MeV. Taking into account the left-hand cut contribution (it was a fitting function), we obtained in [27] the best fit with the σ-meson pole at

$M_\sigma = (430 \pm 150) - i(320 \pm 130)$ MeV. So, the accounting for the left-hand cut and data at $\sqrt{s} > 900$ MeV results in a smaller value of the M_σ.

In the approaches which take into account the left-hand cut as a contribution of some known meson exchanges, the pole positions were obtained at low masses as well. For example, the dispersion relation approach results: $M_\sigma \simeq (470 - i460)$ MeV [28], $M_\sigma \simeq (450 - i375)$ MeV [29], and the meson exchange models give: $M_\sigma \simeq (460 - i450)$ MeV [30], $M_\sigma \simeq (400 - i60)$ MeV [31].

References

[1] S. Mandelstam, Phys. Rev. **112**, 1344 (1958).
[2] S. Mandelstam, Phys. Rev. **115**, 1752 (1959).
[3] G.F. Chew, *The Analytic S-Matrix*, W.A. Benjamin, New York (1966).
[4] E. Salpeter and H.A. Bethe, Phys. Rev. **84**, 1232 (1951).
[5] A.V. Anisovich, V.V. Anisovich, J. Nyiri, V.A. Nikonov, M.A. Matveev and A.V. Sarantsev, *Mesons and Baryons. Systematization and Methods of Analysis*, World Scientific, Singapore (2008).
[6] G.F. Chew and S. Mandestam, Phys. Rev. **119**, 467 (1960).
[7] L. Castillejo, F.J. Dyson, and R.H. Dalitz, Phys. Rev. **101**, 453 (1956).
[8] V.V. Anisovich, M.N. Kobrinsky, D.I. Melikhov and A.V. Sarantsev, Nucl. Phys. A **544**, 747 (1992).
[9] A.V. Anisovich and V.A. Sadovnikova, Yad. Fiz. **55**, 2657 (1992); **57**, 75 (1994); Eur. Phys. J. A **2**, 199 (1998).
[10] V.V. Anisovich, D.I. Melikhov, and V.A. Nikonov, Phys. Rev. D **52**, 5295 (1995); Phys. Rev. D **55**, 2918 (1997).
[11] A.V. Anisovich, V.V. Anisovich and V.A. Nikonov, Eur. Phys. J. A **12**, 103 (2001).
[12] A.V. Anisovich, V.V. Anisovich, M.A. Matveev and V.A. Nikonov, Yad. Fiz. **66**, 946 (2003) [Phys. Atom. Nucl. **66**, 914 (2003)]. exchanges.
[13] A.V. Anisovich, V.V. Anisovich, B.N. Markov, M.A. Matveev, and A. V. Sarantsev, Yad. Fiz. **67**, 794 (2004) [Phys. At. Nucl., **67**, 773 (2004)].
[14] H. Hersbach, Phys. Rev. C **50**, 2562 (1994).
[15] H. Hersbach, Phys. Rev. A **46**, 3657 (1992).
[16] F. Gross and J. Milana, Phys. Rev. D **43**, 2401 (1991).
[17] K.M. Maung, D.E. Kahana, and J.W. Ng, Phys. Rev. A **46**, 3657 (1992).
[18] V.V. Anisovich, L.G. Dakhno, M.A. Matveev, V.A. Nikonov, and A. V. Sarantsev, Yad. Fiz. **70**, 480 (2007) [Phys. Atom. Nucl. **70**, 450 (2007)].
[19] V.V. Anisovich, L.G. Dakhno, M.A. Matveev, V.A. Nikonov, and A. V. Sarantsev, Yad. Fiz. **70**, 68 (2007) [Phys. Atom. Nucl. **70**, 63 (2007)].
[20] V.V. Anisovich, L.G. Dakhno, M.A. Matveev, V.A. Nikonov, and A.V. Sarantsev, Yad. Fiz. **70**, 392 (2007) [Phys. Atom. Nucl. **70**, 364 (2007)].

[21] A.V. Anisovich, Yad. Fiz. **58**, 1467 (1995) [Phys. Atom. Nucl. **58**, 1383 (1995)].
[22] A.V. Anisovich, Yad. Fiz. **66**, 175 (2003) [Phys. Atom. Nucl. **66**, 172 (2003)].
[23] A.V. Anisovich, V.V. Anisovich, V.N. Markov, M.A. Matveev, and A.V. Sarantsev, J. Phys. G: Nucl. Part. Phys. **28**, 15 (2002).
[24] V.V. Anisovich and L.G. Dakhno, Yad. Fiz. **2**, 710 (1966) [Sov. J. Nucl. Phys. **2**, 508 (1966)].
[25] V.V. Anisovich and A.V. Sarantsev, Eur. Phys. J. A **16**, 229 (2003).
[26] S. Pislak, et al. Phys. Rev. Lett., **87**, 221801 (2001).
[27] V.V. Anisovich and V.A. Nikonov, Eur. Phys. J. A**8**, 401 (2000); hep-ph/0008163 (2000).
[28] J.L. Basdevant, C.D. Frogatt and J.L. Petersen, Phys. Lett. **B41**, 178 (1972).
[29] J.L. Basdevant ant J. Zinn-Justin, Phys. Rev. **D3**, 1865 (1971);
D. Iagolnitzer, J. Justin and J.B. Zuber, Nucl. Phys. **B60**, 233 (1973).
[30] B.S. Zou and D.V. Bugg, Phys. Rev. **D48**, R3942 (1994); **D50**, 591 (1994).
[31] G. Janssen, B.C. Pearce, K. Holinde and J. Speth, Phys. Rev. **D52**, 2690 (1995).

Chapter 3

Spectral Integral Equation for the Decay of a Spinless Particle into a Three-Body State

We consider here again a four-point amplitude but, differently from Chapter 2, for a decay process: the transition $4 \to 1 + 2 + 3$. We write the spectral integral equation for a three-particle system using a simple example: the decay of a spinless state into three spinless particles. Final state rescatterings of outgoing particles are taken into account. Two versions of the spectral integral representation are given:
(1) For the case when the spectral integral is written for a two-particle system, and the integral is taken over the energy squared of one of the produced particle pairs, $(k_i + k_j)^2 = s_{ij}$. The integral equation of the decay amplitude $4 \to 1+2+3$ can be obtained by the analytic continuation of the dispersion relation equation of the two-particle amplitude $4 + 3 \to 1 + 2$.
(2) The spectral integral is written for a three-particle system, the integral is taken over the energy squared of all three produced particles, $(k_1 + k_2 + k_3)^2 = s$. In this case we can restrict ourselves to pair interactions, or consider three-particle ones.
Also, two types of representations of the interactions are considered: two-particle interactions written in terms of separable vertices and instantaneous interactions of two particles.

The first steps in accounting for all two-body final state interactions were made in [1, 2, 3, 4] in a non-relativistic approach for three-nucleon systems. In [5] the two-body interactions were successfully considered in the non-relativistic potential approach (the Faddeev equation).

A different approach to handle the three-particle interaction, the dispersion one, is based on finding the relativistic integral equations for the amplitude directly from the two-particle unitarity condition, taking into account, however, only the S-wave interaction in the one-channel problem [6, 7]. However, the transformation [6, 7] of the two-particle unitarity condition into the region of the decay processes did not lead to a result coinciding with those obtained in more elaborate investigations [8, 9].

The relativistic dispersion relation technique was used for the investigation of the final state interactions in three-body systems in [10]. Taking into account two-particle final state interactions, a relativistic dispersion relation equation for the amplitude $\eta \to \pi\pi\pi$ was written in [11]. Later on the method was generalized [12] for the coupled processes $p\bar{p}(\text{at rest}) \to \pi\pi\pi$, $\eta\eta\pi$, $K\bar{K}\pi$: in this way a system of coupled equations for decay amplitudes was written. Following these studies, we explain here the basic points in considering the dispersion relations for a three-particle system. The account of the final state interactions imposes correct unitarity and analyticity constraints on the amplitude.

Some technical details of the calculation procedure are given in Appendices A and B.

3.1 Three-body system in terms of separable interactions: analytic continuation of the four-point scattering amplitude to the decay region

Here we present the dispersion relation N/D-method for a three-body system taking into account final-state two-meson interactions. We consider the decay of a pseudoscalar particle ($J^P_{in} = 0^-$) with the mass M and momentum P into three pseudoscalar particles with masses m_1, m_2, m_3 and momenta k_1, k_2, k_3. There are different contributions to this decay process: those without final state particle interaction (prompt decay, Fig. 3.1a) and decays with subsequent final state interactions (an example is shown in Fig. 3.1b).

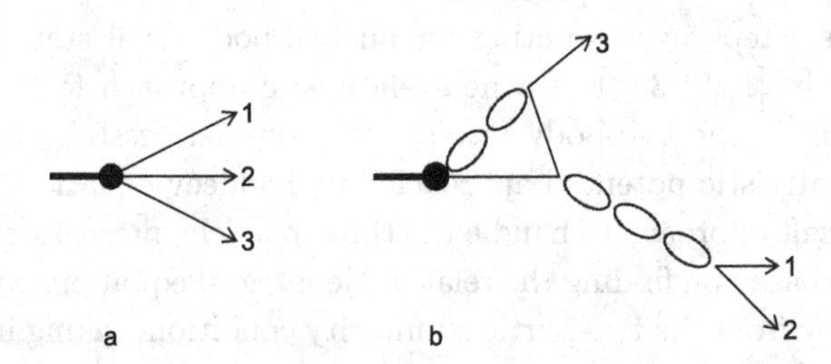

Fig. 3.1 Different types of transitions ($J^P_{in} = 0^-$)-state$\longrightarrow P_1P_2P_3$: a) prompt decay, b) decay with subsequent final state interactions.

For the decay amplitude we consider here an equation which takes into account two-particle final state interactions, such as that shown in Fig. 3.1b. First, we investigate in detail the S-wave rescattering within a separable ansatz. This case clarifies the main points of the dispersion relation approach for the three-particle interaction amplitude. After that we discuss a scheme for generalizing the equations for the case of higher waves.

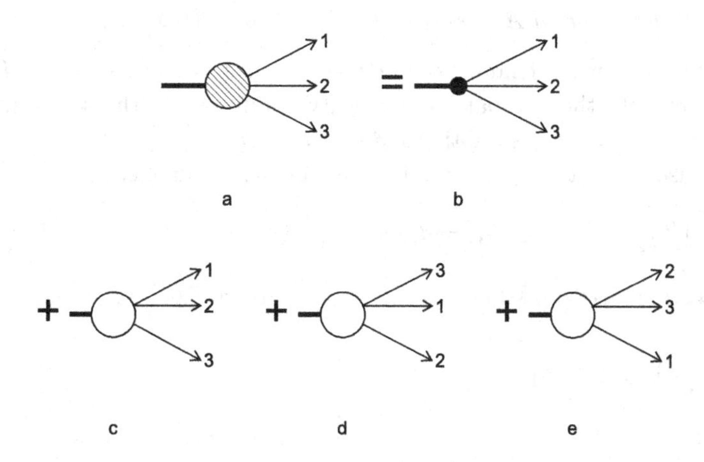

Fig. 3.2 Different terms in the amplitude $A^{(J_{in}=0)}_{P_1P_2P_3}(s_{12}, s_{13}, s_{23})$, Eq. (3.1).

3.1.1 *Final state two-particle S-wave interactions*

Let us begin with the S-wave two-particle interactions.

The decay amplitude is given by

$$A^{(J_{in}=0)}_{P_1P_2P_3}(s_{12}, s_{13}, s_{23}) = \Lambda(s_{12}, s_{13}, s_{23}) + A^{(0)}_{12}(s_{12}) + A^{(0)}_{13}(s_{13}) + A^{(0)}_{23}(s_{23}). \quad (3.1)$$

Different terms in (3.1) are illustrated by Fig. 3.2. We have a prompt production amplitude, Fig. 3.2b, and terms $A^{(0)}_{ij}(s_{ij})$ with particles P_iP_j participating in the final state interactions: P_1P_2 in Fig. 3.2c, P_1P_3 in Fig. 3.2d and P_2P_3 in Fig. 3.2e.

To take into account all S-wave rescatterings in the final state, we should write equations for different terms $A^{(0)}_{ij}(s_{ij})$.

The two-particle unitarity condition is explored to derive the integral equations for the amplitudes $A^{(0)}_{ij}(s_{ij})$ (ij =12, 13, 23). The idea of the approach realized in [8, 13]) is that one should consider the case of a small external mass $M < m_1 + m_2 + m_3$. A standard spectral integral equation (or a dispersion relation equation) is written in this case for the transitions $h_{in}P_\ell \to P_iP_j$. Then the analytic continuation is performed over the mass M back to the decay region: this gives a system of equations for the decay amplitudes $A^{(0)}_{ij}(s_{ij})$.

3.1.1.1 *Calculation of $A^{(0)}_{12}(s_{12})$ in the c.m.s. of particles P_1P_2*

So, let us consider the channel of particles 1 and 2, the transition $h_{in}P_3 \to P_1P_2$. We write the two-particle unitarity condition for the scattering in this channel with the assumption $(M+m_3) \sim (m_1+m_2)$.

The discontinuity of the amplitude in the s_{12}-channel equals

$$\begin{aligned} disc_{12}\, A^{J_{in}=0}_{P_1P_2P_3}(s_{12},s_{13},s_{23}) &= disc_{12}\, A^{(0)}_{12}(s_{12}) \\ = \int d\Phi_2(p_{12};k_1,k_2)\Big(\Lambda(s_{12},s_{13},s_{23}) &+ A^{(0)}_{12}(s_{12}) + A^{(0)}_{13}(s_{13}) + A^{(0)}_{23}(s_{23})\Big) \\ \times\Big(A^{(0)}_{12\to 12}(s_{12})\Big)^* . \end{aligned} \tag{3.2}$$

Recall that

$$d\Phi_2(p_{12};k_1,k_2)=(1/2)(2\pi)^{-2}\delta^4(p_{12}-k_1-k_2)d^4k_1d^4k_2\delta(m_1^2-k_1^2)\delta(m_2^2-k_2^2)$$

is the standard phase volume of particles 1 and 2. In (3.2), we should take into account that only $A^{(0)}_{12}(s_{12})$ has a non-zero discontinuity in the channel 12.

Exploring (3.2), we can write the equation for the decay amplitude $h_{in} \to P_1P_2P_3$.

The full set of rescattering of particles 1 and 2 gives us the factor $(1-B_0(s_{12}))^{-1}$; so, in the case when initial and final energies coincide ($s_{12} = \tilde{s}_{12}$, see Chapter 2, Section 2.2 for more detail), we write:

$$A^{(0)}_{12}(s_{12}) = B^{(0)}_{in}(s_{12})\frac{1}{1-B_0(s_{12})}G^R_0(s_{12}) . \tag{3.3}$$

The first loop diagram $B^{(0)}_{in}(s_{12})$ is determined as

$$B^{(0)}_{in}(s_{12}) = \int\limits_{(m_1+m_2)^2}^{\infty} \frac{ds'_{12}}{\pi}\frac{disc_{12}\, B^{(0)}_{in}(s'_{12})}{s'_{12}-s_{12}-i0} , \tag{3.4}$$

where

$$disc_{12}\, B^{(0)}_{in}(s_{12}) =$$
$$= \int d\Phi_2(p_{12};k_1,k_2)\Big(\Lambda(s_{12},s_{13},s_{23}) + A^{(0)}_{13}(s_{13}) + A^{(0)}_{23}(s_{23})\Big)G^0_L(s_{12})$$
$$\equiv disc_{12}\, B^{(0)}_{\Lambda-12}(s_{12}) + disc_{12}\, B^{(0)}_{13-12}(s_{12}) + disc_{12}\, B^{(0)}_{23-12}(s_{12}). \qquad (3.5)$$

Here we present $disc_{12}\, B^{(0)}_{in}(s_{12})$ as a sum of three terms because each of them needs a special treatment when $M^2 + i\varepsilon$ is increasing.

It is convenient to perform the phase-space integration in equation (3.5) in the centre-of-mass system of particles 1 and 2 where $\mathbf{k}_1 + \mathbf{k}_2 = 0$. In this frame

$$s_{13} = m_1^2 + m_3^2 + 2k_{10}k_{30} - 2z \mid \mathbf{k}_1 \mid\mid \mathbf{k}_3 \mid ,$$
$$s_{23} = m_2^2 + m_3^2 + 2k_{20}k_{30} + 2z \mid \mathbf{k}_2 \mid\mid \mathbf{k}_3 \mid , \qquad (3.6)$$

where $z = \cos\theta_{13}$ and

$$k_{10} = (s_{12} + m_1^2 - m_2^2)/(2\sqrt{s_{12}}),$$
$$k_{20} = (s_{12} + m_2^2 - m_1^2)/(2\sqrt{s_{12}}),$$
$$-k_{30} = (s_{12} + m_3^2 - M^2)/(2\sqrt{s_{12}}),$$
$$\mid \mathbf{k}_j \mid = \sqrt{k_{j\,0}^2 - m_j^2}. \qquad j = 1,2,3. \qquad (3.7)$$

The minus sign in front of k_{30} reflects the fact that in $(\mathbf{k}_1 + \mathbf{k}_2 = 0)$-system the particle P_3 is an outgoing, not an incoming one. In the calculation of $disc_{12}\, B^{(in)}_0(s_{12})$ all integrations are carried out easily except for the contour integral over dz. It can be rewritten in (3.5) as an integral over ds_{13} or ds_{23}:

$$\int\limits_{-1}^{+1} \frac{dz}{2} \to \int\limits_{s_{13}(-)}^{s_{13}(+)} \frac{ds_{13}}{\sqrt{[s_{12} - (m_3+M)^2][s_{12} - (m_3-M)^2]}} \times$$
$$\times \frac{s_{12}}{\sqrt{[s_{12} - (m_1+m_2)^2][s_{12} - (m_1-m_2)^2]}}$$
$$= \int\limits_{s_{23}(-)}^{s_{23}(+)} \frac{ds_{23}}{\sqrt{[s_{12} - (m_3+M)^2][s_{12} - (m_3-M)^2]}} \times$$
$$\times \frac{s_{12}}{\sqrt{[s_{12} - (m_1+m_2)^2][s_{12} - (m_1-m_2)^2]}}, \qquad (3.8)$$

where

$$s_{13}(\pm) = m_1^2 + m_3^2 + 2k_{10}k_{30} \pm 2 \mid \mathbf{k}_1 \mid\mid \mathbf{k}_3 \mid ,$$
$$s_{23}(\pm) = m_2^2 + m_3^2 + 2k_{20}k_{30} \pm 2 \mid \mathbf{k}_2 \mid\mid \mathbf{k}_3 \mid . \qquad (3.9)$$

The relative location of the integration contours (3.8) and amplitude singularities is the determining point for writing the equation.

3.1.1.2 *Analytic continuation of the integration contour over z*

Below we use the notation

$$\int\limits_{s_{i3}(-)}^{s_{i3}(+)} ds_{i3} = \int\limits_{C_{i3}(s_{12})} ds_{i3}. \qquad (3.10)$$

One can see from (3.9) that the integration contours $C_{13}(s_{12})$ and $C_{23}(s_{12})$ depend on M^2 and s_{12}, so we should monitor them when $M^2+i\varepsilon$ increases.

Let us underline again that in this section the idea to consider the decay processes in the dispersion relation approach is the following: we write the equation in the region of the standard scattering *two particles → two particles* (when $m_1 \sim m_2 \sim m_3 \sim M$) with the subsequent analytical continuation (with $M^2+i\varepsilon$ at $\varepsilon > 0$) into the decay region, $M > m_1+m_2+m_3$, and then $\varepsilon \to +0$. In this continuation we need to specify what type of singularities (and corresponding type of processes) we take into account and what type of singularities we neglect. Definitely, we take into account right-hand side and left-hand side singularities of the scattering processes $P_iP_j \to P_iP_j$ (our main aim is to restore the rescattering processes correctly). But singularities of the prompt production amplitude are beyond the field of our interest. In other words, we suppose $\Lambda(s_{12}, s_{13}, s_{23})$ to be an analytical function in the region under consideration.

Other problems with the analyticity of the amplitude $A_{12}^{(0)}(s_{12})$ can appear in choosing $G_0^R(s_{12})$ – this vertex should not violate the analytical properties of the total amplitude, see the Mandelstam plane for reaction $4 \to 1+2+3$, Fig. 6 in Chapter 2. The simplest way to avoid this problem is to use for $G_0^R(s_{12})$ analytical functions or functions with very remote singularities.

Assuming $\Lambda(s_{12}, s_{13}, s_{23})$ and $G_0^R(s_{12})$ to be analytical functions in the considered region, we can easily perform the analytic continuation of the integral over dz, Eq. (3.8), with an increasing $M^2 + i\varepsilon$. In this case the only problems appear in the integrations of $A_{13}^{(0)}(s_{13})$ and $A_{23}^{(0)}(s_{23})$ owing to the threshold singularities in the amplitudes (at $s_{13} = (m_1 + m_3)^2$ and

$s_{23} = (m_2 + m_3)^2$, respectively). However, as it was stressed above, the analytic continuation over $M^2 + i\varepsilon$ resolves these problems easily (see also [14], Chapter 4).

Let us now write the equation for the three-particle production amplitude in more detail. We denote the S-wave projections of $\Lambda(s_{12}, s_{13}, s_{23})$, $A^{(0)}_{13}(s_{23})$ and $A^{(0)}_{23}(s_{23})$ as

$$\Big\langle \Lambda(s_{12}, s_{13}, s_{23}) \Big\rangle^{(0)}_{12} = \int\limits_{-1}^{+1} \frac{dz}{2}\, \Lambda(s_{12}, s_{13}, s_{23}),$$

$$\Big\langle A^{(0)}_{13}(s_{13}) \Big\rangle^{(0)}_{12} = \int\limits_{-1}^{+1} \frac{dz}{2}\, A^{(0)}_{13}(s_{13}) \equiv \int\limits_{C_{13}(s_{12})} \frac{ds_{13}}{4|\mathbf{k}_1||\mathbf{k}_3|} A^{(0)}_{13}(s_{13}),$$

$$\Big\langle A^{(0)}_{23}(s_{23}) \Big\rangle^{(0)}_{12} = \int\limits_{-1}^{+1} \frac{dz}{2}\, A^{(0)}_{23}(s_{23}) \equiv \int\limits_{C_{23}(s_{22})} \frac{ds_{23}}{4|\mathbf{k}_2||\mathbf{k}_3|} A^{(0)}_{23}(s_{23}). \quad (3.11)$$

Recall that the definition of the contours $C_i(s_{12})$ is given in (3.10) while the relative position of the contour $C_2(s_{12})$ and the threshold singularity in the s_{23}-channel is shown in Fig. 3.3.

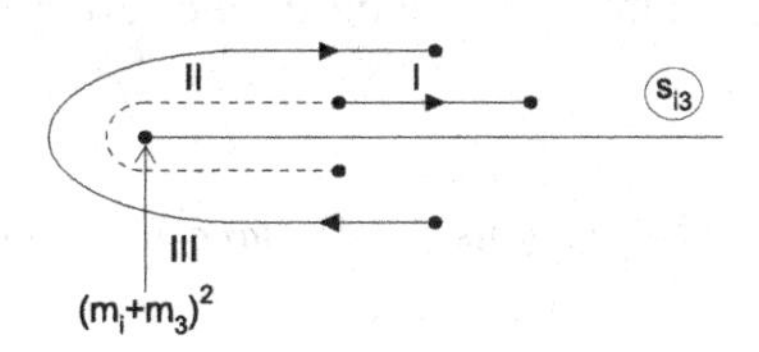

Fig. 3.3 The integration contour $C(s'_{12})$.

3.1.1.3 *Threshold behavior of the three-particle amplitude* $A^{(J_{in}=0)}_{P_1P_2P_3}(s_{12}, s_{13}, s_{23})$

The analytic continuation of the amplitude $A^{(J_{in}=0)}_{P_1P_2P_3}(s_{12}, s_{13}, s_{23})$ results in keeping the threshold relations:

$$A^{(0)}_{P_1P_2P_3}(s_{12}, s_{13}, s_{23})\Big|_{s_{12}\to(m_1+m_2)^2} =$$
$$= \lambda_{12}\Big[1 + i\frac{|\mathbf{k}_{12}|}{8\pi(m_1+m_2)} A_{12\to12}((m_1+m_2)^2)\Big],$$

$$A^{(0)}_{P_1P_2P_3}(s_{12}, s_{13}, s_{23})\Big|_{s_{13}\to(m_1+m_3)^2} =$$
$$= \lambda_{13}\left[1 + i\frac{|\mathbf{k}_{13}|}{8\pi(m_1+m_3)}A_{13\to13}((m_1+m_3)^2)\right],$$
$$A^{(0)}_{P_1P_2P_3}(s_{23}, s_{13}, s_{23})\Big|_{s_{23}\to(m_2+m_3)^2} =$$
$$= \lambda_{23}\left[1 + i\frac{|\mathbf{k}_{23}|}{8\pi(m_2+m_3)}A_{23\to23}((m_2+m_3)^2)\right]. \quad (3.12)$$

Here $|\mathbf{k}_{ij}|$ is the relative momentum of particles and $A_{ij\to ij}((m_i+m_j)^2)$ stands for the threshold value of the scattering amplitude $P_iP_j \to P_iP_j$. The threshold scattering amplitude determines the scattering length:

$$\frac{1}{8\pi(m_i+m_j)}A_{ij\to ij}((m_i+m_j)^2) = a_{ij}\,. \quad (3.13)$$

The constraints (3.12) are easily satisfied by introducing subtractions into spectral integrals for $B^{(0)}_{in}(s_{12})$, Eq. (3.4):

$$\int\limits_{(m_1+m_2)^2}^{\infty}\frac{ds'_{12}}{\pi}\frac{1}{s'_{12}-s_{12}-i0} \to$$
$$\to \int\limits_{(m_1+m_2)^2}^{\infty}\frac{ds'_{12}}{\pi}\frac{\Big(s_{12}-(m_1+m_2)^2\Big)}{\Big(s'_{12}-s_{12}-i0\Big)\Big(s'_{12}-(m_1+m_2)^2\Big)}. \quad (3.14)$$

Analogous subtractions should be made in the integrals for $B^{(0)}_{in}(s_{13})$ and $B^{(0)}_{in}(s_{23})$.

3.1.1.4 *Spectral integral equations for S-wave interactions*

As a result, we have:

$$A^{(0)}_{12}(s_{12}) = \Big(B^{(0)}_{\Lambda-12}(s_{12}) + B^{(0)}_{13-12}(s_{12}) + B^{(0)}_{23-12}(s_{12})\Big)\frac{1}{1-B^{(0)}_{12}(s_{12})}G^R_0(s_{12}) \quad (3.15)$$

where

$$B^{(0)}_{\Lambda-12}(s_{12}) + B^{(0)}_{13-12}(s_{12}) + B^{(0)}_{23-12}(s_{12}) =$$
$$= \int\limits_{(m_1+m_2)^2}^{\infty}\frac{ds'_{12}}{\pi}\frac{\Big(s_{12}-(m_1+m_2)^2\Big)}{\Big(s'_{12}-s_{12}-i0\Big)\Big(s'_{12}-(m_1+m_2)^2\Big)}$$
$$\times\left[\Big\langle\Lambda(s'_{12}, s'_{13}, s'_{23})\Big\rangle^{(0)}_{12} + \Big\langle A^{(0)}_{13}(s'_{13})\Big\rangle^{(0)}_{12} + \Big\langle A^{(0)}_{23}(s'_{23})\Big\rangle^{(0)}_{12}\right]$$
$$\times\rho^{(0)}_{12}(s'_{12})G^L_0(s'_{12}). \quad (3.16)$$

Let us emphasize that in the integrand (3.16) the energy squared is s'_{12} and hence, calculating $\left\langle \Lambda(s'_{12}, s'_{13}, s'_{23}) \right\rangle_{12}^{(0)}$ and $\left\langle A_{i3}^{(0)}(s'_{i3}) \right\rangle_{12}^{(0)}$, we should use Eqs. (3.6) – (3.10) with the replacement $s_{12} \to s'_{12}$.

The equation (3.15) is illustrated by Fig. 3.4.

We have a system of three non-homogeneous equations which determine the amplitudes $A_{ij}^{(0)}(s_{ij})$ when $\Lambda(s_{12}, s_{13}, s_{23})$, $G_0^L(s_{ij})$ and $G_0^R(s_{ij})$ are considered as input functions.

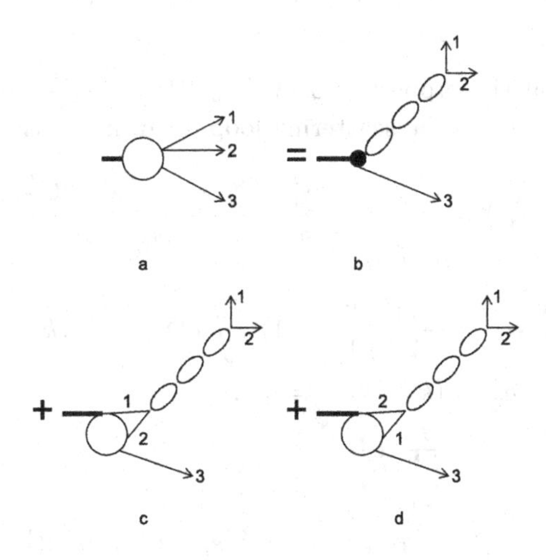

Fig. 3.4 Diagrammatic presentation of Eq. (3.15).

3.1.2 *General case: rescatterings of outgoing particles, $P_iP_j \to P_iP_j$, with arbitrary angular momenta*

We work with three scattering amplitudes and use the following notations:

$$A_{P_1P_2 \to P_1P_2}(s'_{12}, s_{12}) = \sum_{J_{12};\mu_1\ldots\mu_{J_{12}}} X^{(J_{12})}_{\mu_1\ldots\mu_{J_{12}}}(k'^{\perp(p'_1+p'_2)}_{12}) G^L_{J_{12}}(s'_{12})$$
$$\times \frac{1}{1 - B^{(J_{12})}(s_{12})} G^R_{J_{12}}(s_{12}) X^{(J_{12})}_{\mu_1\ldots\mu_{J_{12}}}(k^{\perp(p_1+p_2)}_{12}),$$

$$
\begin{aligned}
A_{P_1P_3\to P_1P_3}(s'_{13},s_{13}) &= \sum_{J_{13};\mu_1\ldots\mu_{J_{13}}} X^{(J_{13})}_{\mu_1\ldots\mu_{J_{13}}}(k'^{\perp(p'_1+p'_3)}_{13})G^L_{J_{13}}(s'_{13}) \\
&\times \frac{1}{1-B^{(J_{13})}(s_{13})}G^R_{J_{13}}(s_{13})X^{(J_{13})}_{\mu_1\ldots\mu_{J_{13}}}(k^{\perp(p_1+p_3)}_{13}), \\
A_{P_2P_3\to P_2P_3}(s'_{23},s_{23}) &= \sum_{J_{23};\mu_1\ldots\mu_{J_{23}}} X^{(J_{23})}_{\mu_1\ldots\mu_{J_{23}}}(k'^{\perp(p'_2+p'_3)}_{23})G^L_{J_{23}}(s'_{23}) \\
&\times \frac{1}{1-B^{(J_{23})}(s_{23})}G^R_{J_{23}}(s_{23})X^{(J_{23})}_{\mu_1\ldots\mu_{J_{23}}}(k^{\perp(p_2+p_3)}_{23}).
\end{aligned}
\tag{3.17}
$$

Let us remind that the properties of $X^{(J_{ij})}_{\mu_1\ldots\mu_{J_{ij}}}(k^{\perp(k_i+k_j)}_{23})$ are given in Appendix A of Chapter 2. The scattering loop diagram reads:

$$
\begin{aligned}
B^{(J)}(s_{ij}) &= \int\limits_{(m_i+m_j)^2}^{\infty} \frac{ds'_{ij}}{\pi}\,\frac{G^R_J(s'_{ij})\rho^{(J)}_{ij}(s'_{ij})G^L_J(s'_{ij})}{s'_{ij}-s_{ij}}\,, \\
\rho^{(J)}_{ij}(s_{ij}) &= \frac{\alpha_J\,|\vec{k}_{ij}|^{2J}}{2J+1}(-1)^J\int d\Phi_2(p_{ij};k_i,k_j), \\
p^2_{ij} &= (k_i+k_j)^2 = s_{ij}\ , \\
\alpha_J &= \prod_{l=1}^{J}\frac{2l-1}{l}, \\
|\vec{k}_{ij}|^2 = -k^{\perp 2}_{ij} &= \frac{[s_{ij}-(m_i+m_j)^2][s_{ij}-(m_i-m_j)^2]}{4s_{ij}}.
\end{aligned}
\tag{3.18}
$$

Here, writing (3.18), we have used the relation

$$
\int\frac{d\Omega}{4\pi}X^{(J)}_{\mu_1\ldots\mu_J}(k^\perp)X^{(J)}_{\nu_1\ldots\nu_J}(k^\perp)=\frac{\alpha_J\,|\vec{k}|^{2J}}{2J+1}(-1)^J O^{\mu_1\ldots\mu_J}_{\nu_1\ldots\nu_J}. \tag{3.19}
$$

In the following we shall transform these equations to equations for the decay amplitude $4\to 1+2+3$ in the same way as it was done for the case of the S-wave final state interactions.

3.1.2.1 *Decay ($J^P_{in}=0^-$)-state $\to P_1P_2P_3$*

Equations for amplitudes which describe the final state interactions in the transition ($J^P_{in}=0^-$)-state$\to P_1P_2P_3$ when the rescattering $P_iP_j\to P_iP_j$ occurs in any state with $J_{ij}\ge 0$ can be written in different forms. The uncertainty is related to the choice of the form of the momentum operator expansion.

In more detail, one can use the momentum expansion totally analogous to that presented in Eq. (3.17), namely, in the rest system of P_iP_j particles:

$$\sum_{J_{ij};\,\mu_1\ldots\mu_{J_{ij}}} X^{(J_{ij})}_{\mu_1\ldots\mu_{J_{ij}}}(k_\ell^{\perp P-k_\ell})\boldsymbol{A}^{(J_{ij})}_{ij}(s_{ij})X^{(J_{ij})}_{\mu_1\ldots\mu_{J_{ij}}}(k_{ij}^{\perp p_{ij}}), \quad i\neq j\neq \ell. \quad (3.20)$$

A modification of the left-hand operator $X^{(J_{ij})}$ in (3.20) leads to a modification of the spectral integral equation. For example, we can write:

$$\sum_{J_{ij};\,\mu_1\ldots\mu_{J_{ij}}} X^{(J_{ij})}_{\mu_1\ldots\mu_{J_{ij}}}(k_\ell^{\perp P})A^{(J_{ij})}_{ij}(s_{ij})X^{(J_{ij})}_{\mu_1\ldots\mu_{J_{ij}}}(k_{ij}^{\perp p_{ij}}). \quad (3.21)$$

This means that we use here the space-like component of k_ℓ in the c.m. system of all outgoing particles, $P_1P_2P_3$.

The amplitudes $\boldsymbol{A}^{(J_{ij})}_{ij}(s_{ij})$ and $A^{(J_{ij})}_{ij}(s_{ij})$ differ in the factor

$$\frac{\left(X^{(J_{ij})}_{\mu_1\ldots\mu_{J_{ij}}}(k_\ell^{\perp P})X^{(J_{ij})}_{\mu_1\ldots\mu_{J_{ij}}}(k_\ell^{\perp P})\right)}{\left(X^{(J_{ij})}_{\mu_1\ldots\mu_{J_{ij}}}(k_\ell^{\perp P})X^{(J_{ij})}_{\mu_1\ldots\mu_{J_{ij}}}(k_\ell^{\perp P-k_\ell})\right)} \quad (3.22)$$

and this factor appears in the comparison $disc\boldsymbol{A}^{(J_{ij})}_{ij}(s_{ij})$ and $discA^{(J_{ij})}_{ij}(s_{ij})$. This results in different non-Landau singularities (kinematic ones) on the unphysical complex-s_{ij} plane.

One can perform other types of the momentum operator decompositions as well.

We use here the decomposition of the type given by Eq. (3.21). Then the amplitude for the decay ($J^P_{in}=0^-$)-state$\to P_1P_2P_3$ reads:

$$\begin{aligned}
A^{(J_{in}=0)}_{P_1P_2P_3}&(s_{12},s_{13},s_{23}) = \Lambda(s_{12},s_{13},s_{23}) + \\
&+ \sum_{J_{12};\,\mu_1\ldots\mu_{J_{12}}} X^{(J_{12})}_{\mu_1\ldots\mu_{J_{12}}}(k_3^{\perp P})A^{(J_{12})}_{12}(s_{12})X^{(J_{12})}_{\mu_1\ldots\mu_{J_{12}}}(k_{12}^{\perp p_{12}}) \\
&+ \sum_{J_{13};\,\mu_1\ldots\mu_{J_{13}}} X^{(J_{13})}_{\mu_1\ldots\mu_{J_{13}}}(k_2^{\perp P})A^{(J_{13})}_{13}(s_{13})X^{(J_{13})}_{\mu_1\ldots\mu_{J_{13}}}(k_{13}^{\perp p_{13}}) \\
&+ \sum_{J_{23};\,\mu_1\ldots\mu_{J_{23}}} X^{(J_{23})}_{\mu_1\ldots\mu_{J_{23}}}(k_1^{\perp P})A^{(J_{23})}_{23}(s_{23})X^{(J_{23})}_{\mu_1\ldots\mu_{J_{23}}}(k_{23}^{\perp p_{23}})\,. \quad (3.23)
\end{aligned}$$

The amplitude term $A^{(J_{ij})}_{ij}(s_{ij})$ manifests that the last interaction of the outgoing particles takes place in the channel ij. One can separate the final state interaction in this channel, then

$$A^{(J_{in}=0)}_{P_1P_2P_3}(s_{12},s_{13},s_{23}) = \Lambda(s_{12},s_{13},s_{23})$$
$$+ \sum_{J_{12};\,\mu_1\ldots\mu_{J_{12}}} X^{(J_{12})}_{\mu_1\ldots\mu_{J_{12}}}(k_3^{\perp P}) B^{(J_{12})}_{12}(s_{12})$$
$$\times \frac{1}{1-B^{(J_{12})}(s_{12})} G^R_{J_{12}}(s_{12}) X^{(J_{12})}_{\mu_1\ldots\mu_{J_{12}}}(k_{12}^{\perp p_{12}})$$
$$+ \sum_{J_{13};\,\mu_1\ldots\mu_{J_{13}}} X^{(J_{13})}_{\mu_1\ldots\mu_{J_{13}}}(k_2^{\perp P}) B^{(J_{13})}_{13}(s_{13})$$
$$\times \frac{1}{1-B^{(J_{13})}(s_{13})} G^R_{J_{13}}(s_{13}) X^{(J_{13})}_{\mu_1\ldots\mu_{J_{13}}}(k_{13}^{\perp p_{13}})$$
$$+ \sum_{J_{23};\,\mu_1\ldots\mu_{J_{23}}} X^{(J_{23})}_{\mu_1\ldots\mu_{J_{23}}}(k_1^{\perp P}) B^{(J_{23})}_{23}(s_{23})$$
$$\times \frac{1}{1-B^{(J_{23})}(s_{23})} G^R_{J_{23}}(s_{23}) X^{(J_{23})}_{\mu_1\ldots\mu_{J_{23}}}(k_{23}^{\perp p_{23}}). \quad (3.24)$$

The loop diagram $B^{(J_{ij})}(s_{ij})$ is just the one which described the two-particle decay, Eq. (3.18); $B^{(J_{ij})}_{ij}(s_{ij})$, however, needs a special consideration.

We have three alternative expansions for the analytic term:

$$\Lambda(s_{12},s_{13},s_{23}) = \sum_{J_{12};\,\mu_1\ldots\mu_{J_{12}}} X^{(J_{12})}_{\mu_1\ldots\mu_{J_{12}}}(k_3^{\perp P}) \Lambda^{(J_{12})}_{12}(s_{12}) X^{(J_{12})}_{\mu_1\ldots\mu_{J_{12}}}(k_{12}^{\perp p_{12}})$$
$$= \sum_{J_{13};\,\mu_1\ldots\mu_{J_{13}}} X^{(J_{13})}_{\mu_1\ldots\mu_{J_{13}}}(k_2^{\perp P}) \Lambda^{(J_{13})}_{13}(s_{13}) X^{(J_{13})}_{\mu_1\ldots\mu_{J_{13}}}(k_{13}^{\perp p_{13}})$$
$$= \sum_{J_{23};\,\mu_1\ldots\mu_{J_{23}}} X^{(J_{23})}_{\mu_1\ldots\mu_{J_{23}}}(k_1^{\perp P}) \Lambda^{(J_{23})}_{23}(s_{23}) X^{(J_{23})}_{\mu_1\ldots\mu_{J_{23}}}(k_{23}^{\perp p_{23}}). \quad (3.25)$$

In the ansatz of the separable vertices we carry out the separabilisation procedure also for the analytic term. We can write

$$\Lambda^{(J_{ij})}_{ij}(s_{ij}) = \sum \Lambda^{(J_{ij}),L}_{ij}(s_{ij}) \Lambda^{(J_{ij}),R}_{ij}(s_{ij}) \quad (3.26)$$

We consider in detail the amplitude with P_1P_2-rescatterings, that is the second term in (3.23):

$$A^{(J_{12})}_{12}(s_{12}) \to A^{(J)}_{12}(s_{12}) = B^{(J)}_{\Lambda-12}(s_{12}) + B^{(J)}_{13-12}(s_{12}) + B^{(J)}_{23-12}(s_{12}). \quad (3.27)$$

Let us first consider the rescattering term initiated by $X^{(J)}_{\mu_1\ldots\mu_J}(k_3^{\perp P})$

$\Lambda^{(J)}_{12}(s_{12})X^{(J)}_{\mu_1\dots\mu_J}(k^{\perp p_{12}}_{12})$:

$$X^{(J)}_{\mu_1\dots\mu_J}(k^{\perp P}_3)\Lambda^{(J),L}_{12}(s_{12})\int\limits_{(m_1+m_2)^2}^{\infty}\frac{ds'_{12}}{\pi}\Lambda^{(J),R}_{12}(s'_{12})$$

$$\times\int d\Phi(p'_{12};k'_1,k'_2)X^{(J)}_{\mu_1\dots\mu_J}(k'^{\perp p'_{12}}_{12})X^{(J)}_{\nu_1\dots\nu_J}(k'^{\perp p'_{12}}_{12})\frac{G^L_J(s'_{12})}{s'_{12}-s_{12}-i0}$$

$$\times G^L_J(s_{12})X^{(J)}_{\nu_1\dots\nu_J}(k^{\perp p_{12}}_{12})\,. \tag{3.28}$$

Not to make complicated expressions even more complicated, we omit provisionally the subtraction term. Due to

$$\int d\Phi(p'_{12};k'_1,k'_2)X^{(J)}_{\mu_1\dots\mu_J}(k'^{\perp p'_{12}}_{12})X^{(J)}_{\nu_1\dots\nu_J}(k'^{\perp p'_{12}}_{12})=\rho^{(J)}_{12}(s'_{12})O^{\mu_1\dots\mu_J}_{\nu_1\dots\nu_J}$$

we write

$$B^{(J)}_{\Lambda-12}(s_{12})=\int\limits_{(m_1+m_2)^2}^{\infty}\frac{ds'_{12}}{\pi}\Lambda^{(J),R}_{12}(s'_{12})\frac{\rho^{(J)}_{12}(s'_{12})}{s'_{12}-s_{12}-i0}G^L_J(s'_{12}) \tag{3.29}$$

with $\rho^{(J)}_{12}(s_{12})$ given Eq. (3.18)

The next two terms in Eq. (3.27) read:

$$i=1,2:\quad B^{(J)}_{i3-12}(s_{12})O^{\mu_1\dots\mu_J}_{\nu_1\dots\nu_J}=\int\limits_{(m_1+m_2)^2}^{\infty}\frac{ds'_{12}}{\pi}$$

$$\times\int d\Phi(p'_{12};k'_1,k'_2)X^{(J)}_{\mu_1\dots\mu_J}(k'^{\perp p'_{12}}_{12})A^{(J)}_{i3}(s'_{i3})X^{(J)}_{\nu_1\dots\nu_J}(k'^{\perp p'_{12}}_{12})\frac{G^L_J(s'_{12})}{s'_{12}-s_{12}-i0}$$

$$=\int\limits_{(m_1+m_2)^2}^{\infty}\frac{ds'_{12}}{\pi}\left\langle A^{(J)}_{i3}(s'_{i3})\right\rangle^{(J)}_{12}O^{\mu_1\dots\mu_J}_{\nu_1\dots\nu_J}\rho^{(J)}_{12}(s'_{12})\frac{G^L_J(s'_{12})}{s'_{12}-s_{12}-i0}\,, \tag{3.30}$$

Finally, we have

$$i=1,2:\quad B^{(J)}_{i3-12}(s_{12})=\int\limits_{(m_1+m_2)^2}^{\infty}\frac{ds'_{12}}{\pi}\left\langle A^{(J)}_{i3}(s'_{i3})\right\rangle^{(J)}_{12}\frac{\rho^{(J)}_{12}(s'_{12})}{s'_{12}-s_{12}-i0}G^L_J(s'_{12}) \tag{3.31}$$

Functions $\left\langle A^{(J)}_{i3}(s'_{i3})\right\rangle^{(J)}_{12}$ are determined by similar integrals as in (3.11) with the obvious replacement $(0)\to(J)$. The reason for this is clear, namely: the factorization of the momentum operators $X^{(J)}_{\mu_1\dots\mu_J}(k^{\perp P}_3)$ and $X^{(J)}_{\mu_1\dots\mu_J}(k^{\perp p_{12}}_{12})$.

Finally, we write the equation for $A_{12}^{(J)}(s_{12})$ in the form

$$A_{12}^{(J)}(s_{12})=\left(B_{\Lambda-12}^{(J)}(s_{12})+B_{13-12}^{(J)}(s_{12})+B_{23-12}^{(J)}(s_{12})\right)\frac{G_J^R(s_{12})}{1-B_{12}^{(J)}(s_{12})}. \tag{3.32}$$

The equations for $A_{23}^{(J)}(s_{23})$ and $A_{13}^{(J)}(s_{13})$ can be written analogously.

In the presented calculation we have omitted the subtraction term. In Eqs. (3.29) – (3.31) it is restored by the replacement

$$\frac{1}{\left(s'_{12}-s_{12}-i0\right)} \to \frac{s_{12}-(m_1+m_2)^2}{\left(s'_{12}-s_{12}-i0\right)\left(s'_{12}-(m_1+m_2)^2\right)}. \tag{3.33}$$

3.1.2.2 *Miniconclusion*

In this section we have presented some characteristic features of the spectral integral equations for three-body systems, $4 \to 1+2+3$, when these equations are obtained by the analytic continuation in the mass m_4 from the equations for two-particle transitions, $4+3 \to 1+2$, $4+2 \to 1+3$ and $4+1 \to 2+3$. Note that the integration contour $C_i(s_{12})$ determined here, see also [13], does not coincide with that of [15].

We have written the equations for arbitrary orbital momenta of the outgoing waves, but for scalar (or pseudoscalar) particles. In principle, the generalization of the equations to the case of particles with any (integer and half-integer) spins is not difficult. The technique of describing two-particle states where the particles have arbitrary spins can be found, *e.g.*, in [14].

The technique can be used both for the determination of levels of compound systems and their wave functions, for instance, by taking into account the leading singularities – this method was applied to the three-nucleon systems, H_3 and He_3 [16]. Another possible topic is the determination of analytical properties of multiparticle production amplitudes when the produced resonances are studied [10]. The interest in the reactions $p\bar{p}(J^{PC}=0^{-+}) \to \pi\pi\pi, \pi\eta\eta, \pi K\bar{K}$ is also due to the resonance production: meson rescatterings in the $p\bar{p}$ annihilation were studied in [12] where a set of equations for the reactions was written.

3.2 Non-relativistic approach and transition of two-particle spectral integral to the three-particle one

Here we transform the equation for $J_{in} = 0$, see Section 1.1, to the non-relativistic limit. First, we write relativistic equations in a compact form:

$$
\begin{aligned}
A^{(J_{in}=0)}_{P_1P_2P_3}(s_{12},s_{13},s_{23}) &= \Lambda(s_{12},s_{13},s_{23})+A^{(0)}_{12}(s_{12})+A^{(0)}_{13}(s_{13})+A^{(0)}_{23}(s_{23}),\\
A^{(0)}_{12}(s_{12}) &= \Big(B^{(0)}_{\Lambda-12}(s_{12})+B^{(0)}_{13-12}(s_{12})+B^{(0)}_{23-12}(s_{12})\Big)\\
&\quad\times\frac{1}{1-B^{(0)}_{12}(s_{12})}G^R_0(s_{12})\,,\\
A^{(0)}_{13}(s_{13}) &= \Big(B^{(0)}_{\Lambda-13}(s_{13})+B^{(0)}_{12-13}(s_{13})+B^{(0)}_{23-13}(s_{13})\Big)\\
&\quad\times\frac{1}{1-B^{(0)}_{13}(s_{13})}G^R_0(s_{13})\,,\\
A^{(0)}_{23}(s_{23}) &= \Big(B^{(0)}_{\Lambda-23}(s_{23})+B^{(0)}_{13-23}(s_{23})+B^{(0)}_{12-23}(s_{23})\Big)\\
&\quad\times\frac{1}{1-B^{(0)}_{23}(s_{23})}G^R_0(s_{23})\,,
\end{aligned}
\tag{3.34}
$$

with

$$
\begin{aligned}
&B^{(0)}_{\Lambda-12}(s_{12})+B^{(0)}_{13-12}(s_{12})+B^{(0)}_{23-12}(s_{12}) =\\
&=\int\limits_{(m_1+m_2)^2}^{\infty}\frac{ds'_{12}}{\pi}\frac{s_{12}-(m_1+m_2)^2}{\Big(s'_{12}-s_{12}-i0\Big)\Big(s'_{12}-(m_1+m_2)^2\Big)}\\
&\times\Big\langle\Lambda(s'_{12},s'_{13},s'_{23})+A^{(0)}_{13}(s'_{13})+A^{(0)}_{23}(s'_{23})\Big\rangle^{(0)}_{12}\rho^{(0)}_{12}(s'_{12})G^L_0(s'_{12}),\\
&B^{(0)}_{\Lambda-13}(s_{13})+B^{(0)}_{12-13}(s_{13})+B^{(0)}_{23-13}(s_{13}) =\\
&=\int\limits_{(m_1+m_3)^2}^{\infty}\frac{ds'_{13}}{\pi}\frac{s_{13}-(m_1+m_3)^2}{\Big(s'_{13}-s_{13}-i0\Big)\Big(s'_{13}-(m_1+m_3)^2\Big)}\\
&\times\Big\langle\Lambda(s'_{12},s'_{13},s'_{23})+A^{(0)}_{12}(s'_{12})+A^{(0)}_{23}(s'_{23})\Big\rangle^{(0)}_{13}\rho^{(0)}_{13}(s'_{13})G^L_0(s'_{13}),
\end{aligned}
$$

$$B^{(0)}_{\Lambda-23}(s_{23})+B^{(0)}_{12-23}(s_{23})+B^{(0)}_{13-23}(s_{23})$$
$$=\int\limits_{(m_2+m_3)^2}^{\infty}\frac{ds'_{23}}{\pi}\frac{s_{23}-(m_2+m_3)^2}{\Big(s'_{23}-s_{23}-i0\Big)\Big(s'_{23}-(m_2+m_3)^2\Big)}$$
$$\times\Big\langle\Lambda(s'_{12},s'_{13},s'_{23})+A^{(0)}_{12}(s'_{12})+A^{(0)}_{13}(s'_{13})\Big\rangle^{(0)}_{23}\rho^{(0)}_{23}(s'_{23})G^L_0(s'_{23})\,. \quad (3.35)$$

For further considerations, let us recall that

$$\frac{s_{12}-(m_1+m_2)^2}{\Big(s'_{12}-s_{12}-i0\Big)\Big(s'_{12}-(m_1+m_2)^2\Big)}=\frac{1}{s'_{12}-s_{12}-i0}-\frac{1}{s'_{12}-(m_1+m_2)^2}\,. \quad (3.36)$$

Below we shall use $G^R_0(s_{ij})=1$, $G^L_0(s_{ij})=N(s_{ij})$.

3.2.1 *Non-relativistic approach*

In the c.m. system, $\mathbf{k}_1+\mathbf{k}_2+\mathbf{k}_3=0$, we write

$$\mathbf{p}_{12}=\mathbf{k}_1+\mathbf{k}_2=-\mathbf{k}_3\,,\quad \mathbf{k}_{12}=\frac{m_2}{m_1+m_2}\mathbf{k}_1-\frac{m_1}{m_1+m_2}\mathbf{k}_2\,.$$
$$\mathbf{k}_1=\mathbf{k}_{12}-\frac{m_1}{m_1+m_2}\mathbf{k}_3\,,\quad \mathbf{k}_2=-\mathbf{k}_{12}-\frac{m_2}{m_1+m_2}\mathbf{k}_3\,,$$
$$\sqrt{s_{12}}\simeq m_1+m_2+E_{12}=m_1+m_2+\frac{\mathbf{k}^2_{12}}{2\mu_{12}},\quad \mu^{-1}_{12}=m^{-1}_1+m^{-1}_2\,,$$
$$\sqrt{s}\simeq m_1+m_2+m_3+E=m_1+m_2+m_3+\frac{\mathbf{k}^2_{12}}{2\mu_{12}}+\frac{\mathbf{k}^2_3}{2\mu_3}\,,$$
$$\mu^{-1}_3=m^{-1}_3+\mu^{-1}_{12}\,. \quad (3.37)$$

For the non-relativistic approach we replace in (3.35):

$$\frac{ds'_{12}}{s'_{12}-s_{12}-i0}\frac{dz}{2}\rho^{(0)}_{12}(s'_{12})\to\frac{dE'_{12}}{E'_{12}-E_{12}-i0}\delta\Big(E'_{12}-\frac{\mathbf{k}'^2_{12}}{2\mu_{12}}\Big)\frac{d^3\mathbf{k}'_{12}}{8m_1m_2(2\pi)^2},$$
$$\frac{ds'_{13}}{s'_{13}-s_{13}-i0}\frac{dz}{2}\rho^{(0)}_{13}(s'_{13})\to\frac{dE'_{13}}{E'_{13}-E_{13}-i0}\delta\Big(E'_{13}-\frac{\mathbf{k}'^2_{13}}{2\mu_{13}}\Big)\frac{d^3\mathbf{k}'_{13}}{8m_1m_3(2\pi)^2},$$
$$\frac{ds'_{23}}{s'_{23}-s_{23}-i0}\frac{dz}{2}\rho^{(0)}_{23}(s'_{23})\to\frac{dE'_{23}}{E'_{23}-E_{23}-i0}\delta\Big(E'_{23}-\frac{\mathbf{k}'^2_{23}}{2\mu_{23}}\Big)\frac{d^3\mathbf{k}'_{23}}{8m_2m_3(2\pi)^2}\,. \quad (3.38)$$

We take into account in (3.38) that $dz/2\;\rho^{(0)}_{ij}(s'_{ij})$ is the two-particle phase space which for the non-relativistic limit reads:

$$d\Phi_2(p'_{ij};k'_i,k'_j)\to d\Phi^{(n-r)}_2(p'_{ij};k'_i,k'_j)=\delta\Big(E'_{ij}-\frac{\mathbf{k}'^2_{ij}}{2\mu_{ij}}\Big)\frac{d^3\mathbf{k}'_{ij}}{8m_im_j(2\pi)^2}\,. \quad (3.39)$$

Analogously, one has for the three particle phase factor:

$$
\begin{aligned}
& d\Phi_3(P;k_1,k_2,k_3) \to d\Phi_3^{(n-r)}(P;k_1,k_2,k_3) = \\
& = \delta\Big(E - \frac{\mathbf{k}_1^2}{2m_1} - \frac{\mathbf{k}_2^2}{2m_2} - \frac{\mathbf{k}_3^2}{2m_3}\Big)\delta^{(3)}\Big(\mathbf{k}_1+\mathbf{k}_2+\mathbf{k}_3\Big)\frac{d^3\mathbf{k}_1 d^3\mathbf{k}_2 d^3\mathbf{k}_3}{16m_1m_2m_3(2\pi)^5} \\
& = \delta\Big(E - \frac{\mathbf{k}_{ij}^2}{2\mu_{ij}} - \frac{\mathbf{k}_\ell^2}{2\mu_\ell}\Big)\frac{d^3\mathbf{k}_{ij} d^3\mathbf{k}_\ell}{16m_1m_2m_3(2\pi)^5} \quad \text{with} \quad i \neq j \neq \ell. \qquad (3.40)
\end{aligned}
$$

Therefore the right-hand side factor in Eq. (3.38) can be re-written as:

$$
\begin{aligned}
& \frac{dE'_{12}}{E'_{12}-E_{12}-i0}\delta\Big(E'_{12} - \frac{\mathbf{k}'^2_{12}}{2\mu_{12}}\Big)\frac{d^3\mathbf{k}'_{12}}{8m_1m_2(2\pi)^2} = \\
& = \frac{dE'}{E'-E-i0}d\Phi_3^{(n-r)}(P';k'_1,k'_2,k'_3)(2\pi)^3 2m_3\delta^{(3)}\Big(\mathbf{k}'_3-\mathbf{k}_3\Big), \\
& \frac{dE'_{13}}{E'_{13}-E_{13}-i0}\delta\Big(E'_{13} - \frac{\mathbf{k}'^2_{13}}{2\mu_{13}}\Big)\frac{d^3\mathbf{k}'_{13}}{8m_1m_3(2\pi)^2} = \\
& = \frac{dE'}{E'-E-i0}d\Phi_3^{(n-r)}(P';k'_1,k'_2,k'_3)(2\pi)^3 2m_2\delta^{(3)}\Big(\mathbf{k}'_2-\mathbf{k}_2\Big), \\
& \frac{dE'_{23}}{E'_{23}-E_{23}-i0}\delta\Big(E'_{23} - \frac{\mathbf{k}'^2_{23}}{2\mu_{23}}\Big)\frac{d^3\mathbf{k}'_{23}}{8m_2m_3(2\pi)^2} = \\
& = \frac{dE'}{E'-E-i0}d\Phi_3^{(n-r)}(P';k'_1,k'_2,k'_3)(2\pi)^3 2m_1\delta^{(3)}\Big(\mathbf{k}'_1-\mathbf{k}_1\Big). \qquad (3.41)
\end{aligned}
$$

Indeed, we have:

$$
E' = E'_{ij} + \frac{\mathbf{k}'^2_\ell}{2m_\ell} = E'_{ij} + \frac{\mathbf{k}^2_\ell}{2m_\ell}, \qquad E'_{ij} - E_{ij} = E' - E \qquad (3.42)
$$

because of the presence of $\delta^{(3)}(\mathbf{k}'_\ell - \mathbf{k}_\ell)$ in the right-hand side of (3.41).

Eq. (3.35) in the non-relativistic approach reads:

$$
\begin{aligned}
& B^{(0)}_{\Lambda-12}(s_{12})+B^{(0)}_{13-12}(s_{12})+B^{(0)}_{23-12}(s_{12}) = \int\limits_0^\infty \frac{dE'_{12}}{\pi}\frac{E_{12}\delta\Big(E'_{12}-\mathbf{k}'^2_{12}/2\mu_{12}\Big)}{E'_{12}(E'_{12}-E_{12}-i0)} \\
& \times\frac{d^3\mathbf{k}'_{12}}{8m_1m_2(2\pi)^2}\Big[\Lambda(s'_{12},s'_{13},s'_{23}) + A^{(0)}_{13}(s'_{13}) + A^{(0)}_{23}(s'_{23})\Big]N_0^{(n-r)}(E'_{12}),
\end{aligned}
$$

$$B^{(0)}_{\Lambda-13}(s_{13})+B^{(0)}_{12-13}(s_{13})+B^{(0)}_{23-13}(s_{13}) = \int\limits_0^\infty \frac{dE'_{13}}{\pi} \frac{E_{13}\delta\left(E'_{13} - \mathbf{k}'^2_{13}/2\mu_{13}\right)}{E'_{13}(E'_{13} - E_{13} - i0)}$$

$$\times \frac{d^3\mathbf{k}'_{13}}{8m_1m_3(2\pi)^2}\left[\Lambda(s'_{12}, s'_{13}, s'_{23}) + A^{(0)}_{12}(s'_{12}) + A^{(0)}_{23}(s'_{23})\right] N_0^{(n-r)}(E'_{13}),$$

$$B^{(0)}_{\Lambda-23}(s_{23})+B^{(0)}_{12-23}(s_{23})+B^{(0)}_{13-23}(s_{23}) = \int\limits_0^\infty \frac{dE'_{23}}{\pi} \frac{E_{23}\delta\left(E'_{23} - \mathbf{k}'^2_{23}/2\mu_{23}\right)}{E'_{23}(E'_{23} - E_{23} - i0)}$$

$$\times \frac{d^3\mathbf{k}'_{23}}{8m_2m_3(2\pi)^2}\left[\Lambda(s'_{12}, s'_{13}, s'_{23}) + A^{(0)}_{12}(s'_{12}) + A^{(0)}_{13}(s'_{13})\right] N_0^{(n-r)}(E'_{23}). \tag{3.43}$$

Deduction of similar formulae starting from the consideration of Feynman diagram amplitudes is given in Appendix C.

3.2.2 *Threshold limit constraint*

It is appropriate to present the threshold limit constraint for the $A^{(J_{in}=0)}_{P_1P_2P_3}(s_{12}, s_{13}, s_{23})$ amplitude.

Let us consider the s_{12}-threshold, here we have

$$s_{12} \to (m_1+m_2)^2 \equiv s_{12}^{(12-thr)},$$
$$s_{13} \to m_1^2 + m_2^2 + \frac{m_2}{m_1+m_2}\left[(m_1+m_2)^2 + m_3^2 - s\right] \equiv s_{13}^{(12-thr)},$$
$$s_{23} \to m_1^2 + m_2^2 + \frac{m_1}{m_1+m_2}\left[(m_1+m_2)^2 + m_3^2 - s\right] \equiv s_{23}^{(12-thr)} \tag{3.44}$$

or, in the non-relativistic limit:

$$\mathbf{k}^2_{12} \to 0,$$
$$\mathbf{k}^2_{13} \to \mathbf{k}^2_{13}(12-thr) = \frac{2m_1^2m_3(m_1+m_2+m_3)}{(m_1+m_2)(m_1+m_3)^2}E,$$
$$\mathbf{k}^2_{23} \to \mathbf{k}^2_{23}(12-thr) = \frac{2m_2^2m_3(m_1+m_2+m_3)}{(m_1+m_2)(m_2+m_3)^2}E. \tag{3.45}$$

Near the 12-threshold, when $s_{12} \to s_{12}^{(12-thr)}$, $s_{13} \to s_{13}^{(12-thr)}$ and $s_{23} \to s_{23}^{(12-thr)}$, we write

$$\left[A^{(J_{in}=0)}_{P_1P_2P_3}(s_{12}, s_{13}, s_{23})\right]_{s_{12}\sim(m_1+m_2)^2} =$$
$$= \left[\Lambda(s_{12}, s_{13}, s_{23}) + A^{(0)}_{12}(s_{12}) + A^{(0)}_{13}(s_{13}) + A^{(0)}_{23}(s_{23})\right]_{s_{12}\sim(m_1+m_2)^2}$$
$$\simeq \lambda_{12} + A^{(0)}_{13}(s_{13}^{(12-thr)}) + A^{(0)}_{23}(s_{23}^{(12-thr)}) + A^{(0)}_{12}(s_{12}\sim(m_1+m_2)^2). \tag{3.46}$$

At $s_{12} \sim (m_1+m_2)^2$, the amplitude $A^{(0)}_{12}(s_{12})$ has a zero real part, while the imaginary part is of the order of $i|\mathbf{k}_{12}|$. This means that we can replace $1/(E'_{12}-E_{12}-i0) \to i\pi\delta(E'_{12}-E_{12})$. Therefore

$$\begin{aligned}
A^{(0)}_{12}(s_{12} \sim (m_1+m_2)^2) \simeq & \int_0^\infty \frac{dE'_{12}}{\pi} \frac{E_{12}\delta\Big(E'_{12} - \mathbf{k}'^2_{12}/2\mu_{12}\Big)}{E'_{12}(E'_{12}-E_{12}-i0)} \\
& \times \frac{d^3\mathbf{k}'_{12}}{8m_1m_2(2\pi)^2}\left[\Lambda(s'_{12},s'_{13},s'_{23}) + A^{(0)}_{13}(s'_{13}) + A^{(0)}_{23}(s'_{23})\right] \\
& \times \frac{N_0^{(n-r)}(E'_{12})}{1-B^{(0)}_{12}(s_{12}=(m_1+m_2)^2)} \\
& \simeq \left[\lambda_{12} + A^{(0)}_{13}(s_{13}^{(12-thr)}) + A^{(0)}_{23}(s_{23}^{(12-thr)})\right] i|\mathbf{k}_{12}|a_{12}\,, \qquad (3.47)
\end{aligned}$$

where $\lambda_{12} = \Big[\Lambda(s'_{12},s'_{13},s'_{23})\Big]_{12-thr}$, and a_{12} is the scattering length for $P_1P_2 \to P_1P_2$:

$$a_{12} = \frac{1}{8\pi(m_1+m_2)} \frac{N_0^{(n-r)}(0)}{1-B^{(0)}_{12}((m_1+m_2)^2)}\,. \qquad (3.48)$$

We see that

$$\begin{aligned}
& \left[A^{(J_{in}=0)}_{P_1P_2P_3}(s_{12},s_{13},s_{23})\right]_{s_{12}\sim(m_1+m_2)^2} \simeq \\
& \simeq \left[\lambda_{12} + A^{(0)}_{13}(s_{13}^{(12-thr)}) + A^{(0)}_{23}(s_{23}^{(12-thr)})\right]\Big(1+i|\mathbf{k}_{12}|a_{12}\Big)\,, \qquad (3.49)
\end{aligned}$$

which means carrying out the threshold requirement.

3.2.3 *Transition of the two-particle spectral integral representation amplitude to the three-particle spectral integral*

Eqs. (3.41) tell us that amplitudes can be represented in terms of the three-particle spectral integrals. Indeed:

$$B^{(0)}_{\Lambda-12}(s_{12})+B^{(0)}_{13-12}(s_{12})+B^{(0)}_{23-12}(s_{12}) = \int\limits_0^\infty \frac{dE'_{12}}{\pi} \frac{E_{12}\delta\left(E'_{12}-\frac{\mathbf{k}'^2_{12}}{2\mu_{12}}\right)}{E'_{12}(E'_{12}-E_{12}-i0)}$$

$$\times \frac{d^3\mathbf{k}'_{12}}{8m_1m_2(2\pi)^2}\left[\Lambda(s'_{12},s'_{13},s'_{23})+A^{(0)}_{13}(s'_{13})+A^{(0)}_{23}(s'_{23})\right]N_0^{(n-r)}(E'_{12})$$

$$= \int\limits_0^\infty \frac{dE'}{\pi}\frac{(E-E_{thr12})}{(E'-E_{thr12})(E'-E-i0)}\delta\left(E'-\frac{\mathbf{k}'^2_{12}}{2\mu_{12}}-\frac{\mathbf{k}'^2_3}{2\mu_3}\right)$$

$$\times \frac{d^3\mathbf{k}'_{12}d^3\mathbf{k}'_3}{16m_1m_2m_3(2\pi)^5}\left[\Lambda(s'_{12},s'_{13},s'_{23})+A^{(0)}_{13}(s'_{13})+A^{(0)}_{23}(s'_{23})\right]$$

$$\times N_0^{(n-r)}(E'_{12})2m_3(2\pi)^3\delta^3(\mathbf{k}'_3-\mathbf{k}_3),$$

$$\text{with} \quad E\Big|_{\text{threshold in 12-channel}} \equiv E_{thr12} = \frac{\mathbf{k}_3^2}{2\mu_3}. \tag{3.50}$$

So, for the amplitude determined in Eq. (3.34) we write:

$$\left[B^{(0)}_{\Lambda-12}(s_{12})+B^{(0)}_{13-12}(s_{12})+B^{(0)}_{23-12}(s_{12})\right]\frac{1}{1-B^{(0)}_{12}(s_{12})} =$$

$$= \int\limits_0^\infty \frac{dE'}{\pi}\frac{(E-E_{thr12})}{(E'-E_{thr12})(E'-E-i0)}\delta\left(E'-\frac{\mathbf{k}'^2_{12}}{2\mu_{12}}-\frac{\mathbf{k}'^2_3}{2\mu_3}\right)\frac{d^3\mathbf{k}'_{12}d^3\mathbf{k}'_3}{16m_1m_2m_3(2\pi)^5}$$

$$\times\left[\Lambda(s'_{12},s'_{13},s'_{23})+A^{(0)}_{13}(s'_{13})+A^{(0)}_{23}(s'_{23})\right]$$

$$\times N_0^{(n-r)}(E'_{12})2m_3(2\pi)^3\delta^3(\mathbf{k}'_3-\mathbf{k}_3)\frac{1}{1-B^{(0)}_{12}(s_{12})}$$

$$\to \int\limits_0^\infty \frac{dE'}{\pi}\frac{1}{E'-E-i0}d\Phi_3^{n-r}(P';k'_1,k'_2,k'_3)$$

$$\times\left[\Lambda(s'_{12},s'_{13},s'_{23})+A^{(0)}_{13}(s'_{13})+A^{(0)}_{23}(s'_{23})+A^{(0)}_{12}(s'_{12})\right]$$

$$\times N_0^{(n-r)}(E'_{12})2m_3(2\pi)^3\delta^3(\mathbf{k}'_3-\mathbf{k}_3) \tag{3.51}$$

In the final lines of (3.51) we omit the subtraction and write down the non-relativistic phase space factor (see Eq. (3.40). We also eliminate $[1-B^{(0)}_{12}(s_{12})]^{-1}$ by introducing an additional term (it is $A^{(0)}_{12}(s'_{12})$) in the spectral integral. That allows to write the interaction block in an universal form. The non-relativistic equation for $A^{(J_{in}=0)}_{P_1P_2P_3}(s_{12},s_{13},s_{23})$ reads:

$$A^{(J_{in}=0)}_{P_1P_2P_3}(s_{12},s_{13},s_{23}) = \Lambda(s_{12},s_{13},s_{23})+A^{(0)}_{12}(s_{12})+A^{(0)}_{13}(s_{13})+A^{(0)}_{23}(s_{23}),$$

$$A_{12}^{(0)}(s_{12}) = \int_0^\infty \frac{dE'}{\pi} \frac{1}{E' - E - i0} d\Phi_3^{(n-r)}(P'; k_1', k_2', k_3')$$
$$\times \Big[\Lambda(s_{12}', s_{13}', s_{23}') + A_{13}^{(0)}(s_{13}') + A_{23}^{(0)}(s_{23}') + A_{12}^{(0)}(s_{12}')\Big]$$
$$\times N_0^{(n-r)}(E_{12}') 2m_3 (2\pi)^3 \delta^3(\mathbf{k}_3' - \mathbf{k}_3)\,,$$
$$A_{13}^{(0)}(s_{13}) = \int_0^\infty \frac{dE'}{\pi} \frac{1}{E' - E - i0} d\Phi_3^{(n-r)}(P'; k_1', k_2', k_3')$$
$$\times \Big[\Lambda(s_{12}', s_{13}', s_{23}') + A_{13}^{(0)}(s_{13}') + A_{23}^{(0)}(s_{23}') + A_{12}^{(0)}(s_{12}')\Big]$$
$$\times N_0^{(n-r)}(E_{13}') 2m_3 (2\pi)^3 \delta^3(\mathbf{k}_2' - \mathbf{k}_2)\,,$$
$$A_{13}^{(0)}(s_{23}) = \int_0^\infty \frac{dE'}{\pi} \frac{1}{E' - E - i0} d\Phi_3^{(n-r)}(P'; k_1', k_2', k_3')$$
$$\times \Big[\Lambda(s_{12}', s_{13}', s_{23}') + A_{13}^{(0)}(s_{13}') + A_{23}^{(0)}(s_{23}') + A_{12}^{(0)}(s_{12}')\Big]$$
$$\times N_0^{(n-r)}(E_{23}') 2m_3 (2\pi)^3 \delta^3(\mathbf{k}_1' - \mathbf{k}_1). \tag{3.52}$$

Introducing the interaction block

$$V\Big(E_{12}', E_{13}', E_{23}'\,;\, (\mathbf{k}_3' - \mathbf{k}_3), (\mathbf{k}_2' - \mathbf{k}_2), (\mathbf{k}_1' - \mathbf{k}_1)\Big)$$
$$= N_0^{(n-r)}(E_{12}') 2m_3 (2\pi)^3 \delta^3(\mathbf{k}_3' - \mathbf{k}_3) + N_0^{(n-r)}(E_{12}') 2m_3 (2\pi)^3 \delta^3(\mathbf{k}_3' - \mathbf{k}_3)$$
$$+ N_0^{(n-r)}(E_{23}') 2m_3 (2\pi)^3 \delta^3(\mathbf{k}_1' - \mathbf{k}_1)\,, \tag{3.53}$$

we read Eq. (3.52) as

$$A_{P_1P_2P_3}^{(J_{in}=0)}(s_{12}, s_{13}, s_{23}) = \Lambda(s_{12}, s_{13}, s_{23}) + \int_0^\infty \frac{dE'}{\pi} d\Phi_3^{(n-r)}(P'; k_1', k_2', k_3')$$
$$\times \frac{A_{P_1P_2P_3}^{(J_{in}=0)}(s_{12}', s_{13}', s_{23}')}{E' - E - i0} V\Big(E_{12}', E_{13}', E_{23}'\,;\, (\mathbf{k}_3' - \mathbf{k}_3), (\mathbf{k}_2' - \mathbf{k}_2), (\mathbf{k}_1' - \mathbf{k}_1)\Big). \tag{3.54}$$

This is the standard non-relativistic three-particle equation. It is reasonable to use here the symmetrized phase space

$$d\Phi_3^{(n-r)}(P'; k_1', k_2', k_3') = \delta\Big(E' - \frac{\mathbf{k}'^2_1}{2m_1} - \frac{\mathbf{k}'^2_2}{2m_2} - \frac{\mathbf{k}'^2_3}{2m_3}\Big)\delta^{(3)}\Big(\mathbf{k}'_1 + \mathbf{k}'_2 + \mathbf{k}'_3\Big)$$
$$\times \frac{d^3\mathbf{k}'_1 d^3\mathbf{k}'_2 d^3\mathbf{k}'_3}{16 m_1 m_2 m_3 (2\pi)^5}\,.$$

Remind that the relation $A^{(J_{in}=0)}_{P_1P_2P_3}(s'_{12}, s'_{13}, s'_{23})/(E' - E - i0)$ is connected to the wave function of the system, it is considered in more detail in the following section.

For the energy E one can use alternatively the expressions

$$E = \frac{\mathbf{k}_j^2}{2\mu_j} + \frac{\mathbf{k}_{i\ell}^2}{2\mu_{i\ell}}, \qquad i \neq j \neq \ell, \tag{3.55}$$

or

$$E = \frac{\mathbf{k}_i^2}{2\mu_{ij}} + \frac{\mathbf{k}_i\mathbf{k}_\ell}{m_j} + \frac{\mathbf{k}_\ell^2}{2\mu_{j\ell}}. \tag{3.56}$$

Recall, here

$$\frac{1}{\mu_{j\ell}} = \frac{1}{m_j} + \frac{1}{m_\ell}, \qquad \frac{1}{\mu_i} = \frac{1}{m_i} + \frac{1}{\mu_{j\ell}}.$$

Writing the standard non-relativistic three-particle equation in the usual for the momentum representation way (3.54), we have to remember that the relative position of the contours of integration over $dz/2$ and the threshold singularities is that presented in Fig. 3.3; the transition to new variables did not affect principally this position.

3.3 Consideration of amplitudes in terms of a three-particle spectral integral

The amplitudes with rescatterings which enter (3.15), i.e. $B^{(0)}_{\lambda-12}$, $B^{(0)}_{13-12}$ and $B^{(0)}_{23-12}$, can be considered in terms of three-particle spectral integrals. In the previous section it was done for the non-relativistic approach, here we present the generalization for relativistic systems. The principal point in this procedure is to write a spectral integral propagator for three particles. The non-relativistic propagation of particles is determined by Eq. (3.54), for the relativistic case we suggest the straightforward generalization:

$$\int\limits_0^\infty \frac{dE'}{\pi}\frac{d\Phi_3^{(n-r)}(P';k'_1,k'_2,k'_3)}{E'-E-i0} \to \int\limits_{(\sum m_a)^2}^\infty \frac{ds'}{\pi}\frac{d\Phi_3(P';k'_1,k'_2,k'_3)}{s'-s-i0}. \tag{3.57}$$

Here $\sum m_a = m_1 + m_2 + m_3$ and, let us recall, the three-particle phase space is determined as

$$d\Phi_3(P;k_1,k_2,k_3) = \frac{1}{2}(2\pi)^4\delta^4(P-k_1-k_2-k_3) \times \frac{d^3k_1}{(2\pi)^3 2k_{10}}\frac{d^3k_2}{(2\pi)^3 2k_{20}}\frac{d^3k_3}{(2\pi)^3 2k_{30}}. \tag{3.58}$$

In the initial state we have the energy squared s, in the intermediate state the total invariant mass squared $P'^2 = (k'_1+k'_2+k'_3)^2 = s'$ and two-particle energies squared $(k'_i+k'_j)^2 = s'_{ij}$. Correspondingly, we re-denote amplitudes which get in Eq. (3.15):

$$
\begin{aligned}
&B^{(0)}_{\lambda-12}(s_{12}) \to B^{(0)}_{\lambda-12}(s,s_{12}),\\
&B^{(0)}_{j3-12}(s_{12}) \to B^{(0)}_{j3-12}(s,s_{12}), \quad j=1,2,\\
&A^{(0)}_{ij}(s_{ij}) \to A^{(0)}_{ij}(s,s_{ij}), \quad i\neq j=1,2,3.
\end{aligned}
\tag{3.59}
$$

Then we write:

$$
\begin{aligned}
A^{(J_{in}=0)}_{P_1P_2P_3}(s,s_{12},s_{13},s_{23}) &= \Lambda(s,s_{12},s_{13},s_{23}) + \\
&+ A^{(0)}_{12}(s,s_{12}) + A^{(0)}_{13}(s,s_{13}) + A^{(0)}_{23}(s,s_{23}),\\
A^{(0)}_{12}(s,s_{12}) &= \Big(B^{(0)}_{\Lambda-12}(s,s_{12})+B^{(0)}_{13-12}(s,s_{12})+B^{(0)}_{23-12}(s,s_{12})\Big)\\
&\times \frac{1}{1-B^{(0)}_{12}(s_{12})}G^R_0(s_{12}),\\
A^{(0)}_{13}(s,s_{13}) &= \Big(B^{(0)}_{\Lambda-13}(s,s_{13})+B^{(0)}_{12-13}(s,s_{13})+B^{(0)}_{23-13}(s,s_{13})\Big)\\
&\times \frac{1}{1-B^{(0)}_{13}(s_{13})}G^R_0(s_{13}),\\
A^{(0)}_{23}(s,s_{23}) &= \Big(B^{(0)}_{\Lambda-23}(s,s_{23})+B^{(0)}_{13-23}(s,s_{23})+B^{(0)}_{12-23}(s,s_{23})\Big)\\
&\times \frac{1}{1-B^{(0)}_{23}(s_{23})}G^R_0(s_{23})
\end{aligned}
\tag{3.60}
$$

with

$$
\begin{aligned}
B^{(0)}_{\Lambda-12}(s,s_{12}) &= \int\limits_{(\sum m_a)^2}^{\infty} \frac{ds'}{\pi}\frac{d\Phi_3(P';k'_1,k'_2,k'_3)}{s'-s-i0} \times\\
&\times \Lambda(s',s'_{12},s'_{13},s'_{23})G^0_L(s'_{12})(2\pi)^3\,2k_{30}\delta(\vec{k}'_3-\vec{k}_3),\\
B^{(0)}_{13-12}(s,s_{12}) &= \int\limits_{(\sum m_a)^2}^{\infty} \frac{ds'}{\pi}\frac{d\Phi_3(P';k'_1,k'_2,k'_3)}{s'-s-i0} \times\\
&\times A^{(0)}_{13}(s',s'_{13})G^0_L(s'_{12})(2\pi)^3\,2k_{30}\delta(\vec{k}'_3-\vec{k}_3),\\
B^{(0)}_{23-12}(s,s_{12}) &= \int\limits_{(\sum m_a)^2}^{\infty} \frac{ds'}{\pi}\frac{d\Phi_3(P';k'_1,k'_2,k'_3)}{s'-s-i0} \times\\
&\times A^{(0)}_{23}(s',s'_{13})G^0_L(s'_{12})(2\pi)^3\,2k_{30}\delta(\vec{k}'_3-\vec{k}_3).
\end{aligned}
\tag{3.61}
$$

for $A^{(0)}_{12}(s,s_{12})$ and analogous formulae for $A^{(0)}_{13}(s,s_{13})$ and $A^{(0)}_{23}(s,s_{23})$. The factor $(2\pi)^3 2k_{30}\delta(\vec{k}'_3-\vec{k}_3)$ in (3.61) arises due to the propagation of the non-interacting particle, P_3, in the final state; we have $k'_{30}=k_{30}$.

3.3.1 *Kinematics of the outgoing particles in the c.m. system*

In the c.m. system of three particles we have for final state momenta:

$$P=k_1+k_2+k_3,\quad P^2=s,\quad p_{12}=k_1+k_2,\quad (k_1+k_2)^2=s_{12},$$
$$P=(\sqrt{s},0,0,0),\quad k_3=(\frac{s+m_3^2-s_{12}}{2\sqrt{s}},\mathbf{n}_3k_3)$$
$$k_3=\frac{1}{2\sqrt{s}}\sqrt{[s-(m_3+\sqrt{s_{12}})^2][s-(m_3-\sqrt{s_{12}})^2]}\,. \qquad (3.62)$$

In the intermediate state:

$$P'=k'_1+k'_2+k_3,\quad P'^2=s',\quad p'_{12}=k'_1+k'_2,\quad (k'_1+k'_2)^2=s'_{12},$$
$$k'_3=k_3\,,\quad P'=(\sqrt{s'},0,0,0),$$
$$k'_3=(\frac{s'+m_3^2-s'_{12}}{2\sqrt{s'}},0,0,\frac{1}{2\sqrt{s'}}\sqrt{[s'-(m_3+\sqrt{s'_{12}})^2][s'-(m_3-\sqrt{s'_{12}})^2]},$$
$$k'_1=\frac{1}{2}p'_{12}+k_{12},\quad k'_2=\frac{1}{2}p'_{12}-k_{12},\quad k_{12}=\frac{1}{2}(k'_1-k'_2),$$
$$p'_{12}=(\sqrt{s'}-\frac{s+m_3^2-s_{12}}{2\sqrt{s}},0,0,-\frac{1}{2\sqrt{s}}\sqrt{[s-(m_3+\sqrt{s_{12}})^2][s-(m_3-\sqrt{s_{12}})^2]}.$$

3.3.2 *Calculation of the block $B^{(0)}_{13-12}(s,s_{12})$*

The block $B^{(0)}_{13-12}(s,s_{12})$ reads:

$$B^{(0)}_{13-12}(s,s_{12})=$$
$$\int\limits_{\sum m_a^2}^{\infty}\frac{ds'}{\pi}\frac{d\Phi_3(P';k'_1,k'_2,k'_3)}{s'-M^2-i0}A^{(0)}_{13}(P'^2,s'_{13})G^0_L(s'_{12})(2\pi)^3\,2k_{30}\delta(\vec{k}'_3-\vec{k}_3)=$$
$$=\int\limits_{\sum m_a^2}^{\infty}\frac{ds'}{\pi}\int ds'_{12}\delta\Big(s'_{12}-(P'-k_3)^2\Big)\frac{d^3k'_1\,d^3k'_2}{k'_{10}k'_{20}}\delta^{(4)}(P'-k'_1-k'_2-k_3)$$
$$\times\frac{1}{8\pi^2}\frac{A^{(0)}_{13}\Big(P'^2,(k'_1+k_3)^2\Big)}{s'-M^2-i0}G^0_L((P'-k_3)^2). \qquad (3.63)$$

We can eliminate some of the delta-functions in (3.63):
(i) The factors

$$d^3k'_2\delta^{(3)}(\vec{P}' - \vec{k}'_1 - \vec{k}'_2 - \vec{k}_3) \tag{3.64}$$

give us

$$\vec{k}'_2 = \vec{P}' - \vec{k}'_1 - \vec{k}_3\,. \tag{3.65}$$

In the c.m. system, $\vec{P}' = 0$, we re-write

$$\vec{k}'_2 = -\vec{k}'_1 - \vec{k}_3\,. \tag{3.66}$$

(ii) The factor

$$\frac{d^3k'_1}{k'_{10}k'_{20}}\delta^{(0)}(P'_0 - \sqrt{m_1^2 + |\vec{k}'_1|^2} - \sqrt{m_2^2 + |\vec{P}' - \vec{k}'_1 - \vec{k}_3|^2} - k_{30}) \tag{3.67}$$

determines $|\vec{k}'_1|$.
(iii) The factor

$$ds'_{12}\delta\Big(s'_{12} - (P' - k_3)^2\Big) \tag{3.68}$$

gives a relation between s'_{12} and s'.

3.4 Three-particle composite systems, their wave functions and form factors

In terms of the three-particle spectral integral technique we introduce the wave function of a composite system and its form factor.

The composite systems appear like poles of amplitudes when summing up the infinite chains of the transitions $1+2+3 \to 1+2+3$. In the framework of the graphical representation of the dispersion diagrams (see Chapter 2), one of them looks like that presented in Fig. 3.5.

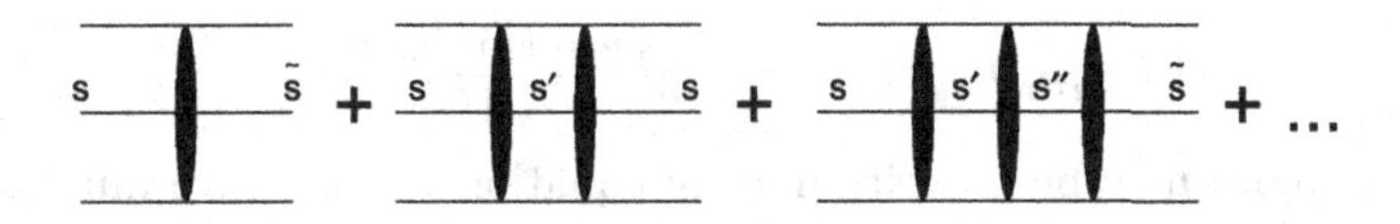

Fig. 3.5 Set of three particle rescattering diagrams in the spectral integral technique.

After the summation, diagrams of the Fig. 3.5 type give a pole amplitude, see Fig. 3.6. From here follows the graphical representation of the spectral integral equation for the pole vertex - it is presented in Fig. 3.7. The analogy between the considered here three-particle case and the two-particle case we have investigated in Chapter 2 is obvious. Following the logics of Chapter 2, we write the equations for the vertex, the wave function, introduce the form-factor and the normalization condition for the wave function.

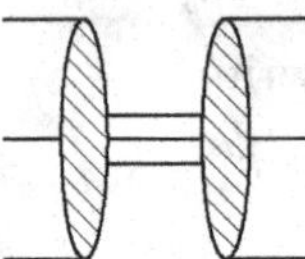

Fig. 3.6 Pole amplitude which corresponds to a three-particle composite state.

3.4.1 *Vertex and wave function*

The equation for the vertex function of a composite system reads:

$$G_{P_1P_2P_3}(k_1,k_2,k_3) = \int\limits_{(\sum m_i)^2}^{\infty} \frac{ds'}{\pi} \int d\Phi_3(P';k_1',k_2',k_3') V_{P_1P_2P_3}(t'^{(12)}_\perp, t'^{(13)}_\perp), t'^{(23)}_\perp) \times \frac{G_{P_1P_2P_3}(k_1',k_2',k_3')}{s'-M^2} \, . \tag{3.69}$$

A specific feature of (3.69) is the homogeneity of the equation.

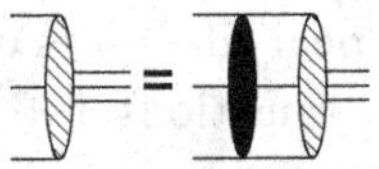

Fig. 3.7 Graphical representation of the equation for the vertex of transition of three particles into composite state, Eq. (3.69).

One can introduce the wave function

$$\Psi_{P_1P_2P_3}(k_1,k_2,k_3) = \frac{G_{P_1P_2P_3}(k_1,k_2,k_3)}{s-M^2} \, , \tag{3.70}$$

then Eq. (3.69) may be re-written as an equation for the wave function:

$$(s-M^2)\,\Psi_{P_1P_2P_3}(k_1,k_2,k_3) = \int\limits_{(\sum m_i)^2}^{\infty} \frac{ds'}{\pi} \int d\Phi_3(P';k_1',k_2',k_3') \times V_{P_1P_2P_3}(t'^{(12)}_\perp, t'^{(13)}_\perp), t'^{(23)}_\perp) \Psi_{P_1P_2P_3}(k_1',k_2',k_3') \, . \tag{3.71}$$

3.4.2 *Three particle composite system form factor*

The vertex function introduced in Eq. (3.69) gives way to study the interaction of the three-particle composite system with the electromagnetic field.

The full amplitude of the interaction of the photon with a composite system, when the charge of the composite system equals unity, is:

$$A_\mu(q^2) = (p_\mu + p'_\mu)F(q^2)\ . \tag{3.72}$$

The form factor of the composite system is an invariant coefficient in front of the transverse part of the amplitude A_μ:

$$(p+p') \perp q\ . \tag{3.73}$$

Consider the dispersion representation of the triangle diagram shown in Fig. 3.8. It can be written in a way similar to the double spectral representation for the triangle diagram in the case of two-particle composite systems. Cuts of this triangle diagram result in the double spectral representation over $P^2 = s$, $P'^2 = s'$ with fixed $(P' - P)^2 = q^2$.

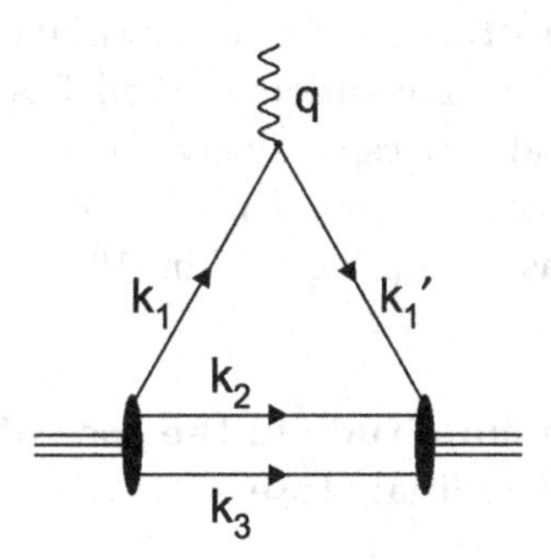

Fig. 3.8 Constituent of the composite system interacts with electromagnetic field.

In this way, the form factor of the composite system reads:

$$F(q^2) = \int\limits_{(\sum m_i)^2}^{\infty} \frac{ds}{\pi} \int\limits_{(\sum m_i)^2}^{\infty} \frac{ds'}{\pi} \frac{disc_s\, disc_{s'} F(s, s', q^2)}{(s - M^2)(s' - M^2)}\ , \tag{3.74}$$

where

$$\begin{aligned} disc_s\, disc_{s'} F(s, s', q^2) &= G_{P_1P_2P_3}(k_1, k_2, k_3) G_{P_1P_2P_3}(k'_1, k_2, k_3) \\ &\times d\Phi_3^{tr}(P, P'; k_1, k'_1; k_2, k_3)\alpha_3(s, s', s_{23}, q^2), \end{aligned} \tag{3.75}$$

with

$$\begin{aligned} d\Phi_3^{tr}(P, P'; k_1, k'_1; k_2, k_3) &= d\Phi_3(P'; k_1, k_2, k_3) d\Phi_3(P'; k'_1, k_2, k_3) \\ &\times (2\pi)^3 2k'_{20}\delta^3(\mathbf{k}_2 - \mathbf{k}'_2)(2\pi)^3 2k'_{30}\delta^3(\mathbf{k}_3 - \mathbf{k}'_3) \end{aligned} \tag{3.76}$$

In the diagram of Fig. 3.8, the gauge invariant vertex for the interaction of a scalar (or pseudoscalar) constituent with a photon is written as $(k_{1\mu} + k'_{1\mu})$, from which one should separate a factor orthogonal to the momentum transfer $P'_\mu - P_\mu$ that results in factor $\alpha_3(s, s', s_{23}, q^2)$:

$$k_{1\mu} + k'_{1\mu} = \alpha_3(s, s', s_{23}, q^2)\left[P_\mu + P'_\mu - \frac{s'-s}{q^2}(P'_\mu - P_\mu)\right] + k_{\perp\mu},$$

$$\alpha_3(s, s', s_{23}, q^2) = -\frac{q^2\left(s + s' + 2m_1^2 - 2s_{23} - q^2\right)}{\lambda(s, s', q^2)},$$

$$\lambda(s, s', q^2) = -2q^2(s+s') + q^4 + (s'-s)^2\,, \qquad (3.77)$$

where $k_{\perp\mu}$ is orthogonal to both $(P_\mu + P'_\mu)$ and $(P_\mu - P'_\mu)$. The term $k_{\perp\mu}$ equals zero after integrating over the phase space.

Formula (3.74) gives a correct normalization of the charge form factor,

$$F(0) = 1\,. \qquad (3.78)$$

The procedure of construction of gauge invariant amplitudes within the framework of the spectral integration method has been realized for the deuteron in [17, 18], and, correspondingly, for the elastic scattering and photodisintegration process. A generalization of the method for the composite quark systems has been performed in [19, 20, 21].

3.5 Equation for an amplitude in the case of instantaneous interactions in the final state

The equation for amplitude reads:

$$A_{P_1P_2P_3}(k_1, k_2, k_3) = \Lambda(k_1, k_2, k_3) + \int\limits_{(\sum m_i)^2}^{\infty} \frac{ds'}{\pi} \frac{d\Phi_3(P'; k'_1, k'_2, k'_3)}{s' - s - i0} A_{P_1P_2P_3}(k'_1, k'_2, k'_3) \times V_{P_1P_2P_3}(k'_1, k'_2, k'_3\,;\, k_1, k_2, k_3). \qquad (3.79)$$

with $\sum m_i = m_1 + m_2 + m_3$. We have changed here the notations, which makes the work with the instantaneous interaction more convenient. Namely, instead of s_{ij} we use four-momenta k_1, k_2, k_3, while the interaction block is denoted as $V_{P_1P_2P_3}(k'_1, k'_2, k'_3\,;\, k_1, k_2, k_3)$. For an instantaneous two-particle interaction we write

$$V_{P_1P_2P_3}(k'_1, k'_2, k'_3\,;\, k_1, k_2, k_3) \to V(t'^{(12)}_\perp)(2\pi)^3\, 2k_{30}\delta(\vec{k}'_3 - \vec{k}_3) + V(t'^{(13)}_\perp)(2\pi)^3\, 2k_{20}\delta(\vec{k}'_2 - \vec{k}_2) + V(t'^{(23)}_\perp)(2\pi)^3\, 2k_{10}\delta(\vec{k}'_1 - \vec{k}_1), \qquad (3.80)$$

where $t'^{(ij)}_{\perp}$ depends on space-like momentum components only (see Chapter 2 also). For a three-particle state we write:

$$t'^{(ij)}_{\perp} = (k_i^{\perp P} - k_i'^{\perp P'})_\mu(-k_j^{\perp P} + k_j'^{\perp P'})_\mu \,. \tag{3.81}$$

In the general case the interaction block may depend on many $t'^{(ij)}_{\perp}$ simultaneously.

3.6 Conclusion

For the description of relativistic three-particle systems we see now two possible approaches:

(1) The development of the system in two-particle channels is considered step by step: in the first, the second, the third etc. channels (this approach is presented in Section 3.1).

(2) The evolution of the system is considered in the three-particle channel (the scheme of the investigation and the equations are given in Sections 3.2, 3.3).

In principle, the second approach seems to be more convenient. Here we have the possibility to introduce three-particle forces, and this may be important for systems of three quarks.

However, the positive features of the first approach are also obvious. In its framework we can make use of the equations for the "two particles → two particles" transitions which occurred to be rather useful in many applied problems. These equations can be continued analytically – by increasing one of the masses – into the "one particle → three particles" decay region. In fact we have here a unique scheme for the description of both the two-particle and three-particle systems.

The two approaches have to be equivalent if we restrict ourselves to the region of two-particle forces. From the point of view of their technical realization, their possible use for applied problems may be, however, quite different. The advantage of one of the methods in a given problem can be verified only by using them in a rather broad region. The application of these technics to some interesting cases will be demonstrated in the following chapters.

We have wrote equations for three spinless particles. The consideration of these cases is definitely sufficient for revealing the principal features of the equations. The transitions to technics describing particles with spins can be found without problems, see, *e.g.*, [14].

3.7 Appendix A. Example: loop diagram with $G^L = G^R = 1$

In the following we shall use chains of loop diagrams in the framework of the diagram technics. To give an idea what this object looks like, let us present an explicit expression of the loop diagram in the simplest case: $G^L = G^R = 1$ and $m_1 = m_2 = m$, and, in order to provide the convergence of the dispersion integral, let us carry out a subtraction in the point $s = 4m^2$:

$$B(s) = (s - 4m^2) \int\limits_{4m^2}^{\infty} \frac{ds'}{\pi} \frac{1}{16\pi} \sqrt{\frac{s' - 4m^2}{s'}} \frac{1}{(s' - 4m^2)(s' - s - i0)}$$
$$= \frac{1}{16\pi} \sqrt{\frac{s - 4m^2}{s}} \left[\frac{1}{\pi} \ln \frac{\sqrt{s} - \sqrt{s - 4m^2}}{\sqrt{s} + \sqrt{s - 4m^2}} + i \right], \quad s > 4m^2. \tag{3.82}$$

The formula (3.82) gives an expression in the physical region, on the first sheet of the complex plane s. In the process of transition to the non-physical region, when $s \gtrsim 0$ is on the first sheet, one should transform $\sqrt{s - 4m^2} \to i\sqrt{4m^2 - s}$. We have:

$$B(s) = \frac{i}{16\pi} \sqrt{\frac{-s + 4m^2}{s}} \left[\frac{1}{\pi} \ln \frac{\sqrt{s} - i\sqrt{-s + 4m^2}}{\sqrt{s} + i\sqrt{-s + 4m^2}} + i \right], \quad 0 < s < 4m^2. \tag{3.83}$$

Here $B(s)$ is real, $\mathrm{Im}B(s) = 0$, and analytical at $s \to 0$:

$$B(s) \simeq \frac{i}{16\pi} \sqrt{\frac{4m^2}{s}} \left[\frac{1}{\pi} \left(- i\pi + 2i\sqrt{\frac{s}{4m^2}} \right) + i \right], \quad s \to 0. \tag{3.84}$$

On the first sheet the only singularity of $B(s)$ is the threshold one, at $s = 4m^2$. The singularity $s = 0$, being absent on the first sheet, is located on the second sheet which can be reached from (3.82) at $\sqrt{s - 4m^2} \to -i\sqrt{4m^2 - s}$ with decreasing s.

In the non-relativistic limit the loop diagram $B(s)$ reads:

$$B(s) \to B^{(n-r)}(k^2) = \frac{1}{16\pi m} ik \quad \text{at} \quad \frac{s}{4} - m^2 = k^2 > 0. \tag{3.85}$$

3.8 Appendix B. Phase space for n-particle state

We write two-particle and three-particle phase spaces as:

$$d\Phi_2(P_{12}; k_1, k_2) = \frac{1}{2}(2\pi)^4 \delta^4(P_{12} - k_1 - k_2)$$
$$\times \frac{d^4k_1}{(2\pi)^3} \delta(k_1^2 - m_1^2) \frac{d^4k_2}{(2\pi)^3} \delta(k_2^2 - m_2^2), \tag{3.86}$$

and

$$d\Phi_3(P_{123};k_1,k_2,k_3) = \frac{1}{2}(2\pi)^4\delta^4(P_{123}-k_1-k_2-k_3) \tag{3.87}$$
$$\times \frac{d^4k_1}{(2\pi)^3}\delta(k_1^2-m_1^2)\frac{d^4k_2}{(2\pi)^3}\delta(k_2^2-m_2^2)\frac{d^4k_3}{(2\pi)^3}\delta(k_3^2-m_3^2),$$

Eq. (3.87) can be rewritten as an integrated product of two two-particle phase spaces:

$$d\Phi_3(P_{123};k_1,k_2,k_3) = \frac{ds_{12}}{\pi}d\Phi_2(P_{123};P_{12},k_3)d\Phi_2(P_{12};k_1,k_2),$$
$$d\Phi_2(P_{123};P_{12},k_3) = \frac{1}{2}(2\pi)^4\delta^4(P_{123}-P_{12}-k_3)\frac{d^4P_{12}}{(2\pi)^3}\delta(P_{12}^2-s_{12})$$
$$\times \frac{d^4k_3}{(2\pi)^3}\delta(k_3^2-m_3^2), \tag{3.88}$$

The general case of the n-particle phase space can be written as:

$$d\Phi_n(P_{1\ldots n};k_1,k_2,\ldots k_n) = \frac{1}{2}(2\pi)^4\delta^4(P_{12\ldots n}-k_1-k_2\cdots-k_n)$$
$$\times\frac{d^4k_1}{(2\pi)^3}\delta(k_1^2-m_1^2)\frac{d^4k_2}{(2\pi)^3}\delta(k_2^2-m_2^2)\ldots\frac{d^4k_n}{(2\pi)^3}\delta(k_n^2-m_n^2)$$
$$=\frac{ds_{12\ldots n-1}}{\pi}d\Phi_2(P_{12\ldots n};P_{12\ldots n-1},k_n)d\Phi_{n-1}(P_{12\ldots n-1};k_1,\ldots,k_{n-1}). \tag{3.89}$$

3.9 Appendix C. Feynman diagram technique and evolution of systems in the positive time-direction

Rather often it is convenient to start the consideration of amplitudes using Feynman diagram technique. Here we consider in this way three-particle non-relativistic amplitudes. The Feynman amplitudes include both positive and negative time-direction evolutions, we transform amplitudes to non-relativistic limit thus eliminating the negative time-direction evolutions.

In terms of the Feynman propagators we write for three particle decay processes $P_4 \to P_1P_2P_3$ in momentum representation:

$$A(P;k_1,k_2,k_3) = A_{12}(k_1,k_2)+A_{13}(k_1,k_3)+A_{23}(k_2,k_3)\,,$$
$$A_{12}(k_1,k_2) = \int\frac{d^4k_2'}{i(2\pi)^4}a_{12}(k_1',k_2')\frac{d^4k_1'\delta^{(4)}(k_1'+k_2'-k_1-k_2)}{(m_1^2-k_1'^2-i0)(m_2^2-k_2'^2-i0)}$$
$$\times\left[A_{12}(k_1',k_2')+A_{13}(k_1',k_3)+A_{23}(k_2',k_3)\right]. \tag{3.90}$$

In (3.90) we do not consider the term with direct production of particles putting $\lambda(k_1,k_2,k_3)=0$.

3.9.1 *The Feynman diagram technique and non-relativistic three particle systems*

The non-relativistic three particle amplitude for the decay process $P_4 \to P_1P_2P_3$ is realized in (3.90) by approximations:

$$k_{j0} \simeq m_j + \varepsilon_j\,, \qquad \left(m_j^2 - k_j^2 - i0\right)^{-1} \simeq \left(\mathbf{k}_j^2 - 2m_j\varepsilon_j - i0\right)^{-1} \quad (3.91)$$

Then Eq. (3.90) reads:

$$A_{12}(\mathbf{k}_1,\mathbf{k}_2) = \int \frac{d^3\mathbf{k}'_2 d\varepsilon'_2}{i(2\pi)^4} a_{12}(\mathbf{k'}_1,\mathbf{k'}_2) \frac{d^3\mathbf{k}'_1\delta^{(3)}(\mathbf{k}'_1+\mathbf{k}'_2-\mathbf{k}_1-\mathbf{k}_2)}{\left(\mathbf{k'}^2_1 - 2m_1\varepsilon'_1 - i0\right)\left(\mathbf{k'}^2_2 - 2m_2\varepsilon'_2 - i0\right)}$$
$$\times d\varepsilon'_1\delta(\varepsilon'_1+\varepsilon'_2-\varepsilon_1-\varepsilon_2)\Big[A_{12}(\mathbf{k}'_1,\mathbf{k}'_2) + A_{13}(\mathbf{k}'_1,\mathbf{k}_3) + A_{23}(\mathbf{k}'_2,\mathbf{k}_3)\Big]. \quad (3.92)$$

After the integration over $d\varepsilon'_1 \cdot d\varepsilon'_2$ one has:

$$A_{12}(\mathbf{k}_1,\mathbf{k}_2) = \int \frac{d^3\mathbf{k}'_2}{(2\pi)^3} \frac{a_{12}(\mathbf{k'}_1,\mathbf{k'}_2)}{4m_1m_2} \frac{d^3\mathbf{k}'_1\delta^{(3)}(\mathbf{k}'_1+\mathbf{k}'_2-\mathbf{k}_1-\mathbf{k}_2)}{\frac{1}{2m_1}\mathbf{k'}^2_1 + \frac{1}{2m_2}\mathbf{k'}^2_2 - \varepsilon_{12} - i0}$$
$$\times \Big[A_{12}(\mathbf{k}'_1,\mathbf{k}'_2) + A_{13}(\mathbf{k}'_1,\mathbf{k}_3) + A_{23}(\mathbf{k}'_2,\mathbf{k}_3)\Big]. \quad (3.93)$$

Here $\varepsilon_{12} = \varepsilon_1 + \varepsilon_2$. Equations for $A_{13}(\mathbf{k}_1,\mathbf{k}_3)$ and $A_{13}(\mathbf{k}_1,\mathbf{k}_3)$ are written analogously.

Denoting

$$A(\mathbf{k}_1,\mathbf{k}_2,\mathbf{k}_3) = A_{12}(\mathbf{k}_1,\mathbf{k}_2) + A_{13}(\mathbf{k}_1,\mathbf{k}_3) + A_{23}(\mathbf{k}_2,\mathbf{k}_3), \quad (3.94)$$

we write the equation for $A(\mathbf{k}_1,\mathbf{k}_2,\mathbf{k}_3)$ in a symmetrical form:

$$A(\mathbf{k}_1,\mathbf{k}_2,\mathbf{k}_3) = \int \frac{d^3\mathbf{k}'_1}{(2\pi)^3}\frac{d^3\mathbf{k}'_2}{(2\pi)^3}\frac{d^3\mathbf{k}'_3}{(2\pi)^3} \frac{(2\pi)^3\delta^{(3)}(\mathbf{k}'_1+\mathbf{k}'_2+\mathbf{k}'_3 - \mathbf{k}_1+\mathbf{k}_2+\mathbf{k}_3)}{\frac{1}{2m_1}\mathbf{k'}^2_2 + \frac{1}{2m_2}\mathbf{k'}^2_2 + \frac{1}{2m_3}\mathbf{k'}^2_3 - E - i0}$$
$$\times A(\mathbf{k}'_1,\mathbf{k}'_2,\mathbf{k}'_3)(2\pi)^3\Big[\delta^{(3)}(\mathbf{k}'_3-\mathbf{k}_3)\frac{a_{12}(\mathbf{k}'_1,\mathbf{k}'_2)}{4m_1m_2} + \delta^{(3)}(\mathbf{k}'_2-\mathbf{k}_2)\frac{a_{13}(\mathbf{k}'_1,\mathbf{k}'_3)}{4m_1m_3}$$
$$+\delta^{(3)}(\mathbf{k}'_1-\mathbf{k}_1)\frac{a_{23}(\mathbf{k}'_2,\mathbf{k}'_3)}{4m_2m_3}\Big]. \quad (3.95)$$

Here $E = \varepsilon_{ij} + \frac{1}{2m_\ell}\mathbf{k}_\ell^2$ with $i \neq j \neq \ell$.

For specifying Eq. (3.95) we may choose two ways of handling the problem; they are presented below. We can work either with ($\mathbf{k}_{ij}$,$\mathbf{k}_\ell$) in considering (ij) in the c.m. system ($\mathbf{k}_i + \mathbf{k}_j = 0$), or with ($\mathbf{k}_1$,$\mathbf{k}_2$,$\mathbf{k}_3$) in the c.m. system of three particles ($\mathbf{k}_1 + \mathbf{k}_2 + \mathbf{k}_3 = 0$).

3.9.1.1 *Two particle c.m. system,* $\mathbf{k}_1 + \mathbf{k}_2 = 0$

In this system we return to dispersion relation formulae presented above. One has in this system $\mathbf{k}_1 = -\mathbf{k}_2 = \mathbf{k}_{12}$ and $\mathbf{k'}_1 = -\mathbf{k'}_2 = \mathbf{k'}_{12}$.

Equation (3.93) reads:

$$A_{12}(\mathbf{k}_1,\mathbf{k}_2) = \int \frac{d^3\mathbf{k}'_{12}}{(2\pi)^3} \frac{a_{12}(\mathbf{k'}_1,\mathbf{k'}_2)}{4m_1m_2} \frac{A_{12}(\mathbf{k}'_1,\mathbf{k}'_2)+A_{13}(\mathbf{k}'_1,\mathbf{k}_3)+A_{23}(\mathbf{k}'_2,\mathbf{k}_3)}{\frac{1}{2\mu_{12}}\mathbf{k'}^2_{12} - \frac{1}{2\mu_{12}}\mathbf{k}^2_{12} - i0}. \tag{3.96}$$

For comparison with dispersion relation formulae it is convenient to re-write (3.96) as

$$\begin{aligned} A_{12}(\mathbf{k}_1,\mathbf{k}_2) &= \int \frac{d^3\mathbf{k}'_{12}}{(2\pi)^3} \frac{a_{12}(\mathbf{k'}_1,\mathbf{k'}_2)}{2(m_1+m_2)} \frac{A_{12}(\mathbf{k}'_1,\mathbf{k}'_2)}{\mathbf{k'}^2_{12} - \mathbf{k}^2_{12} - i0} \\ &+ \int \frac{d^3\mathbf{k}'_{12}}{(2\pi)^3} \frac{a_{12}(\mathbf{k'}_1,\mathbf{k'}_2)}{2(m_1+m_2)} \frac{A_{13}(\mathbf{k}'_1,\mathbf{k}_3)+A_{23}(\mathbf{k}'_2,\mathbf{k}_3)}{\mathbf{k'}^2_{12} - \mathbf{k}^2_{12} - i0}. \end{aligned} \tag{3.97}$$

The first term in the right-hand side of (3.97) is responsible for a set of rescatterings of the (12)-system. For example, if the interaction occurs in the S-wave, it results in the amplitude factor $(1 - B^{(0)}_{12}(\mathbf{k}^2_{12}))^{-1}$ (see the discussion in Section 2.2 of this chapter). The second term describes interactions in the (12)-system when the previous interactions happen in systems (13) or (23). One can re-write (3.97) using notations of Section 2:

$$\begin{aligned} &A_{ij}(\mathbf{k}_i,\mathbf{k}_j) \to A_{ij}(\mathbf{k}_{ij}), \\ &\int \frac{d^3\mathbf{k}'_{ij}}{(2\pi)^3} \frac{a_{ij}(\mathbf{k'}_i,\mathbf{k'}_j)}{2(m_i+m_j)} \to \int\limits_0^\infty \frac{d\mathbf{k'}^2_{ij}}{\pi} \int\limits_{-1}^1 \frac{dz}{2}\rho(\mathbf{k'}^2_{ij}), \end{aligned} \tag{3.98}$$

which leads us to formulae of Eq. (3.43). Equations written in terms of $\mathbf{k}^2_{ij}$ are covariant because $s_{ij} = (m_i+m_j)^2 + 4\mathbf{k}^2_{ij}$. The contour integration over dz is shown on Fig. 3.3.

3.9.1.2 *Three particle c.m. system,* $\mathbf{k}_1 + \mathbf{k}_2 + \mathbf{k}_3 = 0$

In this system one can write:

$$\begin{aligned} &\mathbf{k}_{12} = \frac{m_2}{m_1+m_2}\mathbf{k}_1 - \frac{m_1}{m_1+m_2}\mathbf{k}_2, \qquad \mathbf{k}_1 = -\frac{m_1}{m_1+m_2}\mathbf{k}_3 + \mathbf{k}_{12}, \\ &\frac{1}{2}(\mathbf{k}_1 - \mathbf{k}_2) = \mathbf{k}_{12} + \frac{-m_1+m_2}{2(m_1+m_2)}\mathbf{k}_3, \\ &\mathbf{k}_{13} = \frac{m_1(m_1+m_2+m_3)}{(m_1+m_2)(m_1+m_3)}\mathbf{k}_3 + \frac{m_3}{m_1+m_3}\mathbf{k}_{12} \end{aligned} \tag{3.99}$$

with other similar relations being given by cyclic permutation of indices. Equation (3.95) reads:

$$A_{12}(\mathbf{k}_1,\mathbf{k}_2)=\int\frac{d^3\mathbf{k}'_{12}}{(2\pi)^3}\frac{A_{12}(\mathbf{k}'_1,\mathbf{k}'_2)}{\frac{1}{2\mu_{12}}\mathbf{k'}^2_{12}+\frac{1}{2\mu_3}\mathbf{k}_3^2-E-i0}$$
$$+\int\frac{d^3\mathbf{k}'_2}{(2\pi)^3}\frac{a_{12}(\mathbf{k}'_1,\mathbf{k}'_2)}{4m_1m_2}\frac{A_{13}(\mathbf{k}'_1,\mathbf{k}_3)}{\frac{1}{2\mu_{12}}\mathbf{k'}^2_2+\frac{1}{m_1}(\mathbf{k}'_2\mathbf{k}_3)+\frac{1}{2\mu_{13}}\mathbf{k}_3^2-E-i0}$$
$$+\int\frac{d^3\mathbf{k}'_1}{(2\pi)^3}\frac{a_{12}(\mathbf{k}'_1,\mathbf{k}'_2)}{4m_1m_2}\frac{A_{23}(\mathbf{k}'_2,\mathbf{k}_3)}{\frac{1}{2\mu_{12}}\mathbf{k'}^2_1+\frac{1}{m_2}(\mathbf{k}'_1\mathbf{k}_3)+\frac{1}{2\mu_{23}}\mathbf{k}_3^2-E-i0} \quad (3.100)$$

Let us recall $\mu_{12}^{-1}=m_1^{-1}+m_2^{-1}$ and $\mu_3^{-1}=(m_1+m_2)^{-1}+m_3^{-1}$. In (3.100) we keep $\mathbf{k}'_1+\mathbf{k}'_2+\mathbf{k}_3=0$.

Above we have considered the equation for $A_{12}(\mathbf{k}_1,\mathbf{k}_2)$ – equations for $A_{13}(\mathbf{k}_1,\mathbf{k}_3)$ and $A_{23}(\mathbf{k}_2,\mathbf{k}_3)$ are treated analogously.

At point-like interactions of the outgoing particles, $a_{ij}(\mathbf{k}'_i,\mathbf{k}'_j)\to const$, Eq. (3.100) turns into the Skornyakov–TerMartirosyan equation [1]. The potential interaction, $a_{ij}(\mathbf{k}'_i,\mathbf{k}'_j)\to a_{ij}(|\mathbf{k}'_i-\mathbf{k}_i|)$, gives us the Faddeev equation [5] in the momentum representation. The Skornyakov–TerMartirosyan equation contains divergences which can be eliminated by introducing subtractions [3].

The first term in the right-hand side of Eq. (3.100) is a standard two-particle rescattering amplitude (see, for example, Chapter 2). The second and third terms describe subsequent rescatterings of three particles. For the S-wave interactions amplitudes $a_{12}(\mathbf{k}'_1,\mathbf{k}'_2)$ and $A_{\ell 3}(\mathbf{k}'_\ell,\mathbf{k}_3)$ depend on $\mathbf{k'}^2_{12}$ and $\mathbf{k'}^2_{\ell 3}$, correspondingly, so we can re-denote:

$$a_{12}(\mathbf{k}'_1,\mathbf{k}'_2)\to a_{12}(\mathbf{k'}^2_{12}),$$
$$A_{\ell 3}(\mathbf{k}'_\ell,\mathbf{k}_3)\to A_{\ell 3}(\mathbf{k'}^2_{\ell 3})\to A_{\ell 3}(\mathbf{k'}^2_j)\quad \ell\neq j\neq 3\,. \quad (3.101)$$

Here we take into account that $\frac{1}{2\mu_{ij}}\mathbf{k'}^2_{ij}+\frac{1}{2\mu_\ell}\mathbf{k}^2_\ell=E$. For point-like interactions of the outgoing particles, $a_{ij}(\mathbf{k}'_i,\mathbf{k}'_j)\to const$, the last two terms in (3.100) are determined by two variables under the integral: $\mathbf{k'}^2_\ell$ and $(\mathbf{k}'_\ell\mathbf{k}_3)/|\mathbf{k}'_\ell||\mathbf{k}_3|\equiv\cos\Theta'=z'_\ell$ $(\ell=1,2)$. One can perform the integration $\int dz'_\ell$ thus obtaining the Skornyakov–TerMartirosyan kernel. For example, the last integral in the right-hand side of Eq. (3.100) reads:

$$\ldots+\int\frac{d^3\mathbf{k}'_1}{(2\pi)^3}\frac{a_{12}}{4m_1m_2}\frac{A_{23}(\mathbf{k'}^2_1)}{\frac{1}{2\mu_{12}}\mathbf{k'}^2_1+\frac{1}{m_2}|\mathbf{k}'_1|\,|\mathbf{k}_3|\,z'_1+\frac{1}{2\mu_{23}}\mathbf{k}_3^2-E-i0}=$$
$$\ldots+\int\limits_0^\infty\frac{d\mathbf{k'}^2_1}{(2\pi)^2}\frac{a_{12}A_{23}(\mathbf{k'}^2_1)}{4m_1\,|\mathbf{k}_3|}\ln\frac{\frac{1}{2\mu_{12}}\mathbf{k'}^2_1+\frac{1}{m_2}|\mathbf{k}'_1|\,|\mathbf{k}_3|+\frac{1}{2\mu_{23}}\mathbf{k}_3^2-E-i0}{\frac{1}{2\mu_{12}}\mathbf{k'}^2_1-\frac{1}{m_2}|\mathbf{k}'_1|\,|\mathbf{k}_3|+\frac{1}{2\mu_{23}}\mathbf{k}_3^2-E-i0}$$
$$(3.102)$$

Let us emphasize that the requirement $a_{ij}(\mathbf{k}'_i, \mathbf{k}'_j) \to const$ is essential for performing integration over dz'_ℓ, **i.e.** for getting it from the Skornyakov–TerMartirosyan kernel. But the ansatz $a_{ij}(\mathbf{k}'_i, \mathbf{k}'_j) \to const$ leads to a divergence of the integrals – so we have to use subtraction or cutting procedures. The calculation of the final state rescattering diagrams with point-like interactions is discussed below in Chapter 4.

Another important point in the Skornyakov–TerMartirosyan approach is that of the integration regions in (3.102). The use of $d\mathbf{k}'^2_1$ for performing integration over $A_{23}(\mathbf{k}'^2_1)$ does not require a contour deformation but the integration over $\mathbf{k}'^2_1$ at large values is carried out in the unphysical region. On the contrary, when performing an integration over $d\mathbf{k}'^2_{12}$ (or, ds'^2_{12}), we connect the decay reaction region with regions of isobar production models (see discussion in Chapter 1, around Fig. 1.2).

3.10 Appendix D. Coordinate representation for non-relativistic three-particle wave function

Here we write the Schrödinger equation for a three-particle system in coordinate representation and then demonstrate a transformation of it into a set of non-relativistic momentum representation equations considered in this chapter.

We use here the following coordinate variables for particles i, j, ℓ:

$$\mathbf{r}_{ij} = \mathbf{r}_i - \mathbf{r}_j, \quad \rho_\ell = \mathbf{r}_\ell - \Big(\frac{m_i}{m_i + m_j}\mathbf{r}_i + \frac{m_j}{m_i + m_j}\mathbf{r}_j\Big), \quad i \neq j \neq \ell,$$

$$\mathbf{R} = \frac{1}{m_1 + m_2 + m_3}\Big(m_1\mathbf{r}_1 + m_2\mathbf{r}_2 + m_3\mathbf{r}_3\Big). \tag{3.103}$$

Then the three-particle Laplace operator reads:

$$-\frac{1}{2m_1}\nabla^2_{r_1} - \frac{1}{2m_2}\nabla^2_{r_2} - \frac{1}{2m_3}\nabla^2_{r_3} = -\frac{\nabla^2_R}{2(m_1 + m_2 + m_3)} - \frac{1}{2m_{ij}}\nabla^2_{r_{ij}} - \frac{1}{2\mu_\ell}\nabla^2_{\rho_\ell}\,. \tag{3.104}$$

We write the Schrödinger equation for three-particle system imposing centrally symmetrical two-particle interactions only:

$$V(\mathbf{r}_1, \mathbf{r}_2, \mathbf{r}_3) = V_{12}(r_{12}) + V_{13}(r_{13}) + V_{23}(r_{23}), \tag{3.105}$$

which correspond to two-particle S-wave interactions.

The wave function for three particles, $\Phi(\mathbf{r}_1, \mathbf{r}_2, \mathbf{r}_3)$, can be presented in a factorized form:

$$\Phi(\mathbf{r}_1, \mathbf{r}_2, \mathbf{r}_3) = \phi(\mathbf{R})\Psi(\mathbf{r}_{ij}, \rho_\ell) \tag{3.106}$$

The wave function $\phi(\mathbf{R})$ describes the centre-of-mass motion of a system with mass $(m_1+m_2+m_3)$ while $\Psi(\mathbf{r}_{ij},\rho_\ell)$ describes the relative motion of particles. The wave function of the centre-of-mass motion can be chosen to be equal to unity:

$$\phi(\mathbf{R})=1 \quad \text{at} \quad \mathbf{k}_1+\mathbf{k}_2+\mathbf{k}_3=0\,. \tag{3.107}$$

The Schrödinger equation for a three-particle system is:

$$\begin{aligned}&\Bigg[-\frac{1}{2\mu_{ij}}\nabla^2_{r_{ij}}-\frac{1}{2\mu_\ell}\nabla^2_{\rho_\ell}\\ &\quad+V_{12}(r_{12})+V_{13}(r_{13})+V_{23}(r_{23})-E\Bigg]\Psi_E(\mathbf{r}_{ij},\rho_\ell)=0.\end{aligned} \tag{3.108}$$

The Fourier transformation with using plane waves

$$\begin{aligned}\exp\left(i\mathbf{k}_1\mathbf{r}_1+i\mathbf{k}_2\mathbf{r}_2+i\mathbf{k}_3\mathbf{r}_3\right)&=\exp\left(i\mathbf{r}_{ij}\mathbf{k}_{ij}+i\rho_\ell\mathbf{k}_\ell\right),\\ \mathbf{k}_{ij}&=\frac{m_j}{m_i+m_j}\mathbf{k}_i-\frac{m_i}{m_i+m_j}\mathbf{k}_j\end{aligned} \tag{3.109}$$

can give us the equations in momentum representation. It is convenient to perform a two-step procedure, namely, rewrite Eq. (3.108) in an integral form (using the Green function) and after that perform the Fourier transformation.

We write Eq. (3.108) in the integral form fixing $ij\to 12$ and $\ell\to 3$:

$$\begin{aligned}\Psi_E(\mathbf{r}_{12},\rho_3)&=\int d^3r'_{12}d^3\rho'_3G_E(\mathbf{r}_{12}-\mathbf{r}'_{12},\rho_3-\rho'_3)\\ &\quad\times\Big[V_{12}(r'_{12})+V_{13}(r'_{13})+V_{12}(r'_{12})\Big]\Psi_E(\mathbf{r}'_{12},\rho'_3).\end{aligned} \tag{3.110}$$

The Green function reads:

$$G_E(\mathbf{r}_{23},\rho_1)=-\int\frac{d^3k_1d^3k_{23}}{(2\pi)^6}\frac{\exp\left(i\mathbf{k}_1\rho_1+i\mathbf{k}_{23}\mathbf{r}_{23}\right)}{\mathbf{k}^2_{23}/2\mu_{23}+\mathbf{k}^2_1/2\mu_1-E-i0}. \tag{3.111}$$

To turn to momentum representation one should introduce

$$\Psi_E(\mathbf{k}_{i\ell},\mathbf{k}_j)=\int d^3\rho_jd^3r_{i\ell}\exp\Big(-i\rho_j\mathbf{k}_j-i\mathbf{r}_{i\ell}\mathbf{k}_j\Big)\Psi_E(\mathbf{r}_{i\ell},\rho_j). \tag{3.112}$$

thus converting (3.110) into the non-relativistic momentum representation equations. This wave function is discussed in Appendix C.

References

[1] G.V. Skornyakov and K.A. Ter-Martirosyan, ZhETP **31**, 775 (1956).
[2] V.N. Gribov, ZhETP **38**, 553 (1960).
[3] G.S. Danilov, ZhETP **40**, 498 (1961); **42**, 1449 (1962).
[4] R.A. Minlos and L.D. Faddeev, ZhETP **41**, 1850 (1961).
[5] L.D. Faddeev, ZhETP **39**, 1459 (1960).
[6] N.N. Khuri, S.B. Treiman, Phys. Rev. **119** (1960) 1115.
[7] J.B. Bronzan, C. Kacser, Phys. Rev. **132** (1963) 2703.
[8] V.V. Anisovich, A.A. Anselm, V.N. Gribov, Nucl. Phys. **38** (1962) 132.
[9] V.V. Anisovich, ZhETP **44** 1953, (1963), ZhETP **47** (1964) 240.
[10] V.V. Anisovich and L.G. Dakhno, Phys. Lett. **10**, 221 (1964); Nucl. Phys. **76**, 657 (1966).
[11] A.V. Anisovich, Yad. Fiz. **58**, 1467 (1995) [Phys. Atom. Nucl. **58**, 1383 (1995)].
[12] A.V. Anisovich, Yad. Fiz. **66**, 175 (2003) [Phys. Atom. Nucl. **66**, 172 (2003)].
[13] V.V. Anisovich and A.A. Anselm, UFN **88**, 287 (1966) [Sov. Phys. Usp. **88**, 117 (1966)].
[14] A.V. Anisovich, V.V. Anisovich, J. Nyiri, V.A. Nikonov, M.A. Matveev and A.V. Sarantsev, *Mesons and Baryons. Systematization and Methods of Analysis*, World Scientific, Singapore, 2008.
[15] I.J.R. Aitchison and R. Pasquier, Phys. Rev. **152**, 1274 (1966).
[16] A.V. Anisovich, V.V. Anisovich, Yad. Fiz. **53**, 1485 (1991) [Phys. Atom. Nucl. **53**, 915 (1991)]
[17] V.V. Anisovich, M.N. Kobrinsky, D.I. Melikhov and A.V. Sarantsev, Nucl. Phys. A **544**, 747 (1992).
[18] A.V. Anisovich and V.A. Sadovnikova, Yad. Fiz. **55**, 2657 (1992); **57**, 75 (1994); Eur. Phys. J. A**2**, 199 (1998).
[19] V.V. Anisovich, D.I. Melikhov, V.A. Nikonov, Phys. Rev. D **52**, 5295 (1995); Phys. Rev. D **55**, 2918 (1997).
[20] A.V. Anisovich, V.V. Anisovich and V.A. Nikonov, Eur. Phys. J. A **12**, 103 (2001).
[21] A.V. Anisovich, V.V. Anisovich, M.A. Matveev and V.A. Nikonov, Yad. Fiz. **66**, 946 (2003) [Phys. Atom. Nucl. **66**, 914 (2003)].

Chapter 4

Non-relativistic Three-Body Amplitude

In the present chapter we consider amplitudes when particles in the final state are non-relativistic. Here the following cases are possible:
(1) All particles interact in a non-resonance way in the region of consideration. If so, the amplitude can be expanded into a series over relative momenta of outgoing particles. In the expansion one can distinguish between analytical and singular terms.
(2) One pair of particles interacts strongly. In this case the interaction of two particles are to be taken into account precisely, but the non-resonance interaction terms may be expanded into a series over relative momenta.
(3) Two pairs of particles interact strongly. For the production amplitude this leads to a reduced three-body equation.
(4) All three particles interact strongly – this gives a standard equation for three non-relativistic particles.

4.1 Introduction

Before considering the non-relativistic case, let us present the kinematic relations for the three-body system and formulate the selection rules for the singular diagrams.

4.1.1 *Kinematics*

In the non-relativistic approximation, near the particle production threshold, the values $s_{i\ell} = (k_i + k_\ell)^2$ can be expanded into a series over the momenta $\mathbf{k}^2_{i\ell}$ of the produced particles

$$\sqrt{s_{i\ell}} \cong m_i + m_\ell + \frac{\mathbf{k}^2_{i\ell}}{2\mu_{i\ell}}, \qquad \mu^{-1}_{i\ell} = m_i^{-1} + m_\ell^{-1}, \tag{4.1}$$

where m_i are the masses of particles in the final state.

The kinetic energy E, released in the reaction, is connected to the total energy $\sqrt{s} = [(k_{10} + k_{20} + k_{30})^2 - (\mathbf{k}_1 + \mathbf{k}_2 + \mathbf{k}_3)^2]^{1/2}$ by the conditions

$$\sqrt{s} = m_1 + m_2 + m_3 + E\,,$$

$$E = \frac{m_1 + m_2}{m_1 + m_2 + m_3}\frac{\mathbf{k}_{12}^2}{2\mu_{12}} + \frac{m_1 + m_3}{m_1 + m_2 + m_3}\frac{\mathbf{k}_{13}^2}{2\mu_{13}} + \frac{m_2 + m_3}{m_1 + m_2 + m_3}\frac{\mathbf{k}_{23}^2}{2\mu_{23}}\,,$$

$$E = \frac{\mathbf{k}_{i\ell}^2}{2\mu_{i\ell}} + \frac{\mathbf{k}_j^2}{2\mu_j}\,, \qquad \mu_j^{-1} = m_j^{-1} + (m_i + m_\ell)^{-1}, \quad j \neq i \neq \ell. \tag{4.2}$$

The transition from the momenta $\mathbf{k}_{12}$ and $\mathbf{k}_3$ to, *e.g.*, the momenta $\mathbf{k}_{13}$ and $\mathbf{k}_2$ is carried out via the formulae

$$\mathbf{k}_2 = -\frac{m_2}{m_1 + m_2}\mathbf{k}_3 - \mathbf{k}_{12}\,,$$

$$\mathbf{k}_{13} = -\frac{m_1(m_1 + m_2 + m_3)}{m_1 + m_2)(m_1 + m_3)}\mathbf{k}_3 + \frac{m_3}{m_1 + m_3}\mathbf{k}_{12}\,. \tag{4.3}$$

Other similar relations can be obtained by the cyclic permutation of the indices.

Reaction *two particles* $\to$ *three particles* is characterized by momenta transferred squared $t_1 = (p_1 - k_1)^2$ and $t_2 = (p_1 - k_2)^2$, where p_1 and p_2 refer to momenta of colliding particles. Near the threshold the expressions for the momenta transferred squared can be rewritten up to the linear terms in the momenta of the produced particles, as

$$t_1 \simeq t_1^{(0)} + 2(\mathbf{p}_1^{(0)}\mathbf{k}_1) = t_1^{(0)} + 2|\mathbf{p}_1^{(0)}|\,|\mathbf{k}_1|\,Z_1\,,$$

$$t_2 \simeq t_2^{(0)} + 2(\mathbf{p}_1^{(0)}\mathbf{k}_2)t_2^{(0)} + 2|\mathbf{p}_1^{(0)}|\,|\mathbf{k}_2|\,Z_2\,. \tag{4.4}$$

Here $t_1^{(0)}$, $t_2^{(0)}$ are the threshold values of the invariants t_1 and t_2; $|\mathbf{p}_1^{(0)}|$ is the absolute value of the momentum $\mathbf{p}_1$ at the threshold energy, while Z_1 and Z_2 are the cosines of angles between the vectors $\mathbf{p}_1$, $\mathbf{k}_1$ and $\mathbf{k}_2$.

In the following it may be convenient to use the variables

$$\mathbf{x}_{i\ell} = \frac{\mathbf{k}_{i\ell}}{\sqrt{2\mu_{i\ell}E}} \tag{4.5}$$

instead of $\mathbf{k}_{i\ell}$. We introduce also special notations for combinations of masses which appear frequently:

$$\beta_1 = \frac{m_1(m_1 + m_2 + m_3)}{(m_1 + m_2)(m_1 + m_3)} \tag{4.6}$$

(β_2 and β_3 are defined in a similar way). In these notations useful relations between $\mathbf{x}_{13}$, $\mathbf{x}_{12}$ and z (the cosine of the angle between the momenta $\mathbf{k}_{12}$ and the momentum of the third particle in the c.m. system of particles 1 and 2) are presented as:

$$\mathbf{x}_{13}^2 = (1 - \beta_1)\mathbf{x}_{12}^2 + \beta_1(1 - \mathbf{x}_{12}^2) + 2z\sqrt{\beta_1(1 - \beta_1)\mathbf{x}_{12}^2(1 - \mathbf{x}_{12}^2)}\,. \tag{4.7}$$

The cyclic permutations of indices give us other similar relations.

4.1.2 *Basic principles for selecting the diagrams*

We are considering the amplitude of the transition process of two particles into three near the threshold of the reaction, when the total emerging kinetic energy is much less than the mass of any of the particles. In this case we expand the amplitude of the reaction into a series over the powers $s_{ik} - s_{ik}^{(0)}$ and $t_i - t_i^{(0)}$, where $\sqrt{s_{ik}}$ and $\sqrt{-t_i}$ are the relative energies and the momenta transferred on which the amplitude depends, while $\sqrt{s_{ik}^{(0)}}$ and $\sqrt{-t_i^{(0)}}$ are their values at the thresholds.

It is obvious that the singularities of the amplitude which are close to the threshold values of the invariants will block such an expansion. But after subtracting these singularities, the amplitude can be expanded into a series over the powers of $s_{ik} - s_{ik}^{(0)}$ and $t_i - t_i^{(0)}$.

We consider a singularity to be a close one either if it lies exactly at the threshold value of the invariant or if it is at a distance from the threshold much smaller than the mass squared of any of the particles. Of course, we have to take here into account singularities placed both on the physical and unphysical sheets. The remaining ("far") singularities are at a distance of the order of the particle mass squared ($\sim m^2$), so that, after the subtraction of the close singularities, the expansion is carried out essentially over the powers of $(s_{ik} - s_{ik}^{(0)})/m^2$ and $(t_i - t_i^{(0)})/m^2$. In other words, after subtracting the close singularities, we are facing the expansion over the powers of $(kr_0)^2$, where k is the momentum of any of the produced particles, and r_0 is the interaction radius.

Let us consider first the singularities of the amplitude which correspond exactly to the threshold values of the invariants. In this simple case the position and the type of the singularities can be obtained directly from the unitarity condition in the s- and s_{ik}-channels. This is easier than getting the Landau curves from the unitarity condition. The Landau rules [1] for determining the singularities of the Feynman diagrams are presented in Appendix A, using as an example a triangle diagram. In Appendix B, on the basis of the paper [2], non-Landau type singularities and their influence on the threshold singularities are also considered.

In many hadron reactions the amplitude does not have singularities in momenta transferred at threshold values $t_i = t_i^{(0)}$ (this is so in reactions where two- and many-particle bound states like the deuteron, He_3 etc. are absent), and singularities in s and $s_{i\ell}$ appear at values when the energies squared $s_{i\ell}$ or s are equal to the mass sum squared of all possible intermediate states.

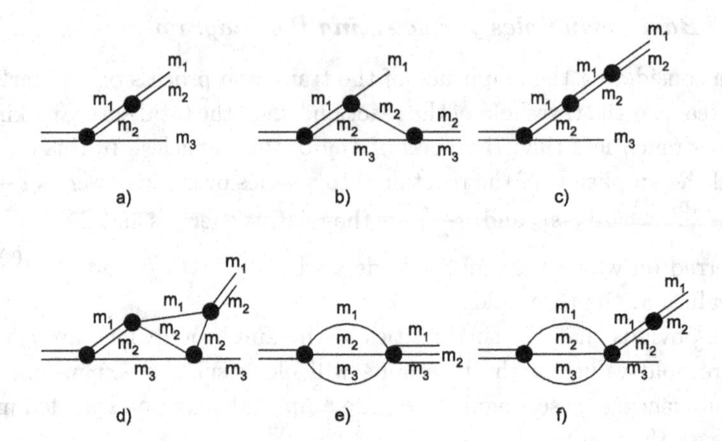

Fig. 4.1 Examples of diagrams with rescatterings of the final state particles – this type of diagrams is responsible for threshold singularities.

If we are interested in singularities at threshold energy values, we have to consider those terms under the unitarity conditions which are connected with final state three-particle scatterings. These terms correspond to Feynman diagrams shown on Fig. 4.1. In all the other diagrams which do not include three-particle decays, singularities at the threshold of the considered reaction are absent (if only the sum of masses of some other particles do not coincide, by accident, with $m_1 + m_2 + m_3$ – but we do not consider this case).

The final state interactions of Fig. 4.1 correspond to transitions of three particles into three ones, including the cases when one of the particles does not interact with the other two (Figs. 4.1a,4.1c).

If all three particles interact, the corresponding Feynman diagram has a three-particle singularity in the total energy squared s at $s = (m_1 + m_2 + m_3)^2$ (or $E = 0$), Fig. 4.1b. In the case if the six-point amplitude is degenerated, as shown in Figs 4.1a,4.1b, a two-particle singularity in s_{ik} appears at $s_{ik} = (m_1 + m_k)^2$ (or $k_{i\ell}^2 = 0$). Many diagrams can have, obviously, both singularities at the same time. Such diagrams are presented, *e.g.*, in Figs. 4.1b,4.1d, 4.1f.

Let us now imagine that we separate from the three particle production amplitude all the possible scatterings of particle pairs in the final state and the processes of three particles turning into three as well. If so, we obtain diagrams shown in Fig. 4.1, where the dashed irreducible blocks do not include transitions of two particles into two or three particles. We get singular terms of the amplitude from these diagrams if all lines correspond to the transitions of real particles ($q_i^2 = m_i^2$) – this is a consequence of the Landau rules. As a consequence, we will expand these blocks into series over the degrees of the difference of the invariants from the threshold values.

Up to now we discussed only singularities which are situated exactly at the threshold values of the invariants. However, other near singularities, being at a distance from the threshold much smaller than the masses squared of any of the particles, may also be interesting. Such singularities appear, *e.g.*, because of the existence of "composite particles" with small binding energies (like the deuteron). In particular, the amplitude of the reaction $\pi + D \to N + N + \pi$ has singularities in the momentum transferred near the threshold, which are not related to diagrams of the type presented in Fig. 4.1. If so, we have to take into account not only diagrams like Fig. 4.1, but also those in the vertices of which "almost real" transitions of deuterons into two nucleons take place. It is clear that the positions of singularities connected with the latter diagrams will be very close to the physical region, since the binding energy of the deuteron is small.

The near singularities can be also on other sheets where they are not defined directly by the unitarity condition. To be situated not far from the physical region, these singularities have to be, of course, under the cuts going either from the point $s_{i\ell} = (m_i + m_\ell)^2$ on the $s_{i\ell}$ plane, or the point $s = (m_1 + m_2 + m_3)^2$ on the s plane. However, these cuts exist only in diagrams presented in Fig. 4.1. Singularities which are placed on other non-physical sheets, determined by other cuts (related to far singularities), do not influence, of course, the expansion of the amplitude near the threshold.

In the following it will be seen that the summation of the diagrams leads to the appearance of singularities in the amplitude the position of which is connected not with the effective interaction radius r_0, but with the value of the scattering amplitude of particle pairs. If these amplitudes are of the order of r_0, $a \sim r_0$ (non-resonant case), the amplitude of the reaction has to be expanded in a power of the degrees of (ka). In fact in this case it turns out to be possible to restrict ourselves to the simplest diagrams, presented in Fig. 4.1. Just such a procedure is carried out when calculating power expansion terms over threshold momenta. For $K \to \pi\pi\pi$ such a power expansion was carried out in [3] (the terms of the order $\sim |k_{i\ell}|$ and $\sim |k_{i\ell}|^2$). The lowest singular terms, $\sim |k_{i\ell}|$, were studied in the general case in [4]. If $a \gg r_0$ (the resonant case, for example, the process of nucleon scattering at small energies) all the diagrams of the type shown in Fig. 4.1 have to be taken into account. In this case, depending on the type of interaction, we arrive at the Skornyakov – Ter-Martirosyan equation [5, 6]) or the Faddeev equation [7].

4.2 Non-resonance interaction of the produced particles

The consideration of the produced particles near the threshold allows one to expand the amplitude in a series over relative momenta of the secondaries. It means that either there are no resonances (and corresponding amplitude poles) near the threshold, or the considered final state total energy is small enough to expand the amplitude over the relative momenta.

4.2.1 *The structure of the amplitude with a total angular momentum $J = 0$*

In the present section we consider the amplitude $A^{(J=0)}(k_{12}^2, k_{13}^2, k_{23}^2)$, corresponding to the zero total angular momentum. As it was said before, this amplitude may be expanded in a series of the relative momenta of the produced particles:

$$
\begin{aligned}
&A^{(J=0)}(k_{12}^2, k_{13}^2, k_{23}^2) = \sum_{n=0}^{\infty} A_n^{(J=0)}(k_{12}^2, k_{13}^2, k_{23}^2)\\
&A_{n=0}^{(J=0)}(k_{12}^2, k_{13}^2, k_{23}^2) \sim const\\
&A_{n=1}^{(J=0)}(k_{12}^2, k_{13}^2, k_{23}^2) \sim |k_{i\ell}|,\\
&A_{n=2}^{(J=0)}(k_{12}^2, k_{13}^2, k_{23}^2) \sim |k_{i\ell}|^2,
\end{aligned}
\tag{4.8}
$$

and so on.

Below we present calculations of terms of the order of $\sim |k_{i\ell}|$ and $\sim |k_{i\ell}|^2$. Terms of the order of $\sim |k_{i\ell}|^3$ were considered in [8], those $\sim |k_{i\ell}|^4$ and $\sim |k_{i\ell}|^5$ are given in [9, 10].

4.2.1.1 *Terms of the order of $\sim$ const and $\sim |k_{i\ell}|$*

Expanding over the relative momenta, the scattering lengths of the pairs of produced particles a_{12}, a_{23} and a_{13} are not considered to be large: $a_{ik} \lesssim r_0$ (r_0 is the radius of the interaction) – this corresponds to the absence of the resonance case in the scattering of the produced particles at zero energy.

The amplitude $A_{n=0}^{(J=0)}(k_{12}^2, k_{13}^2, k_{23}^2)$ does not include final state rescatterings, we put this term to be equal to the threshold value at $k_{12}^2 = k_{13}^2 =$

$k_{23}^2 = 0$:

$$A_{n=0}^{(J=0)}(k_{12}^2, k_{13}^2, k_{23}^2) \equiv \lambda \tag{4.9}$$

The terms $A_{n=1}^{(J=0)}(k_{12}^2, k_{13}^2, k_{23}^2)$ are related to the single particle scattering in the final state, Fig. 4.1a. Calculating the largest singular parts over the threshold momenta of these diagrams, their vertices have to be replaced by the scattering amplitude of particle pairs at zero energy a_{12}, a_{13} and a_{23} and λ – the amplitude of the transformation of the initial particles into three particles at threshold energy. Thus we have to write

$$\begin{aligned} &\int\limits_{(m_i+m_\ell)^2}^{\infty} \frac{ds'}{\pi} A^{(J=0)}(k_{12}'^2, k_{13}'^2, k_{23}'^2) \frac{\rho_{i\ell}(s_{i\ell}')}{s_{i\ell}' - s_{i\ell} - i0} a_{12}(s_{i\ell}') \\ &\to k_{i\ell}^2 \int\limits_0^{\infty} \frac{dk_{i\ell}'^2}{\pi} \lambda \frac{k_{i\ell}'}{k_{i\ell}'^2(k_{i\ell}'^2 - k_{i\ell}^2 - i0)} a_{12} \,. \end{aligned} \tag{4.10}$$

In the right-hand side (4.10) the non-relativistic phase volume

$$\rho_{i\ell}(s_{i\ell}') \to \rho_{i\ell}^{non-rel}(k_{i\ell}'^2) \equiv k_{i\ell}', \tag{4.11}$$

and the non-relativistic dispersion integral with one subtraction at $k_{i\ell}^2 = 0$

$$\int\limits_{(m_i+m_\ell)^2}^{\infty} \frac{ds'}{\pi} \frac{1}{s_{i\ell}' - s_{i\ell} - i0} \to k_{i\ell}^2 \int\limits_0^{\infty} \frac{dk_{i\ell}'^2}{\pi} \frac{1}{k_{i\ell}'^2(k_{i\ell}'^2 - k_{i\ell}^2 - i0)} \tag{4.12}$$

are introduced.

For the first diagram Fig. 4.1a the dispersion relation gives

$$A_{n=1}^{(12)}(k_{12}^2) = \lambda k_{12}^2 \int\limits_0^{\infty} \frac{dk_{12}'^2}{\pi} \frac{k_{12}'}{k_{12}'^2(k_{12}'^2 - k_{12}^2)} a_{12} = \lambda i k_{12} a_{12} \,. \tag{4.13}$$

So, the terms of the order of $\sim |k_{i\ell}|$, see Fig. 4.1a, read:

$$A_{n=1}^{(J=0)}(k_{12}^2, k_{13}^2, k_{23}^2) = i\lambda k_{12} a_{12} + i\lambda k_{13} a_{13} + i\lambda k_{23} a_{23} \,. \tag{4.14}$$

Recall, in this approximation the amplitude is determined by two terms, $A_{n=0}^{(J=0)}(k_{12}^2, k_{13}^2, k_{23}^2)$ and $A_{n=1}^{(J=0)}(k_{12}^2, k_{13}^2, k_{23}^2)$.

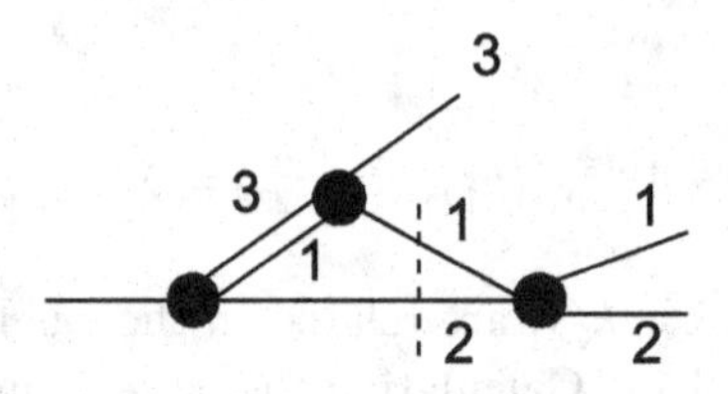

Fig. 4.2 Double rescattering diagram; the dashed line shows the cut in the dispersion integral.

4.2.1.2 *Double rescattering diagram, Fig.* 4.2

Let us turn now our attention to diagrams of the type presented in Fig. 4.2, which is a singular diagram of the order of $\sim |k_{i\ell}|^2$. There are, obviously, six diagrams like that, differing by the permutation of particles in the final and the intermediate states. Again, the vertices of the considered diagrams connect double scattering amplitudes at zero energy and the amplitudes of the production three particles with zero kinetic energy. The diagram Fig. 4.2 depends on two variables – k_{12}^2 and $E = \sqrt{s} - m_1 - m_2 - m_3$. For the calculation of the contribution of Fig. 4.2 which includes singularities in k_{12}^2 at small k_{12}^2 values, it is convenient to make use of the dispersion relations in k_{12}^2. The diagram Fig. 4.2 is, essentially, a three-point amplitude with a decay mass $\sqrt{s}$. Generally speaking, in this case it is difficult to write directly the dispersion relation in k_{12}^2. It can be obtained, however, by consideration of a simple case at small s values, $s \sim m_i^2$, with subsequent analytical continuation to the $s > (m_1 + m_2 + m_3)^2$ region.

At $s \sim m_i^2$ the dispersion relation in k_{12}^2 for the diagram Fig. 4.2 reads:

$$A_{n=2}^{(13\to12)}(k_{12}^2, E) = A_{n=2}^{(13\to12)}(0, E) + k_{12}^2 \int\limits_0^\infty \frac{dk_{12}'^2}{\pi} \langle A_{n=1}^{(13)}(k_{13}'^2, E)\rangle_{12} \frac{k_{12}'}{k_{12}'^2(k_{12}'^2 - k_{12}^2 - i0)} a_{12}\,,$$

$$\langle A_{n=1}^{(13)}(k_{13}'^2, E)\rangle_{12} = \lambda \int\limits_{-1}^{1} \frac{dz_{13}}{2}\, ik_{13}' a_{13} \tag{4.15}$$

Of course, the subtraction constant $A(0, E)$ does not depend on k_{12}^2 and, hence, it is not singular in the k_{12}^2 term. We shall present the calculation of the averaged loop diagram $\langle A_{n=1}^{(13)}(k_{13}'^2, E)\rangle_{12}$ later, now we discuss the subtraction procedure in (4.15).

One would think that Fig. 4.2 can have, in addition to the trivial threshold singularities, also Landau-type singularities [1] which correspond to the case when all Feynman denominators turn to zero (see Appendix A). Let us note that in the considered case the Landau singularities are, in fact, absent. To understand the reason of the absence of these singularities, it is necessary to remember the usual case of the triangle diagrams. At normal values of the two external masses M_1 and M_2 the triangle diagrams have usually two Landau-type singularities on the non-physical sheet in the third mass M_3. These singularities are, at the same time, singularities of the absorption part in M_3. When one of the external masses (for example M_1) is growing, one of the mentioned singularities of the absorption part over the third mass is catching the integration contour. As a result, the dispersion relation obtains a non-trivial form, while the singularity of the amplitude in M_3 itself moves to the physical sheet. In the case of the diagram shown in Fig. 4.2, the positions of the two singularities in M_3 coincide; as a consequence, these singularities cancel each other. We will see this from the exact calculations of the diagram Fig. 4.2. So far let us restrict ourselves to the following comments.

The position of the mentioned singularities are defined by the Mandelstam condition $z_1^2 + z_2^2 + z_3^2 - 2z_1z_2z_2 - 1 = 0$ for three cosines z_1, z_2, z_3 which are related to the masses of the vertices of the triangle diagram (here each cosine equals $(\mu_1^2 + \mu_2^2 - M_3^2)(2\mu_1\mu_2)^{-1}$, where μ_1 and μ_2 are masses of the internal lines, M_3 that of the external line), see Appendix A.

For the diagram presented in Fig. 4.2 the cosine, connected with the vertex where the scattering of particles 1 and 3 takes place, equals zero. Consequently, the point which we consider to be suspicious (usually corresponding to the existence of a Landau singularity) is determined by

$$s_{12} = (m_1 + m_2)^2 + \frac{m_1}{m_1 + m_3}\Big[s - (m_1 + m_2 + m_3)^2\Big]. \qquad (4.16)$$

We see that, independently of the fact whether it is a singular point of the absorption part, this point moves uniformly from left to right when s is growing, and it does not deform the integration contour in the dispersion integral (4.15). Hence, the dispersion relation (4.15) preserves its form when s grows from $s \sim m_i^2$ to $s > (m_1 + m_2 + m_3)^2$.

Calculating the absorption part of Eq. (4.15), we will see that it, indeed, has no singularity in the point determined by the relation (4.16). It turns out, however, that the absorption part has another singularity near the threshold in k_{12}^2 (or s_{12}) which is not of Landau type [2], *i.e.* a singularity not corresponding to $q_i^2 = \mu_i^2$ in the denominator of the Feynman amplitude (non-Landau singularities are discussed in Appendix B).

4.2.1.3 Setting of the relation between s'_{13} and z and calculation $disc_{12} A^{(13\to 12)}_{n=2}(k^2_{12}, E)$

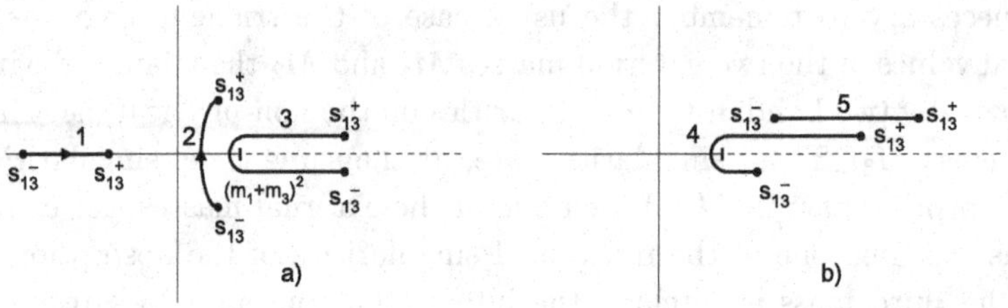

Fig. 4.3 Integration contour over ds'_{13} (or over dz) in the $disc\, A^{(13\to 12)}_{n=2}(k^2_{12}, E)$.

Now let us carry out the analytic continuation of the diagram Fig. 4.2 from $s \sim m_i^2$ to the values $s > (m_1 + m_2 + m_3)^2$ within a relativistic treatment.

For the absorption part we have:

$$disc_{12} A^{(13\to 12)}_{n=2}(k^2_{12}, E) \;=\; \frac{k_{12}}{\sqrt{s_{12}}}\, a_{12}(m_1+m_2) \int\limits_{-1}^{1} \frac{dz}{2}\, A^{(13)}_{n=1}(s'_{13})\,. \quad (4.17)$$

Let us remind that $A^{(13)}_{n=1}(s'_{13})$ is the amplitude drawn in Fig. 4.2 to the left of the dotted line. It depends only on s'_{13}; z is the cosine of the angle between the momentum of the third particle in the final state in the c.m. system of particles 1 and 2 and the relative momentum of particles 1 and 2 in the intermediate state.

The connection between the invariant s'_{13} and the variable z is written as:

$$\begin{aligned} s'_{13} = m_1^2 + m_3^2 &- \frac{(s_{12}+m_3^2-s)(s_{12}+m_1^2-m_2^2)}{2s_{12}} \\ &+ \frac{z}{2s_{12}}\sqrt{[s_{12}-(m_1+m_2)^2][s_{12}-(m_1-m_2)^2]\times} \\ &\overline{\times[s-(\sqrt{s_{12}}+m_3)^2][s-(\sqrt{s_{12}}-m_3)^2]} \end{aligned} \quad (4.18)$$

With the help of (4.18), let us carry out the integration over s'_{13} in (4.17). The integrand $A^{(13)}_{n=1}(s'_{13})$ has, obviously, just one singularity in the s'_{13} plane at $s'_{13} = (m_1+m_3)^2$, see Fig. 4.3. The integration over s'_{13} in the expression (4.17) at small s values ($s < (\sqrt{s_{12}} - m_3)^2$) is carried out along the part of the negative half axis between the values $s^{(-)}_{13}$ and $s^{(+)}_{13}$ (contour 1 in Fig. 4.3. In order to obtain a correct expression at large

s values, we have to perform the analytic continuation over s, adding to s (the external mass squared) a positive imaginary part: $s \to s + i0$. At $s = (\sqrt{s_{12}} - m_3)^2$ the values of s_{13}^- and s_{13}^+ are moving to the complex plane; if $s > (\sqrt{s_{12}} - m_3)^2$, the integration goes between the complex conjugate points (contour 2 in Fig. 4.3. Note that the point $s = (\sqrt{s_{12}} - m_3)^2$ is not a singular one, since the continuation with $s \to s + i0$ leads here to the same result as that with $s - i0$. The reason is that the change in the directions of the integrations is compensated by the change of the sign of the square-root in the multiplication factor which connects dz and ds'_{13}. At $s = (\sqrt{s_{12}} + m_3)^2$ the integration limits s_{13}^+ and s_{13}^- occur on the cut of the function $A_{n=1}^{(13)}(s'_{13})$, going from the point $s'_{13} = (m_1 + m_3)^2$ along the positive part of the real axis. Here s_{13}^+ appears on the upper side of the cut, s_{13}^- on the lower one, while the integration contour encloses the cut with the help of contour 3. It is seen from (4.17) that the point where s_{13}^+ and s_{13}^- coincide on the s'_{13} plane is always on the right-hand side of $s'_{13} = (m_1 + m_3)^2$ if $s_{12} > (m_1 + m_2)^2$. The point $s = (\sqrt{s_{12}} + m_3)^2$ is a singular one, since in the continuation process with $s \to s + i0$, as s is growing, the upper integration limit s_{13}^+ moves to the right, s_{13}^- – to the left along the real axis (Fig. 3, contour 4), while turning around the point $s = \left(\sqrt{s_{12}} + m_3\right)^2$ with a negative addition ($s \to s - i0$), s_{13}^+ moves to the left, s_{13}^- – to the right. Also, at such continuations different signs appear in the multiplication factor which connects dz with ds'_{13} (different signs of the square-roots $\sqrt{s - (\sqrt{s_{12}} + m_3)^2}$). If the cut is absent in the function $A_{n=1}^{(13)}(s'_{13})$, the change in the sign of the square-root could compensate the permutation of the integration limits, as it happened in the point $s = (\sqrt{s_{12}} - m_3)^2$. But due to the presence of the cut in $A_{n=1}^{(13)}(s'_{13})$ the turn around the point $s = (\sqrt{s_{12}} + m_3)^2$ with positive and negative imaginary additions leads to different results, and as a consequence the point $s = (\sqrt{s_{12}} + m_3)^2$ turns out to be a singular one. With the further growth of s the point s_{13}^- turns around the beginning of the cut $(m_1 + m_3)^2$ and occurs on the upper side, while the integration contour obtains the form 5 in Fig. 4.3. The point s_{13}^- coincides with the beginning of the cut just when s and s_{12} satisfy the relation (4.16). Generally speaking, the value of s which corresponds to (4.16) could be a singular point of the absorption part (when the end of the integration contour coincides with the singular point of the integrand). However, as it was said already, this point is in fact not a singular one, what will be seen directly from the explicit expression for the absorption part, to be obtained below.

4.2.1.4 *Calculation of* $A^{(13\to 12)}_{n=2}(k^2_{12}, E)$

Let us turn now to the direct calculation of $A^{(13\to 12)}_{n=2}(k^2_{12}, E)$. The function $A^{(13)}_{n=1}(s'_{13})$ which is included in the integrand in (4.17) may, strictly speaking, contain a constant part, with a contribution linear in k'_{13} and with higher terms in k'_{13}. The constant term in $A^{(13)}_{n=1}(s'_{13})$ leads us again to the expression (4.14); this means that this term can be obtained, practically, by the renormalization of the constant λ. If the latter can be understood as the observable value of the three-particle production amplitude at zero energy, the constant term in $A^{(13)}_{n=1}(s'_{13})$ should be neglected.

Let us consider now the term $A^{(13)}_{n=1}(s'_{13})$ linear in k'_{13}. It is obvious that in the final expression for the amplitude higher order terms in k'_{13} lead to higher order terms in threshold energies. Hence, with the given accuracy, we have to replace $A^{(13)}_{n=1}(s'_{13}) \to i\lambda a_{13}k'_{13}$ and thus obtain the formula (4.15).

Considering the non-relativistic approximation in the expression (4.18), we get a connection between k'^2_{13} and z. For the absorption part we have

$$disc_{12}A^{(13\to 12)}_{n=2}(k^2_{12}, E) = \int\limits_{x^-_{13}}^{x^+_{13}} dx'_{13}x'^2_{13}i\,\lambda\, a_{12}a_{13}\sqrt{\frac{\mu_{12}\mu_{13}}{\beta_1(1-\beta_1)}}\frac{E}{\sqrt{1-x^2_{12}}}$$

$$= i\,\lambda\, a_{12}a_{13}E\,\frac{2\sqrt{\beta_1\mu_{12}\mu_{13}}}{\sqrt{1-x^2_{12}}}\,x_{12}\left(1-\frac{1-4\beta_1}{3\beta_1}\,x^2_{12}\right);$$

$$x^{\pm}_{13} = \sqrt{\beta_1(1-x^2_{12})} \pm \sqrt{(1-\beta_1)\,x^2_{12}},$$

$$x^2_{i\ell} = \frac{k^2_{i\ell}}{2\mu_{i\ell}E} \tag{4.19}$$

At large s values (large E, small x_{12}) both integration limits in x'^2_{13} are placed on the upper part of the cut. According to this, both integration limits in x'_{13} are positive. This condition is satisfied in the integral (4.19).

The expression (4.19) demonstrates explicitly that the absorption part $discA^{(13\to 12)}_{n=2}(k^2_{12}, E)(k^2_{12}, E)$ has singularity at $x^2_{12} = 1$ *i.e.* $k^2_{12} = 2\mu_{12}E$. As it was mentioned already, this singularity is of non-Landau type and corresponds to the case when it appears on the cut of the integration limits. The question is, how to turn around the singularity $k'^2_{12} = 2\mu_{12}E$ (or $x'_{12} = 1$) correctly when integrating over k'^2_{12}. Obviously, the expression (4.19) is correct in both cases: if $E > 0$, and if $E < 0$, when the singularity $k'^2_{12} = 2\mu_{12}E$ is beyond the contour of integration. The right way of analytic continuation from the region $E < 0$ to the region $E > 0$ will be that when a longitudinal imaginary part is added, $E \to E + i0$, since E plays the role of the external mass. Thus the singularity $k'^2_{12} = 2\mu_{12}E$ occurs over the integration contour in k'^2_{12}.

Substituting (4.19) into (4.15), it can be easily seen that the appearing integral is diverging logarithmically. This divergence is a result of expanding the absorption part into a power series of k'^2_{12}. The correct expression would cut the integral at a value k^2_{12} of the order of mass of the particles. It does not make sense to investigate the character of this cut-off by considering the form of the concrete diagram. Indeed, the cut-off may be the result of the decrease of the exact amplitudes which we have substituted by constants in the vertices of the diagrams. On the other hand, the two results obtained using different cuts differ by Ck^2_{12}, where C is a constant, the terms of this type are of analytic character. They are included in many diagrams, and, according to our approach, can not be calculated, but have to be added to the amplitude with arbitrary coefficients. Because of that, we cut off the integral (4.15) at $k'_{12} \sim m$, where m is of the order of the mass of particles appearing in the reaction.

Having this in mind, it turns out to be quite easy to carry out the integration (4.15), which leads to the following expression for the terms singular in k^2_{12}:

$$\begin{aligned} A^{(13\to 12)}_{n=2}(k^2_{12},E) - A^{(13\to 12)}_{n=2}(0,E) &= k^2_{12}\int\limits_0^\infty \frac{dk'^2_{12}}{\pi}\,\frac{discA^{(13\to 12)}_{n=2}(k'^2_{12},E)}{k'^2_{12}(k'^2_{12}-k^2_{12}-i0)} \\ = -2\lambda a_{12}a_{13}E\sqrt{\frac{m_1m_2m_3}{m_1+m_2+m_3}}&\left[\frac{2x_{12}\arccos x_{12}}{\pi\sqrt{1-x^2_{12}}}\left(\beta_1+\frac{1-4\beta_1}{3}x^2_{12}\right)\right. \\ &\left.-\frac{1-4\beta_1}{3\pi}x^2_{12}\left(\ln\frac{m}{E}+i\pi\right)\right]. \qquad (4.20) \end{aligned}$$

As it is seen, the first terms in (4.20) have characteristic square-root threshold singularities at $k^2_{12} = 0$ and a peculiar singularity in k^2_{12} at $k^2_{12} = 2\mu_{12}E$. The latter is, however, on the unphysical sheet, connected with the cut which goes from the point $k^2_{12} = 0$. Indeed, at $x_{12} > 0$ (on the upper part of the cut) at $x^2_{12} \to 1$ $\arccos x_{12}$ behaves like the square-root $\sqrt{(1-x^2_{12}}$ which cancels with the square-root in the denominator. However, on the lower part of the cut where $x_{12} < 0$, at $x^2_{12} \to 1$ $\arccos x_{12} \to \pi$, and we have a singularity of the $(1-x^2_{12})^{-1/2}$ type.

The last term in (4.20) does not contain singularities in k^2_{12}, although there is a singularity of the $k^2_{12}\ln E$ type at $E = 0$. The insertion of this term into (4.20) is somewhat questionable, it is justified by the fact that our next step will be the selection of terms in the diagram Fig. 4.2 which contain singularities in the total energy E. We will see that, except for the last term in (4.20), such a term is included in the subtraction constant $A^{(13\to12)}_{n=2}(0,E)$. The contribution is of $E\ln E$ character, and it reflects the presence of the usual logarithmic singularity in total energy, related with the three-body intermediate state.

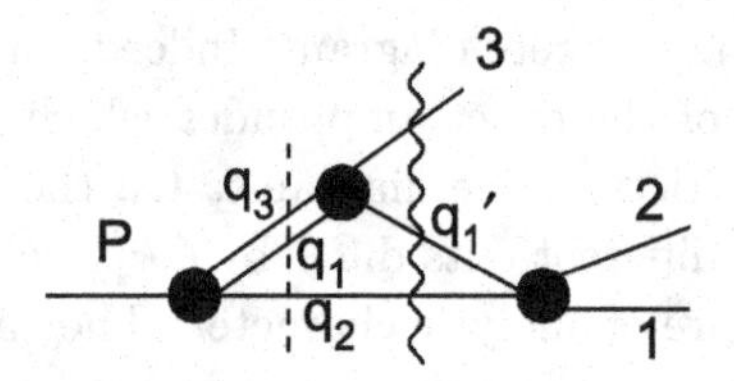

Fig. 4.4 Calculations of $A^{(13\to12)}_{n=2}(0,E)$: cuts in the s-channels.

To calculate singular terms of $A^{(13\to12)}_{n=2}(0,E)$ it will be convenient to make use of the three-particle unitarity condition in the channel where E is the energy (Fig. 4.4). Naturally, for us only the discontinuity on the three-particle cut is of interest.

The discontinuity on the three-particle cut is determined by the expression:

$$disc_s A^{(13\to12)}_{n=2}(0,E) = -\frac{1}{\pi^3}\,\lambda\, a_{12}a_{13}(m_1+m_2)(m_1+m_3) \qquad (4.21)$$
$$\times\int\frac{d^4q_1\,d^4q_2\,d^4q_3}{q_1'^2-m_1^2}\delta^4(q_1+q_2+q_3-P)\delta(q_1^2-m_1^2)\delta(q_2^2-m_2^2)\,\delta(q_3^2-m_3^2).$$

The notations in this expression are evident. The meaning of the 4-momenta q_1,q_2,q_3,q_1' and P is explained in Fig. 4.4. The momentum of the initial state P has in the c.m. system of the three particles only a time component $P_0=\sqrt{s}\simeq m_1+m_2+m_3+E$. In the expression (4.22) only the contribution of the cut shown in Fig. 4.4 by the dotted line is taken into account. The contribution to the three-particle discontinuity due to the division presented by the wavy line is absent since the relative momentum of particles 1 and 2 is zero.

The calculation of the integral (4.22) is quite simple. One gets the following result (making, naturally, the transition to the non-relativistic approximation):

$$disc_s A^{(13\to12)}_{n=2}(0,E) \;=\; \lambda\, a_{12}a_{13}\sqrt{\frac{m_1m_2m_3}{m_1+m_2+m_3}}\,\frac{1}{3}\,(1+2\beta_1)\,E\,. \quad (4.22)$$

The expression (4.22) allows us to obtain the singular part of function $A^{(13\to12)}_{n=2}(0,E)$ using the dispersion relation

$$A^{(13\to12)}_{n=2}(0,E) - A^{(13\to12)}_{n=2}(0,0) \;=\; \frac{E}{\pi}\int\limits_0^\infty dE'\,\frac{discA^{(13\to12)}_{n=2}(0,E')}{E'(E'-E-i0)}\,. \quad (4.23)$$

The subtraction constant $A^{(13\to12)}_{n=2}(0,0)$ should be included into λ. Having this in mind, we take

$$A^{(13\to12)}_{n=2}(0,0) = 0 \quad (4.24)$$

and write the expression for $A^{(13\to12)}_{n=2}(0,E)$, which contains a singularity in E. Inserting (4.22) into (4.23) we obtain a logarithmically divergent integral. We cut it at a value of the order of the mass m of the particle, having in mind the same considerations as in the case when the expression (4.20) was calculated from the dispersion integral (4.15). Hence, we have

$$A^{(13\to12)}_{n=2}(0,E) \;=\; \lambda\, a_{12}a_{13}\sqrt{\frac{m_1m_2m_3}{m_1+m_2+m_3}}\,\frac{1}{3}\,(1+2\beta_1)\frac{E}{\pi}\Big(\ln\frac{m}{E}+i\pi\Big). \quad (4.25)$$

The last term in (4.25) does not contain singularities in E, and is included in (4.25) in a somewhat hypothetical way. But, differently from the real terms in $A^{(13\to12)}_{n=2}(0,E)$, similar imaginary analytic terms occur in the considered case only from diagrams connected with manifold particle scattering, and, because of that, it is unambiguous to write them.

We have calculated linear (Eq. (4.13)) and quadratic (Eqs. (4.20) and (4.25)) non-analytic terms, the terms of the order of $E\ln E$ (or $k^2\ln E$) are considered as quadratic.

If we consider diagrams which correspond to the double scattering of a particle pair (Fig. 4.5a), and select the singular linear terms of both loops, it is seen that their contribution equals $-\lambda\, a_{12}^2k_{12}^2$, which is analytic in k_{12}^2. Because of this, the contribution of such diagrams cannot be separated from other analytic terms.

Concerning diagrams with a large number of scatterings (such as, for example, Fig. 4.5b), the calculations presented in this section tell us that each additional scattering leads to terms with an extra power of the of the threshold momentum. The diagrams Fig. 4.5c give linear terms. So, the diagrams with triple scatterings, Fig. 4.5b, produce cubic terms. Indeed,

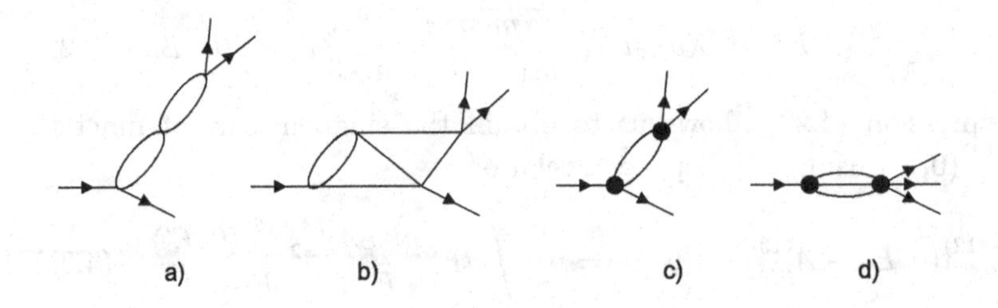

Fig. 4.5 (a,b) Diagrams of two and three rescatterings with point-like interaction blocks. (c,d) Diagrams with re-scattering amplitudes (the black blocks) which are essentially non-point-like.

each additional scattering gives an extra power of the momentum into the absorption part of the diagram. This momentum is, essentially, the phase volume of the scattering particles near the threshold. Reconstructing the singular terms in the amplitude from the absorption part in the dispersion integral, it is always the region of small momenta (or small energies) which is essential, since the large integration momenta give only an analytic contribution. In other words, one can carry out a sufficient number of subtractions in any dispersion integral so that at small external momenta the small integration momenta turn out to be important. The appearing subtraction polynomial gives an analytic dependence. The terms of this type should be included separately in any case, and their normalization is provided by the unitarity condition. Such a subtraction procedure results in the fact that the diagrams with a large number of scatterings (like those of the type Fig. 4.5b) do not contribute to the non-analytic linear and quadratic terms.

Singular terms may appear in diagrams in which the blocks correspond to scatterings or the production of three particles, are not substituted by constants. An example of such a diagram is shown in Fig. 4.5c. Let us expand the blocks contained in the diagram into a series over the difference between the respective invariants and the threshold values. The constant terms in these expansions lead just to one of the diagrams we have taken into account, the linear terms. The next addends give us quadratic, cubic and other terms in the absorption part.

Finally, let us turn our attention to the diagram shown in Fig. 4.5d which contains the block of transforming three particles into three. Note that diagrams of this type include non-analytic terms near the threshold independently of the fact whether there is a particle pair scattering.

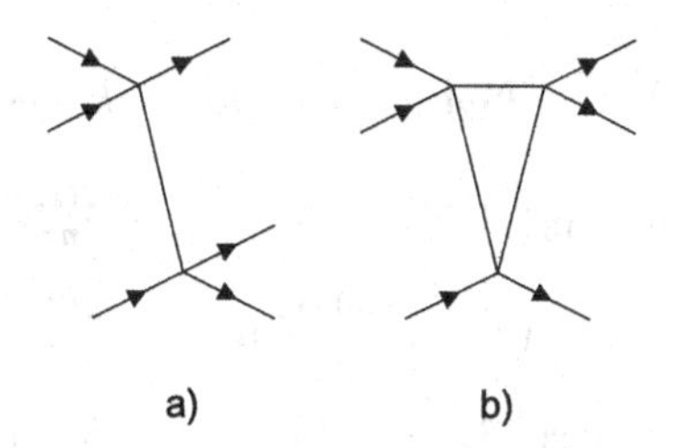

Fig. 4.6 Diagrams for the $3 \to 3$ transition of the order of $1/E$ (a), and $1/\sqrt{E}$ (b).

If the block of the transition of three particles into three is a constant, the absorption part of such a diagram is of the order of E^2 (the phase volume of three particles near the threshold), and the amplitude contains a non-analytic term of the order of $E^2 \ln E$. In other words, in this case the diagram contributes only to terms of the fourth order in threshold momenta. The block of the transition of three particles into three can, however, turn into infinity at small particle momenta. This can happen, for example, when the amplitude bears the character of a pole (Fig. 4.6a). In this case we come to the diagram Fig. 4.2, investigated before. Since the pole-type six-point amplitude is of the order of E^{-1}, the value of the absorption part turns out to be of the order of E, as it was seen in Eqs. (4.19) and (4.22). The total amplitude occurs to be also of the order of E (or $E \ln E$).

The six-point amplitude, presented in Fig. 4.6b, is of the order of $E^{-1/2}$ near the threshold. The corresponding contribution of the diagram Fig. 4.5b to the amplitude, containing such a six-point amplitude, is of the order of $E^{3/2}$ near threshold.

4.2.1.5 *Amplitude and total cross section up to terms* $\sim E$

Now we present the amplitude with the accuracy up to quadratic terms in threshold momenta. From what was discussed above it follows that this expression can be obtained from the contributions of three diagrams shown in Fig. 4.1a (linear terms), six diagrams of the type presented in Fig. 4.1b (quadratic non-analytic terms) and analytic terms of the form $\lambda + \alpha_3 k_{12}^2 + \alpha_2 k_{13}^2 + \alpha_1 k_{23}^2$. The analytic quadratic terms have to be added to the amplitude with some unknown coefficients. Since these terms characterize the contributions of far singularities, the coefficients are of the order of λ/m^2, where m is a value of the order of the low-mass hadrons, $m \sim 200 - 500$ MeV. According to this, we take $\alpha_i = \lambda C_i$, where $C_i \sim 1/m^2$.

Hence, we obtain:

$$\begin{aligned}A(k_{12}^2, k_{13}^2, k_{23}^2) = \lambda\Big[&1 + ik_{12}a_{12} + ik_{13}a_{23} + ik_{23}a_{23} \\ &+ a_{12}a_{13}\Big(A_{n=2}^{(13\to12)}(k_{12}^2, E) + A_{n=2}^{(12\to13)}(k_{13}^2, E)\Big) \\ &+ a_{12}a_{23}\Big(A_{n=2}^{(12\to23)}(k_{23}^2, E) + A_{n=2}^{(23\to12)}(k_{12}^2, E)\Big) \\ &+ a_{13}a_{23}\Big(A_{n=2}^{(23\to13)}(k_{13}^2, E) + A_{n=2}^{(13\to23)}(k_{23}^2, E)\Big) \\ &+ C_1k_{23}^2 + C_2k_{13}^2 + C_3k_{12}^2\Big], \end{aligned} \tag{4.26}$$

where, for example,

$$\begin{aligned}A_{n=2}^{(13\to12)}(k_{12}^2, E) = &-2E\sqrt{\frac{m_1m_2m_3}{m_1+m_2+m_3}} \\ &\times\left[\frac{2x_{12}\arccos x_{12}}{\pi\sqrt{1-x_{12}^2}}\left(\beta_1 + x_{12}^2\frac{1-4\beta_1}{3}\right)\right. \\ &\left.-\frac{1}{\pi}\left(\ln\frac{m}{E} + i\pi\right)\left(\frac{1}{6}(1+2\beta_1) + x_{12}^2\frac{1-4\beta_1}{3}\right)\right], \\ x_{12} = \frac{k_{12}}{\sqrt{2\mu_{12}E}},\quad &\beta_1 = \frac{m_1(m_1+m_2+m_3)}{(m_1+m_2)(m_1+m_3)}, \end{aligned} \tag{4.27}$$

while other $A_{n=2}^{(i\ell\to ij)}(k_{ij}^2, E)$ can be obtained by index permutations.

In the expression (4.26) there are three undetermined constants C_1, C_2, C_3, and, besides, a term of the type of $E\ln m$ is included so that $\ln m$ turns out to be also an undetermined constant. Since, however, E is the sum of k_{12}^2, k_{13}^2 and k_{23}^2, the expression (4.26) contains, essentially, just three undetermined constants.

Let us underline that at $k_{12}\to 0$, the cross section which is calculated according to Eq. (4.26), does not contain a term linear in k_{12}. This is the consequence of the fact that in the physical region of the reaction at $k_{12}\to 0$ we have $x_{13}^2 \to \beta_1$, *i.e.* $k_{13}\to\sqrt{2\beta_1\mu_{13}E}$. On the other hand, the function $A_{n=2}^{(13\to12)}(k_{12}^2, E)$ at small k_{12} contains addends of the type $-k_{12}\sqrt{2\beta_1\mu_{13}E}$, which cancels exactly the term linear in k_{12}. This behaviour of the cross section bears a rather general character due to the Landau threshold relation:

$$A\Big(k_{12}^2, k_{13}^2, k_{23}^2\Big) = A\Big(0, k_{13}^2, k_{23}^2\Big)(1 + ik_{12}a_{12}). \tag{4.28}$$

In the physical region there is only one point corresponding to the value $k_{12}^2 = 0$, and thus the variables k_{13}^2 and k_{23}^2 in (4.28) are expressed via the total energy E.

Equation (4.26) gives us the total cross section:

$$\sigma = \sigma_0 E^2 \left[1 + \alpha E \ln \frac{\mu}{E}\right],$$
$$\alpha = \frac{8}{3\pi}\sqrt{\frac{m_1 m_2 m_3}{m_1 + m_2 + m_3}}$$
$$\times \Big[a_{12}a_{13}(1-\beta_1) + a_{12}a_{23}(1-\beta_2) + a_{13}a_{23}(1-\beta_3)\Big]. \quad (4.29)$$

where σ_0 and α are parameters related to λ and C_i while α is determined on scattering lengths a_{ij}.

4.2.1.6 *Miniconclusion*

The presented investigation allows us to give a qualitative picture for the structure of the expansion of the amplitude with $J = 0$ near the threshold. We have demonstrated that the amplitude can be written in the form of the sum (4.8) and that each of the $A_n^{J=0}$ values is of the order of $(E/m)^{n/2}$. At $n = 0$ we have $A_0^{J=0} = \lambda$, the amplitude of the production of three particles at zero energy. At $n = 1$ $A_1(k_{12}^2, k_{13}^2, k_{23}^2) \sim |k_{i\ell}|$. The linear terms, if considered as functions of the complex variables $k_{12}^2, k_{13}^2, k_{23}^2$, are non-analytic. The quadratic terms in the expansion of the amplitude ($n = 2$ in eq. (4.8)) include new unknown real parameters C_1, C_2, C_3. The physical meaning of these parameters is obvious: they characterize the interaction of particles at large energies. The inclusion of the constants C_i in the expression for the amplitude is equivalent to the introduction of the interaction radius in the particle rescattering amplitudes.

In the quadratic terms we face singularities at $x_{i\ell}^2 = 1$ (or at $k_{i\ell}^2 = 2\mu_{i\ell}E$). Such singularities can not be present on the physical sheet and have to be "hidden" under the cuts of the threshold singularities.

As we have seen already, cubic terms have to appear in diagrams corresponding to the case when these particles scatter not more than three times in the final state. The absorption parts of the diagrams concerned is expressed, obviously, in terms of pair scattering amplitudes and the production amplitude of three particles calculated up to quadratic terms. In addition to the pair amplitudes, the latter contains also the constants C_1, C_2, C_3 which, consequently, enter the final expression for the cubic terms. There are, however, no new unknown constants – this is a proof of the fact that all the cubic terms are non-analytic.

The fourth order terms contain a large number of new unknown parameters like $B_1 k_{23}^4$, $B_2 k_{13}^4$,..., $D_1 k_{12}^2 k_{13}^2$, $D_2 k_{12}^2 k_{23}^2$ etc. Besides, as it was shown in the previous section, in this order one has to take into account the diagram presented in Fig. 4.6d. This diagram leads to the term $E^2 \ln E$, non-analytic in the total energy, with a coefficient proportional to the constant part of the amplitude of three particles transforming into three; such a quantity does not appear in lower order terms.

Going to higher order terms, the structure of the expansion (4.8) is obvious. For example, terms of the fifth order do not contain new unknown parameters compared to the first to fourth terms.

4.2.2 *Production of three particles in a state with $J = 1$*

Up to now we have investigated the amplitude of the production of three spinless particles with a total angular momentum $J = 0$. In the present section we consider also spinless particles which have, however, a angular momentum $J = 1$ in the final state. This means that we either have one particle with $J = 1$ in the initial state (*i.e.* a decay process), or two particles. Let us consider the latter case, the transition of two particles into three. We carry out an expansion of the amplitude with the accuracy up to quadratic terms in the momenta.

We define the amplitude of the transition of two particles into three with the total angular momentum $J = 1$ in the c.m. system as:

$$A^{(J=1)}(k_{12}^2, k_{13}^2, k_{23}^2) = T_I^{(J=1)}(k_{12}^2, k_{13}^2, k_{23}^2)k_1Z_1 + T_{II}^{(J=1)}(k_{12}^2, k_{13}^2, k_{23}^2)k_2Z_2 \,, \tag{4.30}$$

where

$$k_iZ_i = (\mathbf{nk_i}). \tag{4.31}$$

Here $\mathbf{n}$ is a unit vector for the spin of the initial state. For example, if we consider the decay of a vector particle, then $\mathbf{n}$ is the polarization vector of the decaying particle.

Due to $\mathbf{k}_1 + \mathbf{k}_2 + \mathbf{k}_3 = 0$ in the c.m. system, we come to a relation

$$k_1Z_1 + k_2Z_2 + k_3Z_3 \ = \ 0 \,. \tag{4.32}$$

This means that the expression (4.30) can be rewritten in the form of the linear combination of k_1Z_1 and k_2Z_2 or k_2Z_2 and k_3Z_3. To be definite, in the beginning we use the cosines of the angles Z_1 and Z_2. In the end of the section we will demonstrate how the obtained result can be rewritten in a symmetric form.

The terms linear in the momenta can be obtained from (4.30) if $T_I^{(J=1)}(k_{12}^2,k_{13}^2,k_{23}^2)$ and $T_{II}^{(J=1)}(k_{12}^2,k_{13}^2,k_{23}^2)$ are substituted by their threshold values

$$T_I^{(J=1)}(k_{12}^2,k_{13}^2,k_{23}^2)\to\alpha_1\,,\qquad T_{II}^{(J=1)}(k_{12}^2,k_{13}^2,k_{23}^2)\to\alpha_2 \quad (4.33)$$

We can get the quadratic terms because the amplitudes $T_I^{(J=1)}(k_{12}^2,k_{13}^2,k_{23}^2)$ and $T_{II}^{(J=1)}(k_{12}^2,k_{13}^2,k_{23}^2)$ contain non-analytic terms linear in $k_{i\ell}$. Naturally, these terms appear from diagrams shown Fig. 4.7. The object of the present section is just the calculation of these linear corrections in $T_I^{(J=1)}(k_{12}^2,k_{13}^2,k_{23}^2)$ and $T_{II}^{(J=1)}(k_{12}^2,k_{13}^2,k_{23}^2)$.

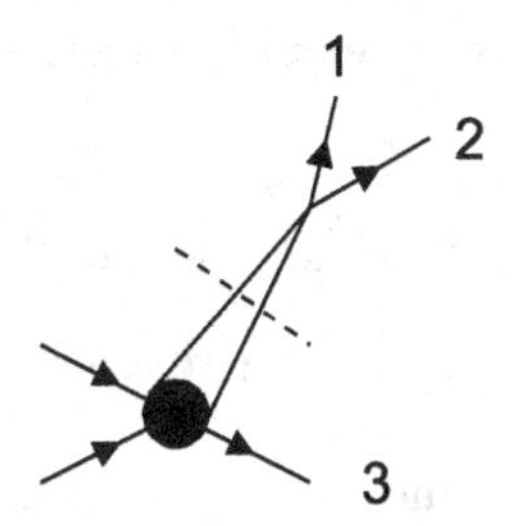

Fig. 4.7 Diagrams of this type in the amplitude with $J=1$ result in non-analytical terms of the order of $|k_{i\ell}^2|$.

For example, in Fig. 4.7 the block to the left of the dotted line has to be replaced by $\alpha_1k_1Z_1+\alpha_2k_2Z_2$. The diagram Fig. 4.7 can be calculated in the framework of the method discussed in the previous section. The linear corrections of interest in this diagram are determined by a dispersion integral over k_{12}^2:

$$A^{(J=1)}(k_{12}'^2,k_{13}'^2,k_{23}'^2)=\frac{k_{12}^2}{\pi}\int\limits_0^\infty dk_{12}'^2\,\frac{disc_{12}A^{(J=1)}(k_{12}'^2,k_{13}'^2,k_{23}'^2)}{k_{12}'^2(k_{12}'^2-k_{12}^2-i0)}\,,$$

$$disc_{12}A^{(J=1)}(k_{12}'^2,k_{13}'^2,k_{23}'^2)=k_{12}'a_{12}\int\limits_{-1}^{1}\frac{dz}{2}\Big(\alpha_1k_1'Z_1'+\alpha_2k_2'Z_2'\Big). \quad (4.34)$$

Let us remind that $\mathbf{k}_1'$ and $\mathbf{k}_2'$ are the momenta of the intermediate particles 1 and 2 in the centre-of-mass system, z is the cosine of the angle between the relative momentum of particles 1 and 2 in the intermediate state ($\mathbf{k}_{12}'$) and the momentum of one of the incident particles ($\mathbf{n}$), while Z_1' and Z_2' are the cosines of the angles between $\mathbf{k}_1'$ and $\mathbf{k}_2'$ and the unit vector $\mathbf{n}$. We obtain

$$k'_1 Z'_1 = -\frac{m_1}{m_1+m_2}\, k_3 Z_3 + k'_{12} z\,, \quad k'_2 Z'_2 = -\frac{m_2}{m_1+m_2}\, k_3 Z_3 - k'_{12} z\,, \quad (4.35)$$

what allows us to carry out the integration over z in (4.34) immediately. We have

$$disc_{12} A^{(J=1)}(k'^2_{12}, k'^2_{13}, k'^2_{23}) = k_3 Z_3 a_{12} k'_{12} \left(-\frac{m_1}{m_1+m_2}\, a_1 - \frac{m_2}{m_1+m_2}\, a_2 \right). \quad (4.36)$$

From here it follows that the integral (4.34) equals $i\; disc_{12} A^{(J=1)}(k^2_{12}, k^2_{13}, k^2_{23})$. Considering in the same way corrections related with the scattering of other particle pairs, we get for the quadratic terms

$$i \Big(disc_{12} A^{(J=1)}(k^2_{12}, k^2_{13}, k^2_{23}) + disc_{13} A^{(J=1)}(k^2_{12}, k^2_{13}, k^2_{23}) + disc_{23} A^{(J=1)}(k^2_{12}, k^2_{13}, k^2_{23}) \Big). \quad (4.37)$$

So, taking into account the linear and quadratic terms in the momenta, we obtain for the amplitude with $J = 1$:

$$\begin{aligned} A^{(J=1)}(k^2_{12}, k^2_{13}, k^2_{23}) &= \quad (4.38) \\ = k_1 Z_1 &\left[\alpha_1 + i k_{12} a_{12} \frac{m_1\alpha_1 + m_2\alpha_2}{m_1+m_2} + i k_{23} a_{23} \left(\alpha_1 - \frac{m_2}{m_2+m_3}\alpha_2 \right) \right] \\ + k_2 Z_2 &\left[\alpha_2 + i k_{12} a_{12} \frac{m_1\alpha_1 + m_2\alpha_2}{m_1+m_2} + i k_{13} a_{13} \left(\alpha_2 - \frac{m_1}{m_1+m_3}\alpha_1 \right) \right]. \end{aligned}$$

The structure of the production amplitude of three particles with $J = 1$ is analogous to the structure of the amplitude with $J = 0$. The first terms of this amplitude are also defined by the unknown constants (amplitudes at threshold energies) α_i; the corrective terms are determined by the same constants and the scattering lengths of the produced particles. The unitarity condition connects the imaginary and real parts of the amplitude.

The expression (4.38) is asymmetric in the indices 1, 2 and 3. This is due to the special choice of the amplitude in the form (4.30). If we use from the very beginning the the symmetric from of the amplitude

$$A^{(J=1)}(k^2_{12}, k^2_{13}, k^2_{23}) = T^{(J=1)}_I k_1 Z_1 + T^{(J=1)}_{II} k_2 Z_2 + T^{(J=1)}_{III} k_3 Z_3\,, \quad (4.39)$$

we can obtain instead of (4.38) the following expression, symmetric in the indices:

$$A^{(J=1)}(k_{12}^2, k_{13}^2, k_{23}^2) = \sum_{n=1,2} A_n^{(J=1)}(k_{12}^2, k_{13}^2, k_{23}^2)$$

$$= k_1 Z_1 \left[\tilde{\alpha}_1 + ik_{23}a_{23}\left(\tilde{\alpha}_1 - \frac{m_2}{m_2+m_3}\tilde{\alpha}_2 - \frac{m_3}{m_2+_3}\tilde{\alpha}_3\right)\right]$$
$$+ k_2 Z_2 \left[\tilde{\alpha}_2 + ik_{13}a_{13}\left(\tilde{\alpha}_2 - \frac{m_1}{m_1+m_3}\tilde{\alpha}_1 - \frac{m_3}{m_1+m_3}\tilde{\alpha}_3\right)\right]$$
$$+ k_3 Z_3 \left[\tilde{\alpha}_3 + ik_{12}a_{12}\left(\tilde{\alpha}_3 - \frac{m_1}{m_1+m_2}\tilde{\alpha}_1 - \frac{m_2}{m_1+m_2}\tilde{\alpha}_2\right)\right], \quad (4.40)$$

where $\tilde{\alpha}_1, \tilde{\alpha}_2$ and $\tilde{\alpha}_3$ are the threshold values of $T_I^{(J=1)}$, $T_{II}^{(J=1)}$ and $T_{III}^{(J=1)}$. Due to the relation (4.32) between k_1Z_1, k_2Z_2 and k_3Z_3, the function $\tilde{\alpha}_i$ in Eq. (4.35) can not be determined unambiguously. Essentially, one of the $\tilde{\alpha}_i$ values is arbitrary; Eq. (4.38) corresponds to the choice $\tilde{\alpha}_3 = 0$.

4.3 The production of three particles near the threshold when two particles interact strongly

Many reactions exist in which pairs of the produced particles interact strongly near the threshold, or even in a resonant way. For example, in the reaction $\pi + N \to N + \pi + \pi$ the $\pi + N$ interaction is not large near the threshold. Two pions have, however, a low-energy resonance σ, which should be taken into account precisely near $\sqrt{s_{\pi\pi}} \sim 2\mu_\pi$. The two low-energy baryons in the reactions $N+N \to N+N+\pi$ or $N+N \to N+\Lambda+K$ are interacting strongly: the amplitude of the pn system has a deuteron pole on the first sheet of the complex-s_{pn} plane while amplitudes of the nn and pp systems have analogous poles on the second sheets though close to the physical region.

The given reactions near the threshold can be described in a rough approximation (in terms of the order of $n = 0$) applying the Watson–Migdal formalism. But a more correct investigation requires to take into account the corrections to the Watson–Migdal formula which appear when the non-resonant interactions of the produced particles are considered.

In this section we present in the framework of the dispersion technique the scheme for calculating the corrections of the order of $\sqrt{E}$ to the Migdal–Watson formula. The corrections depend both on the relative momenta of the particles produced and on the total kinetic energy. These corrections can be calculated explicitly when $a_{i\ell} \sim const$, see, for example [11, 12]. However, for a realistic description of the low energy interactions we have to use the energy-dependent $a_{i\ell}(k_{i\ell}^2)$ both in the $\pi\pi$ and NN systems.

4.3.1 *The production of three spinless particles*

We consider here the production amplitude of three particles with $J = 0$ in the case when the interaction of two particles (*e.g.*, particles 1 and 2) is large, i.e. $a_{12}(k_{12}^2)k_{12} \sim 1$. We shall presume that, at the same time, the interaction of particles $1+2$ does not lead to the appearance of real bound states. This means that we consider reactions of the type $\pi + N \to N + \pi\pi$ or $pp \to pp + \pi$ (strong interactions in the systems $\pi\pi$ and pp), but in a simplified version without taking into account the spin and the isospin.

The amplitudes a_{13} and a_{23} are considered to be of the order of m^{-1}. This means that it is necessary to take into account manifold scatterings of the particles 1, 2 which do not lead to extra smallness (see Fig. 4.8). An additional smallness appears only from scatterings of the first two particles on the third one.

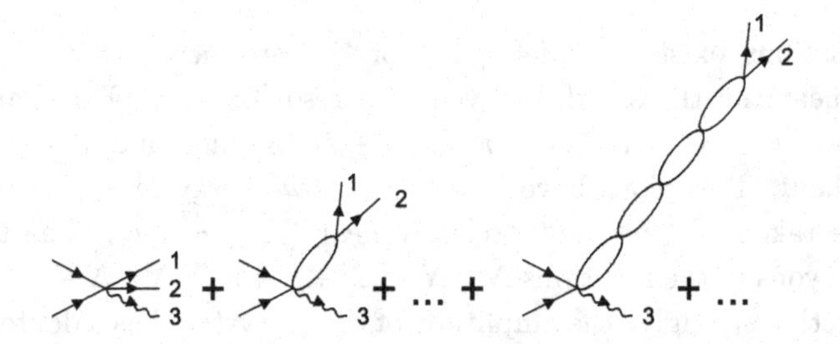

Fig. 4.8 Diagrams up to the order $n = 0$ in the case when two particles, $(1+2)$, interact strongly.

The diagrams which are essential in the investigation of the amplitude taking into account terms of the order of $\sqrt{E}$ are presented in Figs. 4.9 and 4.10.

Let us return now to the diagrams of the order of $n = 0$. We identify the first diagram – in Fig. 4.8 – with the threshold amplitude λ; the second diagram (the loop diagram) gives us $\lambda i k_{12} a_{12}$. Each new loop leads to the appearance of an extra factor $ik_{12}a_{12}(k_{12}^2)$ which is not small. Consequently, the sum of the diagrams in Fig. 4.8 equals

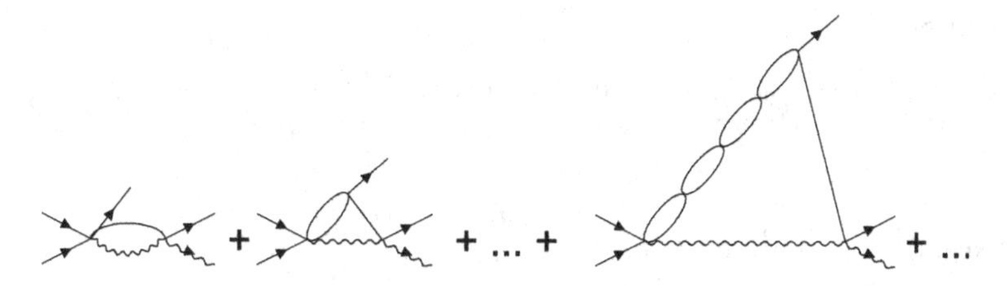

Fig. 4.9 Diagrams of the order $n = 1$ (or $\sqrt{E}$) when the third particle (wave line) participates in the "last interaction" in the final state.

$$\lambda \frac{1}{1 - ik_{12}\, a_{12}(k_{12}^2)} \equiv A_{n=0}^{(12)}(k_{12}^2)\,. \tag{4.41}$$

The summation of the infinite series of diagrams leaves us always with an unpleasant feeling. Nevertheless, there is a justification for this procedure: Eq. (4.41) is the Watson–Migdal formula. In terms of the expansion of Eq. (4.8) it is term of the order $n = 0$.

The corrections to (4.41) of the order of $\sqrt{E}$ appear as the result of taking into account the interaction of the two first particles with the third one. Diagrams of this type are shown in Figs. 4.9 and 4.10. The diagrams Figs. 4.9a, 4.9b were considered in detail. The method for calculating the diagrams Figs. 4.9c and 4.10 is quite similar, but technically much more complicated. Let us demonstrate the scheme of these calculations.

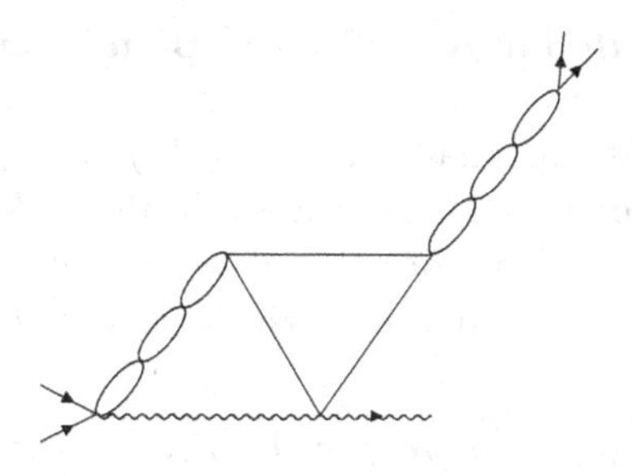

Fig. 4.10 Diagrams of the order $n = 1$ when the "last interaction" is in the $(1 + 2)$-channel.

The diagrams Fig. 4.9 depend on two variables: the relative momenta of the produced particles $k_{\ell 3}^2$ and on the total kinetic energy E. Consequently, they can be calculated with the help of the dispersion relations in these variables:

$$A^{(12\to\ell 3)}_{n=1}(k^2_{\ell 3},E)=\int\limits_0^\infty \frac{dk'^2_{\ell 3}}{\pi}\frac{disc_{\ell 3}\,A^{(12\to\ell 3)}_{n=1}(k'^2_{\ell 3},E)}{k'^2_{\ell 3}-k^2_{\ell 3}-i0}\,,\qquad \ell=1,2\,,$$

$$disc_{\ell 3}\,A^{(12\to\ell 3)}_{n=1}(k^2_{\ell 3},E)=k_{\ell 3}a_{\ell 3}\int\limits_{-1}^{1}\frac{dz}{2}\frac{\lambda(k'^2_{12})}{1-ik'_{12}a_{12}(k'^2_{12})}$$

$$\equiv k_{\ell 3}a_{\ell 3}\langle A^{(12)}_{n=0}(k'^2_{12})\rangle_{\ell 3}. \tag{4.42}$$

Here $disc_{\ell 3}\,A^{(12\to\ell 3)}_{n=1}(k^2_{\ell 3},E)$ is the absorption part in the $\ell 3$ channel.

The assumption that the integral (4.42) is convergent means that we suppose that this is ensured by the behaviour of $a_{12}(k'^2_{12})$ and $\lambda(k'^2_{12})$.

For diagram of Fig. 4.10 we write

$$A^{(\ell 3\to 12)}_{n=1}(k^2_{12},E)=\frac{1}{1-ik_{12}a_{12}(k^2_{12})}\int\limits_0^\infty \frac{dk'^2_{12}}{\pi}\frac{disc_{12}\,A^{(\ell 3)\to 12}_{n=1}(k'^2_{12},E)}{k'^2_{12}-k^2_{12}-i0}\,,$$

$$disc_{12}\,A^{(\ell 3\to 12)}_{n=1}(k^2_{12},E)=k_{12}a_{12}(k^2_{12})\int\limits_{-1}^{1}\frac{dz}{2}\,A^{(12\to\ell 3)}_{n=1}(k'^2_{\ell 3}),\qquad \ell=1,2. \tag{4.43}$$

Here, as previously, the convergence of the integral is insured by the behaviour of $a_{12}(k^2_{12})$.

4.4 Decay amplitude for $K\to 3\pi$ and pion interaction

Pion scattering lengths are small but $a_{I=0,2}(k^2)$ promptly increase with increasing k^2, therefore in the amplitudes of the decay $K\to 3\pi$ the full set of the pion rescatterings should be taken into account. Thus, the $K\to 3\pi$ amplitudes obey the integral equations which are presented below.

4.4.1 *The dispersion relation for the decay amplitude*

In the $K\to 3\pi$ decay process the $\Delta I=1/2$ rule is satisfied, and the produced pions are in a state with an isospin $I=1$. Hence, the amplitude of the decay $K\to 3\pi$ can be written in the form

$$M^d_{abc}(k_{12},k_{13},k_{23})=A(k_{12},k_{13},k_{23})\delta_{da}\delta_{bc}+B(k_{12},k_{13},k_{23})\delta_{db}\delta_{ac}+C(k_{12},k_{13},k_{23})\delta_{dc}\delta_{ab}\,, \tag{4.44}$$

where δ_{ab} is the Kronecker symbol, while the isotopic indices (a, b, c) refer to pions and (d) to the total isospin of the system . If the pion interaction at small energies takes place only in the S-state, then

$$\begin{aligned} A(k_{12}, k_{13}, k_{23}) &= \lambda + A(k_{12}) + A(k_{13}) + D(k_{23}), \\ B(k_{12}, k_{13}, k_{23}) &= A(k_{13}, k_{12}, k_{23}), \\ C(k_{12}, k_{13}, k_{23}) &= A(k_{23}, k_{13}, k_{12}). \end{aligned} \tag{4.45}$$

We use $\delta_{00} = \delta_{1-1} = \delta_{-11} = 1$ with $(1, 0, -1) = (\pi^+, \pi^0, \pi^-)$, and put the other Kronecker symbols equal to zero. It is convenient to choose the constant λ so that

$$A(0) = D(0) = 0. \tag{4.46}$$

The amplitude $M^d_{abc}(k_{12}, k_{13}, k_{23})$ satisfies the following dispersion relation:

$$\begin{aligned} M^d_{abc}(k_{12}, k_{13}, k_{23}) &= \lambda\Big(\delta_{da}\delta_{bc} + \delta_{db}\delta_{ac} + \delta_{dc}\delta_{ab}\Big) + \\ &+ k_{12}^2 \int_0^\infty \frac{dk_{12}'^2}{\pi} \frac{k_{12}' f^*_{a'b'ab}(k_{12}')}{k_{12}'^2(k_{12}'^2 - k_{12}^2 - i0)} \langle M^d_{a'b'c}(k_{12}', k_{13}'', k_{23}'') \rangle_{12} + \\ &+ k_{13}^2 \int_0^\infty \frac{dk_{13}'^2}{\pi} \frac{k_{13}' f^*_{a'c'ac}(k_{13}')}{k_{13}'^2(k_{13}'^2 - k_{13}^2 - i0)} \langle M^d_{a'bc'}(k_{12}'', k_{13}', k_{23}'') \rangle_{13} + \\ &+ k_{23}^2 \int_0^\infty \frac{dk_{23}'^2}{\pi} \frac{k_{23}' f^*_{b'c'bc}(k_{23}')}{k_{23}'^2(k_{23}'^2 - k_{23}^2 - i0)} \langle M^d_{ab'c'}(k_{12}'', k_{13}'', k_{23}') \rangle_{23}. \end{aligned} \tag{4.47}$$

In the above formula the subtractions are carried out according to (4.46).

Let us remind that the symbol $\langle\ \rangle_{i\ell}$ means that we take the average of the quantity over the cosine of the angle between the direction $\vec{k}_{i\ell}$ and that of the momentum of the third particle. For example,

$$\langle M(k_{12}k_{13}''k_{23}'') \rangle_{12} = \int_{-1}^{1} \frac{dz}{2} M(k_{12}k_{13}''k_{23}''), \tag{4.48}$$

with

$$k_{13}''^2 = \frac{1}{4}k_{12}^2 + \frac{3}{4}(E\mu - k_{12}^2) + \frac{\sqrt{3}}{2} z\, k_{12}\sqrt{E\mu - k_{12}^2}. \tag{4.49}$$

and the integration is carried out taking into account the threshold singularities at $k_{13}''^2 = 0$ around the contour shown in Fig. 4.3.

In the case of the pion scatterings in S-wave

$$f_{a'b'ab}(k) = \frac{1}{3}\, f_0(k)\delta_{a'b'}\delta_{ab} + \frac{1}{2}\, f_2(k)\Big(\delta_{a'a}\delta_{b'b} + \delta_{a'b}\delta_{b'a} - \frac{2}{3}\delta_{a'b'}\delta_{ab}\Big),$$

$$f_I(k) = \frac{1}{k}e^{i\delta_I}\sin\delta_I = \frac{a_I(k^2)}{1-ika_I(k^2)}\,. \tag{4.50}$$

For further investigations it is convenient to write the dispersion relation (4.47) in a different form. First, we carry out the corresponding transformation in the case of $a_2 = 0$ and $a_0 \sim const.$ If so, $A(k) = 0$, and $D(k)$ satisfies the equation

$$D(k) = \frac{k^2}{1-ika_0}\int\limits_0^\infty \frac{dk'^2}{\pi}\frac{k'a_0}{k'^2(k'^2-k^2-i0)}\left[\frac{5}{3}\lambda + D(k') + \frac{2}{3}\langle D(k'')\rangle\right]. \tag{4.51}$$

Further simplification of the dispersion relation (4.51) is possible by introducing the function $d(k)$

$$D(k) \;=\; \frac{d(k)}{1-i\,k\,a_0}. \tag{4.52}$$

It follows from (4.51) that the discontinuity of the function $D(k)$ on the threshold cut is equal to

$$\left[\frac{5}{3}\lambda + D(k) + \frac{2}{3}\langle D(k'')\rangle\right]\frac{ik\,a_0}{1-ika_0}\,. \tag{4.53}$$

Carrying out this calculation, we have to take into account that the dispersion integral in (4.51) consists of two addends: the principal part (P) and the half-residue $(i\pi\delta(k'^2-k^2))$ in the pole:

$$\int\limits_0^\infty \frac{dk'^2}{\pi}\frac{1}{k'^2-k^2-i0} = P\int\limits_0^\infty \frac{dk'^2}{\pi}\frac{1}{k'^2-k^2} + \int\limits_0^\infty \frac{dk'^2}{\pi}\, i\pi\,\delta(k'^2-k^2). \tag{4.54}$$

The principal part has a jump due to the factor $(1-ika_0)^{-1}$; the half-residue changes its sign on the lower edge.

On the other hand, it is seen from expression (4.52) that

$$\Delta D(k) \;\equiv\; D(k) - D(-k) \;=\; \frac{\Delta\, d(k)}{1-ika_0} + d(k)\,\frac{2ika_0}{1+k^2a_0^2}\,. \tag{4.55}$$

Comparing (4.53) and (4.55), we get

$$\Delta d(k) \;=\; \left(\frac{5}{3}\lambda + \frac{2}{3}\langle D(k'')\rangle\right)2ika_0\,; \tag{4.56}$$

from here it follows that

$$d(k) = k^2 \int\limits_0^\infty \frac{dk'^2}{\pi} \frac{k' a_0}{k'^2(k'^2 - k^2 - i0)} \left[\frac{5}{3}\lambda + \frac{2}{3}\langle D(k'')\rangle\right]. \tag{4.57}$$

In (4.57) it is taken into account that $d(0) = 0$. The amplitude $D(k)$, as can be seen from formulae (4.52) and (4.57), satisfies the dispersion relation

$$D(k) = \frac{5}{3}\lambda \frac{i k a_0}{1 - ika_0} + \frac{2}{3}\frac{k^2}{1 - ika_0}\int\limits_0^\infty \frac{dk'^2}{\pi}\frac{k' a_0\langle D(k'')\rangle}{k'^2(k'^2 - k^2 - i0)}. \tag{4.58}$$

The dispersion relation for the decay amplitude in the general case, when a_2 and a_0 are arbitrary, can be easily obtained in a similar way. We can write

$$\begin{aligned}
A(k) &= \lambda \frac{ika_2(k^2)}{1 - ika_2(k^2)} \\
&+ \frac{k^2}{1 - ika_2(k^2)} \int\limits_0^\infty \frac{dk'^2}{\pi} \frac{k' a_2(k'^2)\left[\langle A(k'')\rangle + \langle D(k'')\rangle\right]}{k'^2(k'^2 - k^2 - i0)}, \\
D(k) &= -\frac{2}{3}\lambda \frac{ika_2(k^2)}{1 - ika_2(k^2)} \\
&- \frac{2}{3}\frac{k^2}{1 - ika_2(k^2)} \int\limits_0^\infty \frac{dk'^2}{\pi} \frac{k' a_2(k'^2)\left[\langle A(k'')\rangle + \langle D(k'')\rangle\right]}{k'^2(k'^2 - k^2 - i0)} + \\
&+ \frac{5}{3}\lambda \frac{ika_0(k^2)}{1 - ika_0(k^2)} \\
&+ \frac{2}{3}\frac{k^2}{1 - ika_0(k^2)} \int\limits_0^\infty \frac{dk'^2}{\pi} \frac{k' a_0(k'^2)\left[4\langle A(k'')\rangle + \langle D(k'')\rangle\right]}{k'^2(k'^2 - k^2 - i0)}.
\end{aligned} \tag{4.59}$$

The dispersion relations for the non-relativistic decay amplitudes $A(k)$ and $D(k)$ were written in [13] taking into account properly the integration over z (the status of the z-integration is discussed when considering Fig. 4.3).

In terms of $D(k)$ and $A(k)$ the amplitudes of the $K \to \pi\pi\pi$ decays have the form:

$$\begin{aligned}
\pi^+\pi^-\pi^0 &: \lambda + D(k_{12}) + A(k_{13}) + A(k_{23}) \\
\pi^0\pi^0\pi^0 &: -\big[3\lambda + D(k_{12}) + A(k_{13}) + A(k_{23}) \\
&\qquad + D(k_{13}) + A(k_{23}) + A(k_{12}) + D(k_{23}) + A(k_{12}) + A(k_{13})\big] \\
\pi^+\pi^+\pi^- &: 2\lambda + D(k_{13}) + A(k_{23}) + A(k_{12}) + D(k_{23}) + A(k_{12}) + A(k_{13}) \\
\pi^0\pi^0\pi^+ &: -\big[\lambda + D(k_{12}) + A(k_{13}) + A(k_{23})\big]
\end{aligned} \tag{4.60}$$

Let us underline that these formulae are possible if the produced pions have a total isospin $I = 1$, and the pion masses are equal. The effect of the mass differences of kaons and pions is considered in the next section.

4.4.2 Pion spectra and decay ratios in $K \to \pi\pi\pi$ within taking into account mass differences of kaons and pions

In the previous section we presented the decay amplitudes of $K \to \pi\pi\pi$ assuming that the pion masses are equal, $\Delta_\pi^2 = \mu_\pm^2 - \mu_0^2 \to 0$. The mass differences ($\Delta_\pi^2$ and $\Delta_K^2 = m_0^2 - m_\pm^2$ for kaons) change the relations between the probabilities of the decays. It is relevant also for the pion spectra in the region of small relative momenta, leading to cusps in the spectra. The problem of the influence of the mass differences was raised and investigated long ago, see [14]. Later the question was considered in [15]. The observation of a cusp in the $\pi^0\pi^0$-spectrum of $K^+ \to \pi^0\pi^0\pi^+$ resulted in a huge interest in the problem, see, first of all, [16, 17].

4.4.2.1 *Decay ratios in $K \to \pi\pi\pi$*

The decay ratios are determined, first of all:
(i) by the relation of the amplitudes squared in the average points of the Dalitz plots $k_{23}^2 = k_{12}^2 = k_{13}^2 = \mu E/2$,
(ii) and by the ratios of the phase spaces for three-pion states (*i.e.* the ratios of the released energies squared, for example in the K^+ decays, $\Phi_{++-}/\Phi_{00+} \simeq E_{++-}^2/E_{00+}^2$).

This is the main contribution to the ratios of the decay probabilities. The latter, however, depend also on the details of the behaviour of the amplitudes on the peripheries of the Dalitz plots which depend on the charge exchange amplitude $\frac{2}{3}(a_2 - a_0)$. This is so if we suppose that the main contribution to the violation of the isotopic invariance in the $\pi\pi$-amplitudes near the threshold give kinematic factors conditioned by the difference of the pion masses, while the relations between the scattering lengths are preserved (this hypothesis was discussed in Chapter 2, Appendix C).

Let us consider the processes $K^+ \to \pi^+\pi^+\pi^-$ and $K^+ \to \pi^0\pi^0\pi^+$ in more detail. Direct calculations with the use of (4.59) result in

$$\frac{W_{00+}}{W_{++-}} \simeq \frac{\Phi_{00+}}{4\Phi_{++-}}\left[1+\frac{1}{2}\left(\overline{|D(k)|^2}-\overline{D(k)D^*(k')}\right)\right.$$
$$+\frac{1}{2}\left(\overline{|A(k)|^2}-\overline{A(k)A^*(k')}\right)$$
$$\left.-\operatorname{Re}\left(\overline{D(k)A^*(k)}-\overline{D(k)A^*(k')}\right)\right]. \quad (4.61)$$

Here the overline means averaging over the Dalitz-plot. In this way we have

$$\frac{W_{00+}}{W_{++-}} \simeq \frac{E^2_{00+}}{4E^2_{++-}}[1+0.042(a_2-a_0)^2], \quad (4.62)$$
$$\frac{W_{000}}{W_{+-0}} \simeq \frac{3E^2_{000}}{2E^2_{+-0}}[1-0.056(a_2-a_0)^2].$$

Having in mind that $(a_2-a_0)^2 \simeq 1/16$, this means that the contributions of the scattering lengths in the probabilities of the decays are of the order of 0.4%. But in (4.62) the cusps near the thresholds which are essential and have to be considered, are not taken into account.

4.4.2.2 *Cusps in pion spectra at small relative momenta*

In order to observe cusps in the spectra, it is sufficient to calculate the linear singularities near the threshold in the channels $\pi^+\pi^-$ and $\pi^0\pi^0$. The singularities linear in $|k|$ were calculated in [14].

For the K^+-decays near the thresholds $\pi^0\pi^0$ and $\pi^+\pi^-$ we write in the linear approximation the following amplitudes:

$$K^+ \to \left(\pi^0\pi^0\pi^+\right)_{\sim\pi^0\pi^0\ threshold}:$$
$$-\tilde{\lambda}_{00+}\left[1+(\frac{1}{3}a_0+\frac{2}{3}a_2)ik_{00}+\frac{\tilde{\lambda}_{-++}}{\tilde{\lambda}_{00+}}\frac{4}{3}(a_0-a_2)ik_{+-}\right]$$
$$=-\tilde{\lambda}_{00+}\left[1+(\frac{1}{3}a_0+\frac{2}{3}a_2)ik_{00}+\frac{\tilde{\lambda}_{-++}}{\tilde{\lambda}_{00+}}\frac{4}{3}(a_0-a_2)i\sqrt{k^2_{00}-\Delta^2_\mu}\right],$$
$$\text{at} \quad k^2_{00}>\Delta^2_\mu$$
$$\sqrt{k^2_{00}-\Delta^2_\mu} \to i\sqrt{\Delta^2_\mu-k^2_{00}} \quad \text{at} \quad k^2_{00}<\Delta^2_\mu. \quad (4.63)$$

Recall, $\Delta^2_\mu = \mu^2_\pm - \mu^2_0$. Analogously we write:

$$K^+ \to (\pi^-\pi^+\pi^+)_{\sim\pi^-\pi^+ \, threshold}:$$

$$2\tilde{\lambda}_{-++}\Big[1 + (\frac{2}{3}a_0 + \frac{1}{3}a_2)ik_{+-} + \frac{\tilde{\lambda}_{00+}}{\tilde{\lambda}_{-++}}\frac{1}{6}(a_0 - a_2)ik_{00}\Big]$$

$$= 2\tilde{\lambda}_{-++}\Big[1 + (\frac{2}{3}a_0 + \frac{1}{3}a_2)ik_{+-} + \frac{\tilde{\lambda}_{00+}}{\tilde{\lambda}_{-++}}\frac{1}{6}(a_0 - a_2)i\sqrt{k^2_{+-} + \Delta^2_\mu}\Big],$$

$$\text{at} \quad k^2_{+-} > 0. \tag{4.64}$$

Let us discuss Eq. (4.63) in more detail. First, we underline that the amplitudes (4.63) satisfy the Landau threshold singularities (4.28), which exist for pions at $a_0 = a_2 \equiv a$:

$$\pi^0\pi^0 - \text{threshold}: A_{00+} \sim 1 + iak_{00} \quad \text{at } k^2_{00} \to 0,\ k^2_{0+} \to \frac{3}{4}\mu E, \tag{4.65}$$

$$\pi^+\pi^- - \text{threshold}: A_{-++} \sim 1 + iak_{+-} \quad \text{at } k^2_{+-} \to 0,\ k^2_{++} \to \frac{3}{4}\mu E.$$

In the amplitude A_{00+} the singularity at the $\pi^0\pi^0$ threshold appears from the direct transition $\pi^0\pi^0 \to \pi^0\pi^0$ which, if written part by part, consists of the factors

$$-\tilde{\lambda}_{00+} \cdot ik_{00}\frac{1}{2} \cdot (\frac{2}{3}a_0 + \frac{4}{3}a_2), \tag{4.66}$$

namely, the A_{00+} amplitude in zero approximation $-\tilde{\lambda}_{00+}$, the loop diagram contribution $ik_{00}\frac{1}{2}$ with pion identity factor $\frac{1}{2}$, and the $\pi^0\pi^0$ scattering amplitude $2(\frac{1}{3}a_0 + \frac{2}{3}a_2)$. The charge exchange amplitude contains the amplitude of the $\pi^-\pi^+\pi^+$ production, the $\pi^-\pi^+$ loop diagram and the charge exchange amplitude $(-\frac{2}{3}a_0 + \frac{2}{3}a_2)$:

$$2\tilde{\lambda}_{-++} \cdot ik_{+-} \cdot (-\frac{2}{3}a_0 + \frac{2}{3}a_2). \tag{4.67}$$

The threshold amplitudes are determined, according to (4.60), as

$$2\tilde{\lambda}_{-++} = 2\lambda_0 + D_0(0) + A_0(0) + D_0(\frac{3}{4}\mu E) + 3A_0(\frac{3}{4}\mu E),$$

$$-\tilde{\lambda}_{00+} = \lambda_0 + D_0(0) + 2A_0(\frac{3}{4}\mu E) \tag{4.68}$$

We carry out here the re-definition

$$\lambda \to \lambda_0, \quad D(k) \to D_0(k^2), \quad A(k) \to A_0(k^2) \tag{4.69}$$

with the requirement

$$D_0(\frac{1}{2}\mu E) = A_0(\frac{1}{2}\mu E) = 0. \tag{4.70}$$

For the decays $K_L \to \pi^+\pi^-\pi^0$ and $K_L \to \pi^0\pi^0\pi^0$ we have near the thresholds:

$$K_L \to \left(\pi^+\pi^-\pi^0\right)_{\sim\pi^+\pi^-\ threshold}:$$

$$\tilde{\lambda}_{+-0}\Big[1+(\frac{2}{3}a_0+\frac{1}{3}a_2)ik_{+-}+\frac{\tilde{\lambda}_{000}}{\tilde{\lambda}_{+-0}}\,(a_0-a_2)ik_{00}\Big]$$

$$=\tilde{\lambda}_{+-0}\Big[1+(\frac{2}{3}a_0+\frac{1}{3}a_2)ik_{+-}+\frac{\tilde{\lambda}_{000}}{\tilde{\lambda}_{+-0}}\,(a_0-a_2)i\sqrt{k^2_{+-}+\Delta^2_\mu}\Big]$$

$$\text{at}\quad k^2_{+-}>0\,, \tag{4.71}$$

and

$$\left(K_L \to \pi^0\pi^0\pi^0\right)_{\sim\pi^0\pi^0\ threshold}:$$

$$-3\tilde{\lambda}_{000}\Big[1+(\frac{1}{3}a_0+\frac{2}{3}a_2)ik_{00}+\frac{\tilde{\lambda}_{+-0}}{\tilde{\lambda}_{000}}\,\frac{2}{9}(a_0-a_2)ik_{+-}\Big]$$

$$=-3\tilde{\lambda}_{000}\Big[1+(\frac{1}{3}a_0+\frac{2}{3}a_2)ik_{00}+\frac{\tilde{\lambda}_{+-0}}{\tilde{\lambda}_{000}}\,\frac{2}{9}(a_0-a_2)i\sqrt{k^2_{00}-\Delta^2_\mu}\Big]$$

$$\text{at}\quad k^2_{00}>\Delta^2_\mu\,;$$

$$\sqrt{k^2_{00}-\Delta^2_\mu}\to i\sqrt{\Delta^2_\mu-k^2_{00}}\quad\text{at}\quad k^2_{00}<\Delta^2_\mu\,. \tag{4.72}$$

In these threshold regions the amplitudes are determined, according to (4.60), as

$$\tilde{\lambda}_{+-0}=\lambda_0+D_0(0)+2A_0(\frac{3}{4}\mu E),$$

$$-3\tilde{\lambda}_{000}=-\Big[3\lambda_0+D_0(0)+2A_0(0)+2D_0(\frac{3}{4}\mu E)+4A_0(\frac{3}{4}\mu E)\Big]\,. \tag{4.73}$$

In a very rough approximation one can accept

$$\tilde{\lambda}_{+-0}\simeq\tilde{\lambda}_{000}\simeq\tilde{\lambda}_{00+}\simeq\tilde{\lambda}_{++-}\simeq\lambda_0\,. \tag{4.74}$$

For a more precise estimation we neglect the $I=2$ contribution and use for $D_0(k^2_{ij})$ the linear approximation:

$$A_0(k^2_{ij})=0,$$

$$\frac{1}{\lambda_0}D_0(k^2_{ij})=\delta\,(\frac{1}{2}-\epsilon)\,,\quad \epsilon=1-\frac{k^2_{ij}}{\mu E}\quad\text{with}\quad \delta\simeq 0.6\,, \tag{4.75}$$

so $D_0(0)=\frac{1}{2}\,\delta$ and $D_0(\frac{3}{4}\mu E)=-\frac{1}{4}\,\delta$.

The cusp contributions into the ratios $\frac{W_{00+}}{W_{++-}}$, $\frac{W_{000}}{W_{+-0}}$, as it is shown by the estimation [14], are not large, being just of the order of less than a percent. Equation (4.75) gives a good possibility to see that.

For $K^\pm \to \pi^0\pi^0\pi^\pm$ the cusp region amplitude, given by Eq. (4.63) and that of [14], coincides with the amplitude proposed in [16] for $M_{00}^2 \equiv s < 4\mu_+^2$ and written as:

$$\mathbf{M}_0 - 2a_x\mathbf{M}_+\sqrt{\mu_+^2 - \frac{M_{00}^2}{4}}, \qquad a_x \simeq \frac{1}{3}(a_0 - a_2) \tag{4.76}$$

where $\mathbf{M}_0$ is the "unperturbed amplitude" for the $K^\pm \to \pi^0\pi^0\pi^\pm$ process, and $\mathbf{M}_+$ is the contribution of the $K^\pm \to \pi^\pm\pi^+\pi^-$ decay amplitude. The cusp in $K^\pm \to \pi^0\pi^0\pi^\pm$ was observed experimentally and results in $a_0 - a_2 = 0.264\ ^{+0.033}_{-0.020}$ [18].

4.4.3 *Transformation of the dispersion relation for the $K \to \pi\pi\pi$ amplitude to a single integral equation*

Two integrations are included in the dispersion relations for $D(k)$ and $A(k)$, Eq. (4.59, namely, the averaging over dz and the dispersive integration over $dk_{i\ell}^2$. Hypotheses about the point-like $\pi\pi$ structure of the interaction in the low energy region allow us to transform the equation to the Skornyakov–TerMartirosyan kernel, which means to work with a single integral equation. The way of transformation of the dispersion relations into single integral equations will be discussed below.

For simplifying the calculation, let us consider the case when $a_2 = 0$ and $A(k) = 0$, *i.e.* we begin with the equation (4.58). Similarly to the investigation carried out in the previous section, let us re-write the integral in (4.58) selecting the subtraction term:

$$k^2\int_0^\infty dk'^2\,\frac{k'\langle D(k'')\rangle}{k'^2(k'^2-k^2-i0)} = \int_0^\infty dk'^2\,\frac{k'\langle D(k'')\rangle}{k'^2-k^2-i0} - \int_0^\infty dk'^2\,\frac{k'\langle D(k'')\rangle}{k'^2}\,. \tag{4.77}$$

Consider now the first term in the right-hand side of (4.77):

$$\int_0^\infty dk'^2\,\frac{k'\langle D(k'')\rangle}{k'^2-k^2-i0} \to \int_0^{\Lambda^2}\frac{dk'^2\,k'}{k'^2-k^2-i0}\,\frac{1}{\sqrt{3\,k'^2(\mu E-k'^2)}}\int_{k'^2_-}^{k'^2_+} dk''^2 D(k''). \tag{4.78}$$

In order to guaranty the convergence in the dispersion integrals of the separate terms, we introduce here a cut-off Λ^2 in the right-hand side. In the final expression we shall take $\Lambda^2 \to \infty$ having in mind that we started from a formula with a subtraction, see (4.77). Further, in (4.78) an exchange $z \to k''^2 = \frac{3}{4}\mu E - \frac{1}{2}k'^2 + \frac{1}{2}z\sqrt{3k'^2(\mu E - k'^2)}$ was made, namely:

$$\int\limits_{-1}^{1} \frac{dz}{2} \to \frac{1}{\sqrt{3\,k'^2(\mu E - k'^2)}} \int\limits_{k'^2_-}^{k'^2_+} dk''^2$$

where $k'^2_\pm = \frac{3}{4}\mu E - \frac{1}{2}k'^2 \pm \frac{1}{2}\sqrt{3k'^2(\mu E - k'^2)}$ at $0 < k'^2 < \frac{3}{4}\mu E$ and the analytic continuation into other regions with a positive imaginary addition $E + i\epsilon$, $\epsilon > 0$ is assumed.

The right-hand side (4.78) can be re-written in the form

$$\int\limits_0^{\Lambda^2} dk'^2 \frac{k'}{k'^2 - k^2 - i0} \frac{1}{\sqrt{3\,k'^2(\mu E - k'^2)}} \int\limits_{k'^2_-}^{k'^2_+} dk''^2 \oint\limits_{C(\kappa^2)} \frac{d\kappa^2}{2\pi i} D(\kappa) \frac{1}{\kappa^2 - k''^2} \tag{4.79}$$

where the integration over κ^2 is taken along a contour around the point $\kappa^2 = k''^2$ anti-clockwise. After that the integrations over k'^2 and k''^2 can be carried out explicitly. The integration over k''^2 gives

$$\begin{aligned} &\oint\limits_{C(\kappa^2)} \frac{d\kappa^2}{2\pi i} D(\kappa) \int\limits_0^{\Lambda^2} dk'^2 \frac{k'}{k'^2 - k^2 - i0} \frac{1}{\sqrt{3\,k'^2(\mu E - k'^2)}} \ln\frac{k'^2_- - \kappa^2}{k'^2_+ - \kappa^2} \\ &= \oint\limits_{C(\kappa^2)} \frac{d\kappa^2}{2\pi i} D(\kappa) \oint\limits_{C(k'^2)} \frac{dk'^2}{2} \frac{k'}{k'^2 - k^2 - i0} \frac{1}{\sqrt{3\,k'^2(\mu E - k'^2)}} \ln\frac{k'^2_- - \kappa^2}{k'^2_+ - \kappa^2} \end{aligned} \tag{4.80}$$

Here the integration is first taken over k'^2 the upper edge of the cut at $0 < k'^2 < \Lambda^2$. After the exchange $dk'^2 \to \frac{1}{2}dk'^2$ we introduce the integration dk'^2 over both parts of the cut $0 < k'^2 < \Lambda^2$, *i.e.* we have a closed contour $C(k'^2)$ compensated by the factor 1/2.

The other closed contour which corresponds to the integration over κ^2 is also changed. It was first a small contour around the point $\kappa^2 = k''^2$. But we can deform it, and have to do that, since the integration is carried out over k''^2 in the region $k''^2 < \Lambda^2$. If so, the contour over $d\kappa^2$ has to be widened. Hence, the integration around a big circle is taken in such a way that $|\kappa^2| > \Lambda^2$. The dotted line shows the cut going from the singularity $D(\kappa)$ in the point $\kappa^2 = 0$.

Let us now begin to deform the contour $C(k'^2)$, widening it. After a while it will catch the singularities of the integrand in the complex plane

k'^2. This is a pole singularity at $k'^2 = k^2$, it results in:

$$\oint \frac{d\kappa^2}{2\pi i} D(\kappa) \cdot i\pi \left[\frac{k'}{\sqrt{3\,k'^2(\mu E - k'^2)}} \ln \frac{k'^2_- - \kappa^2}{k'^2_+ - \kappa^2} \right]_{k'^2 = k^2} \tag{4.81}$$

and logarithmic singularities at $k'^2_\pm = \kappa^2$ which result in the contribution:

$$\oint \frac{d\kappa^2}{2\pi i} D(\kappa) \cdot i\pi \left[\int\limits_{k^2_-(\kappa^2)}^{k^2_+(\kappa^2)} dk'^2 \frac{k'}{k'^2 - k^2 - i0} \frac{1}{\sqrt{3\,k'^2(\mu E - k'^2)}} \right],$$

$$k^2_\pm(\kappa^2) = \frac{3}{4}\mu E - \frac{1}{2}\kappa^2 \pm \frac{1}{2}\sqrt{3\kappa^2(\mu E - \kappa^2)} \tag{4.82}$$

The expression in the square brackets can be calculated explicitly. Summing up the terms (4.80) and (4.80), we obtain a single integral which equals

$$\oint\limits_{C(\kappa^2)} d\kappa^2 \frac{D(\kappa)}{\sqrt{3(\mu E - k^2)}} \ln \left(\frac{\left[\left(\frac{1}{2}k - \frac{\sqrt{3}}{2}\sqrt{\mu E - k^2}\right)^2 - \kappa^2 \right]}{\left[\left(\frac{1}{2}k + \frac{\sqrt{3}}{2}\sqrt{\mu E - k^2}\right)^2 - \kappa^2 \right]} \right.$$
$$\left. \times \frac{\left[\left(\frac{1}{2}\kappa + \frac{\sqrt{3}}{2}\sqrt{\mu E - \kappa^2}\right)^2 - k^2 \right] \left[\frac{\sqrt{3}}{2}\kappa + \frac{1}{2}\sqrt{\mu E - \kappa^2} + \sqrt{\mu E - k^2} \right]^2}{\left[\left(\frac{1}{2}\kappa - \frac{\sqrt{3}}{2}\sqrt{\mu E - \kappa^2}\right)^2 - k^2 \right] \left[\frac{\sqrt{3}}{2}\kappa - \frac{1}{2}\sqrt{\mu E - \kappa^2} - \sqrt{\mu E - k^2} \right]^2} \right) \tag{4.83}$$

We got here an integral equation for $D(k)$ – it remains only to give a representation of the integral which can be satisfied. Let us go first to the limit $\Lambda^2 \to \infty$. We see that Λ^2 is not present explicitly, but it is necessary for the convergence of the integral, in order to carry out the integration with the help of the contour integral $C(k'^2)$. That's why we have first made the subtraction Eq. (4.77).

After that we begin to narrow the contour of integration over $d\kappa^2$. The contour has to catch the singularities of the kernel of the equation – we have:

$$\frac{1}{\sqrt{3}} \int\limits_{\mu E}^{-\infty} d\kappa^2\, D(\kappa^2) \left[\frac{1}{\sqrt{\mu\, E - k^2}} \ln \frac{(\sqrt{\mu\, E - k^2} - \frac{1}{2}\sqrt{\mu\, E - \kappa^2})^2 - \frac{3}{4}\kappa^2}{(\sqrt{\mu\, E - k^2} + \frac{1}{2}\sqrt{\mu\, E - \kappa^2})^2 - \frac{3}{4}\kappa^2} \right.$$
$$\left. - \frac{1}{\sqrt{\mu E}} \ln \frac{(\sqrt{\mu E} - \frac{1}{2}\sqrt{\mu\, E - \kappa^2})^2 - \frac{3}{4}\kappa^2}{(\sqrt{\mu E} + \frac{1}{2}\sqrt{\mu\, E - \kappa^2})^2 - \frac{3}{4}\kappa^2} \right]. \tag{4.84}$$

The analytic continuation from the $(E - \kappa^2 > 0)$-region to $E < \kappa^2$ is performed with $E + i0$. Let us remind that the second addend in (4.84) is a subtraction term which is responsible for the convergence of the equation.

The introduction of

$$\mu E - \kappa^2 = \frac{3}{4}p'^2, \quad \mu E - k^2 = \frac{3}{4}p^2,$$

allows us to write (4.58) in a form

$$D(\mu E - \frac{3}{4}p^2) = \frac{5}{3}\frac{ia_0\sqrt{\mu E - \frac{3}{4}p^2}}{1 - i\,a_0\sqrt{\mu E - \frac{3}{4}p^2}} +$$

$$+ \frac{a_0}{1 - i\,a_0\sqrt{\mu E - \frac{3}{4}p^2}} \int\limits_0^\infty \frac{dp'^2}{\pi} D(\mu E - \frac{3}{4}p'^2)$$

$$\times \left[\frac{1}{\sqrt{3}p}\ln\frac{p'^2 + p^2 + pp' - \mu E - i0}{p'^2 + p^2 - pp' - \mu E - i0} - \frac{1}{2\sqrt{\mu E}}\ln\frac{p'^2 + \frac{1}{3}\mu E + p'\sqrt{4\mu E/3}}{p'^2 + \frac{1}{3}\mu E - p'\sqrt{4\mu E/3}}\right] \tag{4.85}$$

We discussed the Skornyakov–Ter-Martirosyan kernel in Chapter 3, here we introduce the subtraction procedure to eliminate divergences inherent when using this kernel.

In Eq. (4.85) the integration is performed over the region $0 \le p'^2 < +\infty$ so that $k'^2 = \mu E - \frac{3}{4}p'^2$ is positive (physical region) and negative (non-physical region). In the non-physical region

$$\sqrt{\mu(E + i0) - \frac{3}{4}p'^2} \to i\sqrt{\frac{3}{4}p'^2 - \mu E}$$

which means that we act on the first plane of variables k'^2, k^2. But the integration in the region $k'^2 < 0$ tells that for obtaining a reliable solution one need to take into account the left-hand-side singularities which are located exactly at $k'^2 < 0$.

4.5 Equation for the three-nucleon amplitude

Lively interest in three-body problem was risen over a long period of time by the intention to develop an approach for the description of light nuclei, first of all H_3 and He_3. That was the Skornyakov–Ter-Martirosyan equation [5] for short-range NN interaction, $r_{NN} \to 0$, with the subsequent developing it to a solvable level [6, 19, 20]. The generalization of the Skornyakov–Ter-Martirosyan equation to the two-particle potential interactions was made by Faddeev [7]. Versions of application of the Faddeev equation to NNN systems that time were surveyed in [21].

Non-relativistic NNN-systems were treated in terms of other approaches as well. Here, following to [22], we perform the construction of the tritium/helium-3 wave function by means of extraction of the leading singularities. We use the non-relativistic three-particle dispersion relation equation with a subtraction and the S-wave NN interaction amplitudes. In terms of this method we find the wave function of tritium and helium-3 states and their form factors. The calculated form factors agree qualitatively with experimental data at $q^2 < 2\,\mathrm{fm}^{-2}$. The advantage of this method is its easy generalization to the case of relativistic description of the nucleon systems.

4.5.1 *Method of extraction of the leading singularities*

Here we present an approximate solution of the three-particle equation in a method based on the extraction of the leading singularities in the amplitude of the transition of a bound state into three nucleons. This method looks as rather universal and so it might be extended to composite systems with more constituents.

The discussion below is guided to a large extent by the ideas that
(i) singular parts of the three-particle amplitude give the leading contribution at large distances,
(ii) the short-range region leads to a contribution which can be taken into account by introducing effective cuttings in amplitude integrals at large momenta.

4.5.1.1 *Amplitude of production of three spinless particles*

Let us discuss a computational method for the simple example of three spinless particles. We construct the amplitude for the interaction of the three particles, taking only two-particle interactions into account.

We consider the standard situation discussed previously, namely, when some current produces three constituents. The S-wave binary interactions lead to the diagrams with separable interaction of the outgoing particles. So, the diagrams can be classified on the basis of which of the three pairs of particles is the last to interact. It means we can present the total amplitude as the sum:

$$A(E, k_{12}^2, k_{13}^2, k_{23}^2) = \lambda + A_{12}(E, k_{12}^2) + A_{13}(E, k_{13}^2) + A_{23}(E, k_{23}^2)\,. \quad (4.86)$$

As usually, the k_{ij} are the relative momenta of the produced particles (1,2,3) and E is the total energy of the system.

In general, the functions $A_{12}(E,k_{12}^2)$, $A_{13}(E,k_{13}^2)$, and $A_{23}(E,k_{23}^2)$ can be different. For simplicity we assume here that the masses of all the produced particles are equal and their binary interactions are equal as well. In other words, we set $A_{12}(E,k^2) = A_{13}(E,k^2) = A_{23}(E,k^2) \equiv A(E,k^2)$.

For the specification of the equation we have to know the amplitude $a(k_{ij}^2)$ of the binary interaction of the produced particles. In order not to overload the presentation we use here the simplest case: the scattering-length approximation, $a(k_{ij}^2) = const \equiv a$, with the two-particle scattering amplitude $a(1-ik_{ij}a)^{-1}$.

The equation for the amplitude $A(E,k_{12}^2)$ reads

$$\begin{aligned} A(E,k_{12}^2) &= \lambda \frac{ik_{12}a}{1-ik_{12}a} \\ &+ \frac{k_{12}^2}{1-ik_{12}a} \int\limits_C \frac{dk_{12}'^2}{\pi} \frac{k_{12}'a}{k_{12}'^2(k_{12}'^2-k_{12}^2-i0)} \int\limits_{-1}^{1} \frac{dz}{2} \Big[A(E,k_{13}'^2) + A(E,k_{23}'^2)\Big]. \end{aligned} \tag{4.87}$$

The integration contour C depends on the sign of the scattering length a.

(i) Interaction with the absence of deuteron-like composite states, $a > 0$

If the interacting particles do not form bound states, the integration in the dispersion integral (4.87) (contour C) is carried out from 0 to ∞:

$$\int\limits_C \frac{dk_{12}'^2}{\pi} \to \int\limits_0^\infty \frac{dk_{12}'^2}{\pi} \tag{4.88}$$

Recall that the values $dz/2$, $k_{12}'^2$ and $k_{\ell 3}'^2$ with $\ell = 1,2$ are related in the equal-mass case by

$$k_{\ell 3}'^2 = \frac{3}{4}mE - \frac{1}{2}k_{12}'^2 - 2(-1)^\ell z\sqrt{\frac{3}{16}k_{12}'^2(Em-k_{12}'^2)}. \tag{4.89}$$

(ii) Two-particle deuteron-like composite state, $a < 0$

If a two-particle system forms a deuteron-like composite state, one has to take it into account.

On the first (physical) sheet a pole of the partial amplitude $a(1-ika)^{-1}$ at $k^2 < 0$ corresponds to the two-particle bound state; this is possible if $a <$

0. If so, one can disjoint the pole and continuous spectrum contributions:

$$\frac{a}{1-ika} = A_{pole}(k) + A_{non-pole}(k)$$
$$A_{pole}(k) = \frac{2/a}{1/a^2+k^2},$$
$$A_{non-pole}(k) = \frac{a}{1-ika}\left(1-\frac{2}{1+ika}\right) - \frac{a}{1+ika}. \tag{4.90}$$

Let us remind that Eq. (4.90) is written for the first sheet where

$$\sqrt{k^2} = i|k| \quad \text{at} \quad k^2 < 0, \quad \text{and} \quad \frac{1}{a^2} = m\varepsilon\,. \tag{4.91}$$

Here ε is the binding energy of the deuteron-like composite system. The non-pole contribution is described by standard formulae with integrations carried out according to (4.89). But the pole amplitude needs a special treatment. The matter is that the diagrams with deuteron-like composite states have anomalous singularities, and the intermediate state integration is carried out with the counter shown in Fig. 4.11. The values of $k^2_\pm$ are given by Eq. (4.89) with replacing $k'^2_{\ell 3} \to m\varepsilon$.

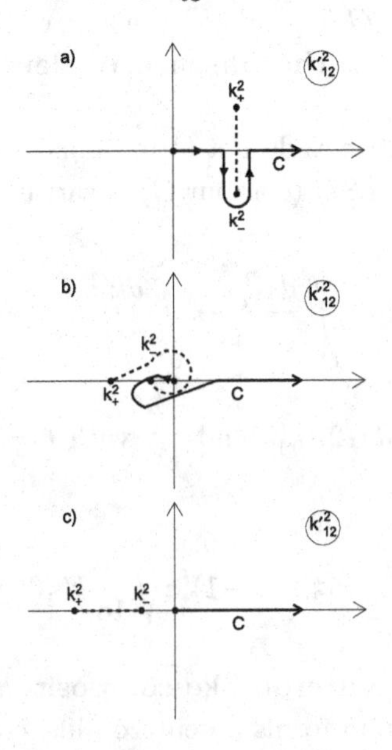

Fig. 4.11 The integration contour C for $A_{pole}(k'_{12})$ in the following cases: (a) $E > -1/ma^2$; (b) $-\frac{4}{3}(1/ma^2) < E < -1/ma^2$; (c) $E < -\frac{4}{3}(1/ma^2)$.

(iii) Pole in the three-nucleon channel

The appearance of a pole (*i.e.* of a bound state) in the E-channel can also be expected due to two-particle rescatterings. Singling out this pole, we can write:

$$A_{12}(E, k_{12}^2) = g_{123}(E)\frac{1}{E+\epsilon}G(E, k_{12}^2) \tag{4.92}$$

where ϵ is the binding energy of three particles; the total energy $E = k_{12}^2/m + 3k_1^2/(4m)$ may be expressed both over $(k_{13},\, k_2)$ and $(k_{23},\, k_1)$.

In the following our problem will be just to extract the pole $1/(E+\epsilon)$ from $A_{i\ell}(E, k_{i\ell}^2)$.

4.5.1.2 *Separation of the threshold singularities and the cutoff procedure*

It is convenient to extract from the amplitude $A(E, k_{12}^2)$ the two-nucleon threshold singularities:

$$A(E, k_{12}^2) \;=\; \lambda\frac{ik_{12}a}{1-ik_{12}a}\,b(E, k_{12}^2)\,. \tag{4.93}$$

The equation for the amplitude $b(E, k_{12}^2)$ is written in the following form

$$b(E, k_{12}^2) = 1 + \frac{1}{ik_{12}}\int\limits_C \frac{dk_{12}'^2}{\pi}\frac{k_{12}'}{k_{12}'^2 - k_{12}^2 - i0}\int\limits_{-1}^{1}\frac{dz}{2}\,\frac{2b(E, k_{13}'^2)}{1-ik_{13}'a}\,ik_{13}'a\,. \tag{4.94}$$

Here we replace $[A(E, k_{13}'^2) + A(E, k_{23}'^2]$ by $2A(E, k_{13}'^2)$.

The next step is to introduce a cutoff at large $k_{12}'^2$ in (4.94). As we mentioned earlier, this cutoff is made in order to approximate the contribution from the short-range interaction. Accordingly, we write integration contour in Eq. (4.94) as

$$\int\limits_C \frac{dk_{12}'^2}{\pi} \to \int\limits_C \frac{dk_{12}'^2}{\pi}\,C_X(k'^{\,2}_{12}) \tag{4.95}$$

Here $C_X(k'^{\,2}_{12})$ is the cutoff function. In calculating the tritium/helium-3 wave function below, we use cutoffs of two types: a "hard" cutoff corresponding to vanishing of the integrand at $k'^{\,2}_{12} > \Lambda_h^2$, and a "soft" cutoff:

$$\begin{aligned} \text{hard cutoff}: \qquad C_h(k_{12}'^2) \;&=\; \theta(\Lambda_h^2 - k'^{\,2}_{12})\,, \\ \text{soft cutoff}: \qquad C_s(k'^{\,2}_{12}) \;&=\; \frac{\Lambda_s^2}{k_{12}'^2 + \Lambda_s^2}\,. \end{aligned} \tag{4.96}$$

4.5.1.3 *Extraction of the leading singularities*

An approximate solution of Eq. (4.95) is constructed by extracting the leading singularities in the region $k_{ij}^2 \simeq 0$. The strongest singularities in k_{ij}^2 arise from the binary rescatterings of nucleons: root singularities corresponding to thresholds and pole singularities corresponding top bound states (on the first sheet in the case of real bound states; on the second in the case of virtual bound states). In addition there are some specific triangle singularities which appear in diagrams with subsequent rescatterings of all three particles. The specific singularities which correspond to a large number of rescatterings become progressively weaker.

We can seek an approximate solution of Eq. (4.94) which incorporates a definite number of leading singularities ignoring the weaker singularities. Thus one can solve Eq. (4.94) by an iterative procedure where each iteration incorporates a definite and distinct set of singularities.

Several iterative approximations are given below for $b(E, k^2)$.

(i) Zeroth approximation

This is the crudest approximation. It involves replacement of the function in the integral over z by a constant (*i.e.* by the value of the integrand at some average point):

$$\frac{b(E, k'^{\,2}_{13})}{1 - ik'_{13}a}\, ik'_{13}a \;\to\; \frac{b(E, k_0^2)}{1 - ik_0 a}\, ik_0 a\,, \tag{4.97}$$

where k_0 is an arbitrary point on the contour of the integration over z. At fixed values of E and $k'^{\,2}_{12}$, the integration is carried out over a region of the variable $k'^{\,2}_{13}$ corresponding to the physical transition of the current into the three nucleons (the physical region of the Dalitz plot). It is convenient to take k_0 to be the midpoint of this region, which corresponds to $z = 0$. As can be seen from (4.89), at this point of the Dalitz plot we have

$$k_0^2 \;=\; \frac{1}{2} mE\,. \tag{4.98}$$

This choice of the point k_0 allows us to go over from the integral equation (4.95) to the equation

$$b(E, k^2) \;=\; 1 + I_0(E, k^2)\, \frac{b(E, k_0^2)}{1 - ik_0 a}\, 2ik_0 a\,, \tag{4.99}$$

where

$$I_0(E, k^2) \;=\; \frac{1}{ik} \int\limits_C \frac{dk'^{\,2}}{\pi}\, C_X(k'^{\,2}) \frac{k'}{k'^{\,2} - k^2 - i0}\,.$$

The value of $b(E, k_0^2)$ is found from (4.99) with $k^2 = k_0^2$:

$$b(E, k_0^2) = \frac{1 - ik_0 a}{1 - ik_0 a\Big(1 + 2I_0(E, k_0^2)\Big)}. \tag{4.100}$$

The right-hand side of (4.100) can have a pole in E, which corresponds to a bound state of the three particles. According to the computational procedure outlined above, the cutoff in (4.96) must be chosen in such a way that the pole in (4.100) corresponds to the physical value of the binding energy.

The function $I_0(E, k^2)$ contains only two-particle singularities, so that the zeroth approximation corresponds to the case in which only two-particle singularities are retained in the function $b(E, k^2)$.

(ii) First approximation

This is the approximation is which we incorporate the singularity which corresponds to the interaction of all three particles. Correspondingly, in Eq. (4.95) we need to retain only the two-particle rescatterings in the integrand of the integral over z, and we take the function $b(E, k^2)$ to be constant over k^2:

$$\int\limits_{-1}^{1} \frac{dz}{2} \frac{2b(E, k_{13}'^2)}{1 - ik_{13}' a} ik_{13}' a \to b(E, k_0^2) \int\limits_{-1}^{1} \frac{dz}{2} \frac{2}{1 - ik_{13}' a} ik_{13}' a. \tag{4.101}$$

where k_0^2 is an arbitrary point on the contour of the integration over z. It is convenient to take k_0^2 to be in the region $k_0^2 \sim 0$.

We find

$$b(E, k^2) = 1 + I_1(E, k^2)\, 2b(E, k_0^2)\,,$$

$$I_1(E, k^2) = \frac{1}{ik} \int\limits_C \frac{dk'^2}{\pi} C_X(k'^2) \frac{k'}{k'^2 - k^2 - i0} \int\limits_{-1}^{1} \frac{dz}{2} \frac{ik_{13}' a}{1 - ik_{13}' a}. \tag{4.102}$$

The amplitude $b(E, k_0^2)$ reads:

$$b(E, k_0^2) = \frac{1}{1 - 2I_1(E, k_0^2)}. \tag{4.103}$$

The function $I_1(E, k^2)$ correctly incorporates the singularities corresponding to the vanishing of the propagators of the triangle diagrams in rescattering processes.

(iii) Second approximation

In this case we include weaker singularities of the amplitude:

$$b(E,k_{12}^2)=1+\frac{i}{ik_{12}}\int\limits_C\frac{dk_{12}'^2}{\pi}C_X(k_{12}'^2)\frac{k_{12}'}{k_{12}'^2-k_{12}^2-i0}\int\limits_{-1}^{1}\frac{dz}{2}\frac{2ik_{13}'}{1-ik_{13}'a}$$
$$\times\left[1+\frac{1}{ik_{13}'}\int\limits_C\frac{dk_{12}''^2}{\pi}C_X(k_{12}''^2)\frac{k_{12}''}{k_{12}''^2-k_{13}'^2-i0}\int\limits_{-1}^{1}\frac{dz}{2}\frac{2b(E,k_{13}''^2)}{1-ik_{13}''a}ik_{13}''a\right]. \tag{4.104}$$

This second approximation corresponds to the following substitution in (4.104):

$$b(E,k_{13}''^2)\longrightarrow b(E,k_0^2)\,. \tag{4.105}$$

The function $b(E,k_0^2)$ is found by a method like that used above – by working from Eq. (4.104) with $k_{13}''=k_0$. We find

$$b(E,k_{12}^2)=1+2I_1(E,k_{12}^2)+4I_2(E,k_{12}^2)b(E,k_0^2)\,,$$
$$b(E,k_0^2)=\frac{1+I_1(E,k_0^2)}{1-4I_2(E,k_0^2)}\,, \tag{4.106}$$

where the function $I_2(E,k_{12}^2)$ is given by

$$I_2(E,k_{12}^2)\;=\;1+\frac{1}{ik}\int\limits_C\frac{dk_{12}'^2}{\pi}C_X(k_{12}'^2)\frac{k_{12}'}{k_{12}'^2-k_{12}^2-i0}\int\limits_{-1}^{1}\frac{dz}{2}\frac{ik_{13}'}{1-ik_{13}'a}$$
$$\times\left[1+\frac{1}{ik_{13}'}\int\limits_C\frac{dk_{12}''^2}{\pi}C_X(k_{12}''^2)\frac{k_{12}''}{k_{12}''^2-k_{13}'^2-i0}\int\limits_{-1}^{1}\frac{dz}{2}\frac{ik_{13}''a}{1-ik_{13}''a}\right]. \tag{4.107}$$

(iv) Iteration procedure

The principle for constructing the successive iterations is obvious:

$$b(E,k_{12}^2)=1+2I_1(E,k_{12}^2)+4I_2(E,k_{12}^2)+\ldots$$
$$+\,2^{n-1}I_{n-1}(E,k_{12}^2)+2^nI_n(E,k_{12}^2)b(E,k_0^2)$$
$$b(E,k_0^2)=\left[1+\sum_{\ell=1}^{n-1}2^\ell I_\ell(E,k_0^2)\right]\frac{1}{1-2^nI_n(E,k_0^2)}\,, \tag{4.108}$$

The bound state corresponds to the pole in the function $b(E,k_0^2)$:

$$b(E,k_0^2)\sim\frac{1}{E+\varepsilon},\qquad \text{over}\quad 1-2^nI_n(E,k_0^2)\;=\;0. \tag{4.109}$$

The terms of the amplitude which contain a pole at $E = -\varepsilon$ determine the amplitude of the transition (*bound state*) $\to$ (*three particle state*):

$$A(E,k_{12}^2,k_{13}^2,k_{23}^2) = G\,\frac{1}{E+\varepsilon}\,g_{123}(E)\left[\frac{I_n(E,k_{12}^2)}{1-ik_{12}a}\,ik_{12}a + \frac{I_n(E,k_{13}^2)}{1-ik_{13}a}\,ik_{13}a + \frac{I_n(E,k_{23}^2)}{1-ik_{23}a}\,ik_{23}a\right], \qquad (4.110)$$

and the corresponding wave function and vertex in the momentum representation:

$$\psi(k_1,k_2,k_3) = \frac{N}{E+\varepsilon}\,g_{123}(E)\left[\frac{I_n(E,k_{12}^2)}{1-ik_{12}a}\,ik_{12}a + \frac{I_n(E,k_{13}^2)}{1-ik_{13}a}\,ik_{13}a + \frac{I_n(E,k_{23}^2)}{1-ik_{23}a}\,ik_{23}a\right] \equiv \frac{G(E,k_{12}^2,k_{13}^2,k_{23}^2)}{E-H_0}\,. \qquad (4.111)$$

Using the wave function (4.111), we can calculate the form factor $F(q)$ of the three-particle system, which is defined by

$$F(q) = \int \frac{dk_1^3 dk_2^3 dk_3^3}{(2\pi)^9}\,\psi(k_1,k_2,k_3)\psi^*(k_1+q,k_2,k_3)\delta^3(k_1+k_2+k_3-P),$$
$$F(0) = 1 \qquad (4.112)$$

where P is the momentum of the tritium. The condition $F(0) = 1$ determines N^2 and thus the normalization of the wave function.

4.5.2 *Helium-3/tritium wave function*

Using the amplitude for the rescattering of three nucleons in the scattering-length approximation, we can construct the wave function of the system and calculate the form factor. We go through the calculations with the goal of testing the discussed method. In these calculations we use the first approximation in accordance with the classification of the preceding section.

To be definite, we first consider the helium-3 ground state with isospin $(I=\frac{1}{2},\ I_3=\frac{1}{2})$ and spin $J^P=\frac{1}{2}^+,\ J_3=\frac{1}{2}$. It is a mixture of two waves, ${}^2S_{1/2}$ and ${}^4D_{1/2}$. Below we neglect the D-wave admixture.

To construct the helium-3 wave function, we introduce a current for which each pair of nucleons is in the S wave:

$$\frac{1}{\sqrt{6}}\Big[p_\uparrow p_\downarrow n_\uparrow - p_\downarrow p_\uparrow n_\uparrow - p_\uparrow n_\uparrow p_\downarrow + p_\downarrow n_\uparrow p_\uparrow - n_\uparrow p_\downarrow p_\uparrow + n_\uparrow p_\uparrow p_\downarrow\Big]$$
$$\equiv \frac{1}{\sqrt{6}}\,P_a\big[p_\uparrow p_\downarrow n_\uparrow\big]\,. \qquad (4.113)$$

Here we have introduced an antisymmetrizing operator P_a. This operator makes it possible to write the antisymmetric combinations of amplitudes in a compact way.

We consider the interaction of two nucleons in the states $J^P = 0^+, 1^+$. The S-wave amplitudes are characterized by a singlet scattering length, a_s, and a triplet one, a_t. The amplitude of the two-nucleon interaction is then

$$\begin{aligned} \mathbf{a}_{NN} &= \mathbf{a}_s P_s + \mathbf{a}_t P_t\,, \\ \mathbf{a}_s &= \frac{a_s}{1-ika_s}\,, \quad \mathbf{a}_t = \frac{a_t}{1-ika_t}\,, \end{aligned} \tag{4.114}$$

where the operators P_s and P_t project onto the 0^+ and 1^+ states:

$$\begin{aligned} P_s &= \frac{1}{16}\Big[1-\sigma(1)\sigma(2)\Big]\Big[3+\tau(1)\tau(2)\Big]\,, \\ P_t &= \frac{1}{16}\Big[3+\sigma(1)\sigma(2)\Big]\Big[1-\tau(1)\tau(2)\Big]\,. \end{aligned} \tag{4.115}$$

Using (4.114), we find the amplitudes for the scattering of two nucleons in various states:

$$\begin{aligned} &\langle p_\uparrow p_\downarrow|\mathbf{a}_{NN}|p_\uparrow p_\downarrow\rangle = \mathbf{a}_s/2\,, \qquad \langle p_\uparrow n_\uparrow|\mathbf{a}_{NN}|p_\uparrow n_\uparrow\rangle = \mathbf{a}_t/2\,, \\ &\langle p_\uparrow p_\downarrow|\mathbf{a}_{NN}|p_\downarrow p_\uparrow\rangle = -\mathbf{a}_s/2\,, \qquad \langle p_\uparrow n_\uparrow|\mathbf{a}_{NN}|n_\uparrow p_\uparrow\rangle = -\mathbf{a}_t/2\,, \\ &\langle p_\uparrow n_\downarrow|\mathbf{a}_{NN}|p_\uparrow n_\downarrow\rangle = (\mathbf{a}_s+\mathbf{a}_t)/4\,, \quad \langle p_\uparrow n_\uparrow|\mathbf{a}_{NN}|n_\uparrow p_\uparrow\rangle = (\mathbf{a}_s-\mathbf{a}_t)/4\,, \\ &\langle p_\uparrow n_\downarrow|\mathbf{a}_{NN}|n_\downarrow p_\uparrow\rangle = -(\mathbf{a}_s+\mathbf{a}_t)/4\,, \quad \langle p_\uparrow n_\downarrow|\mathbf{a}_{NN}|p_\downarrow n_\uparrow\rangle = (\mathbf{a}_t-\mathbf{a}_s)/4\,. \end{aligned} \tag{4.116}$$

We have to consider all possible rescatterings for each term in (4.113). Grouping the terms by the final state of the particles, we find the amplitude:

$$\begin{aligned} A(^3\mathrm{He}) = P_a[p_\uparrow p_\downarrow n_\uparrow]\Bigg[&\frac{\lambda}{\sqrt{6}} + A_s(E,k_{12}^2) + A_t(E,k_{13}^2) \\ &+\frac{1}{2}\Big(A_s(E,k_{23}^2) + A_t(E,k_{23}^2)\Big)\Bigg] \\ + P_a[p_\uparrow p_\uparrow n_\downarrow]\frac{1}{2}&\Big[A_s(E,k_{13}) - A_t(E,k_{13})\Big]\,. \end{aligned} \tag{4.117}$$

Let us remind that the first nucleon in the factor $p_\uparrow p_\downarrow n_\uparrow$ is $p_\uparrow$; $p_\uparrow$ is the second, and $n_\uparrow$ is the third one. The operator P_a permits both these nucleons and the indices in the amplitudes $A_s(E,k_{ij}^2)$ and $A_t(E,k_{ij}^2)$. The functions $A_s(E,k_{ij}^2)$ and $A_t(E,k_{ij}^2)$ in (4.117) satisfy the system of integral equations:

$$A_s(E,k_{12}^2) = \frac{\lambda i k_{12} a_s}{\sqrt{6}(1-ik_{12}a_s)} + \frac{a_s}{1-ik_{12}a_s} \int\limits_C \frac{dk_{12}'^2}{\pi} \frac{k_{12}'}{k_{12}'^2 - k_{12}^2}$$

$$\times \int\limits_{-1}^{1} \frac{dz}{2} \left[\frac{3}{4} A_t(E,k_{13}'^2) + \frac{1}{4} A_s(E,k_{13}'^2) + \frac{3}{4} A_t(E,k_{23}'^2) + \frac{1}{4} A_s(E,k_{23}'^2) \right],$$

$$A_t(E,k_{12}^2) = \left[A_s(E,k_{12}^2) \right]_{s \rightleftharpoons t}, \tag{4.118}$$

where $\left[... \right]_{s \rightleftharpoons t}$ means the expression in the previous formula with the permutation $s \rightleftharpoons t$.

Expressions (4.116) and (4.117) determine a three-nucleon system in a pure S-wave state, $^2S_{1/2}$. Recall that there is no D-wave admixture in this case.

Let us separate two-particle threshold singularities:

$$A_s(E,k_{ij}^2) = \frac{1}{\sqrt{6}} \frac{ik_{ij}a_s}{1-ik_{ij}a_s} b_s(E,k_{ij}^2),$$

$$A_t(E,k_{ij}^2) = \frac{1}{\sqrt{6}} \frac{ik_{ij}a_t}{1-ik_{ij}a_t} b_t(E,k_{ij}^2). \tag{4.119}$$

The equations for b_s and b_t read:

$$b_s(E,k_{12}^2) = 1 + \frac{1}{ik_{12}} \int\limits_C \frac{dk_{12}'^2}{\pi} C_X(k_{12}'^2) \frac{k_{12}'}{k_{12}'^2 - k_{12}^2 - i0}$$

$$\times \int\limits_{-1}^{1} \frac{dz}{2} \left[\frac{3}{2} \frac{ik_{13}'a_t}{1-ik_{13}'a_t} b_t(E,k_{13}'^2) + \frac{1}{2} \frac{ik_{13}'a_s}{1-ik_{13}'a_s} b_s(E,k_{13}'^2) \right],$$

$$b_t(E,k_{12}^2) = \left[b_s(E,k_{12}^2) \right]_{s \rightleftharpoons t}. \tag{4.120}$$

The first approximation corresponds to the following replacements on the right-hand side of (4.120): $b_s(E,k_{13}'^2) \to b_s(E,k_0^2)$ and $b_t(E,k_{13}'^2) \to b_t(E,k_0^2)$. We then find:

$$b_s(E,k_{12}^2) = 1 + \frac{3}{2} I_{t1}(E,k_{12}^2)\, b_t(E,k_0^2) + \frac{1}{2} I_{s1}(E,k_{12}^2)\, b_s(E,k_0^2),$$

$$b_t(E,k_{12}^2) = 1 + \frac{1}{2} I_{t1}(E,k_{12}^2)\, b_t(E,k_0^2) + \frac{3}{2} I_{s1}(E,k_{12}^2)\, b_s(E,k_0^2). \tag{4.121}$$

The functions $b_s(E,k_0^2)$ and $b_t(E,k_0^2)$ are given by

$$b_s(E,k_0^2) = i\frac{1 - iI_{t1}(E,k_0^2)}{\Delta(E)}, \quad b_t(E,k_0^2) = i\frac{1 - iI_{s1}(E,k_0^2)}{\Delta(E)}, \tag{4.122}$$

$$\Delta(E) = \left(1 - \frac{1}{2} I_s(E,k_0^2)\right)\left(1 - \frac{1}{2} I_t(E,k_0^2)\right) - \frac{9}{4} I_t(E,k_0^2)\, I_s(E,k_0^2).$$

The pole at $E = -\varepsilon$ (the bound state of tritium, $\varepsilon = 8.5\,\mathrm{MeV}$), is determined by the condition

$$\Delta(E) = 0\,. \tag{4.123}$$

As we mentioned above, this condition fixes the cutoff parameter Λ in (4.96).

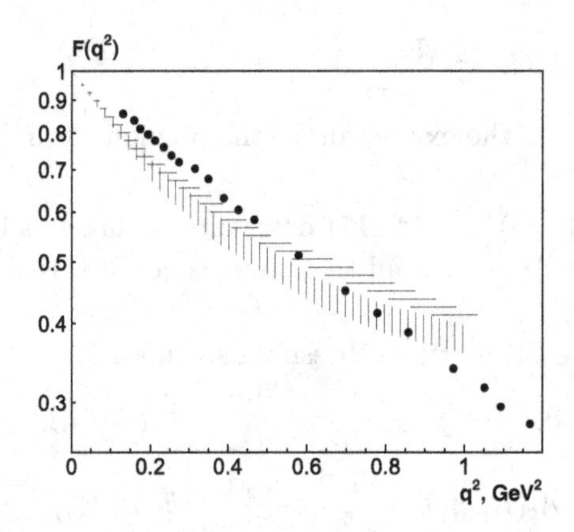

Fig. 4.12 Results calculated for the tritium form factor in the case of a "hard" cutoff (vertical hatching) and in the case of a "soft" cutoff (horizontal hatching), in comparison with the experimental data of Ref. [23]. The length of the hatching lines corresponds to the numerical error in the calculations.

The wave function is found by the prescription given in the previous section. Singling out the pole terms of the amplitude (4.117), we find the following expression for the wave function:

$$\begin{aligned}\psi(k_1,k_2,k_3) = P_a[p_\uparrow p_\downarrow n_\uparrow]\frac{1}{\Delta(E)}\Big[G_s(E,k_{12}^2)+G_t(E,k_{13}^2)\\ +\frac{1}{2}\Big(G_s(E,k_{23}^2)+G_t(E,k_{23}^2)\Big)\Big]\\ +\ P_a[p_\uparrow p_\uparrow n_\downarrow]\frac{1}{2\Delta(E)}\Big[G_s(E,k_{13}^2)-G_t(E,k_{13}^2)\Big],\end{aligned} \tag{4.124}$$

where

$$\begin{aligned}G_s(E,k_{ij}^2) &= G\frac{ik_{ij}a_s}{1-ik_{ij}a_s}\\ &\times\Big[3I_{t1}(E,k_{ij}^2)\Big(1+2I_{t1}(E,k_0^2)\Big)+I_{s1}(E,k_{ij}^2)\Big(1+2I_{t1}(E,k_0^2)\Big)\Big],\\ G_t(E,k_{ij}^2) &= \Big[G_s(E,k_{ij}^2)\Big]_{s\rightleftarrows t},\end{aligned} \tag{4.125}$$

and G is a normalization constant. Here, in contrast with (4.110), the wave function contains spin and isospin components.

To find the tritium form factor, we consider the simplest case, in which only the proton interacts with the external electric field. We set the vertex of this interaction equal to one, *i.e.* the nucleon form factors are excluded from consideration.

Results of calculations of the charge form factor are shown in Fig. 4.12, where they are compared with experimental data of [23] on ^{3}He (in Fig. 4.12, experimental data on ^{3}He have been divided by the proton form factor). It can be seen from this diagram that the wave functions which we have found give a fairly good description of the form factor up to $q^2 = 0.1\,\mathrm{GeV}^2$. We want to stress here that in this region the q^2 dependence of the form factor is basically linear, and this allows to guess the value of the form factor which is not determined exclusively by its charge radius.

4.5.2.1 *Miniconclusion*

The presented here method for solving three particle equation by the extraction of the leading singularities has been tested in the particular case of the rather rough approximation of the low-energy NN amplitudes. Nevertheless, the wave functions which are found give a comparatively satisfactory description of the tritium form factor at small values of q^2.

Further approbation of the method should include a use of more precise two-nucleon amplitudes, $a_s(k_{ij}^2)$ and $a_t(k_{ij}^2)$, and testing a stability of the iteration procedure.

Several ways for developing of this method are seen:
(*i*) Extending of the method for relativistic systems. Then we can consider a three-nucleon systems in a wide energy range (up to $s \simeq (2.2\,\mathrm{GeV})^2$), and with higher partial waves.
(*ii*) Application of this method to equations for four and more particles. For example, in the four-nucleon case efforts to carry out calculations by the standard methods run into serious technical problems. The binding energies calculated for the ^4He particle with the help of most of the realistic NN potentials for ^4He differ from the experimental value by 6–9 MeV. One can hope that the method of extraction of leading singularities, and its relativistic generalization, will permit some progress toward the construction of a realistic wave function for ^4He. Or, give a valuable information about multinucleon (multiquark) bags.

4.6 Appendix A. Landau rules for finding the singularities of the diagram

In this appendix we present a simple deduction of the Landau rules [1] for finding the singularities of the Feynman diagrams. To be definite, we shall consider the triangle diagram shown in Fig. 4.13; however, the considerations presented below will bear a rather general character.

Let the external particles have momenta p_1, p_2, p_3 and masses m_1, m_2, m_3; by definition, $p_i^2 = m_i^2$ $(i = 1, 2, 3)$. It follows from the momentum conservation law that $p_1 + p_2 + p_3 = 0$. Let us denote the momenta of the internal (virtual) particles as q_1, q_2, q_3, and their masses as μ_1, μ_2, μ_3. Generally speaking, here $q_i^2 \neq \mu_i^2$.

The Feynman integral for the diagram in Fig. 4.13 is, with the accuracy of a multiplying factor:

$$\int \frac{d^4 q_1}{(q_1^2 - \mu_1^2)(q_2^2 - \mu_2^2)(q_3^2 - \mu_3^2)}, \quad q_2 = q_1 - p_3, \quad q_3 = q_1 + p_2. \quad (4.126)$$

Let us first consider the integration over dq_{10}. In the complex q_{10} plane the integrand has poles at

$$q_{10} = \pm\sqrt{\mathbf{q}_1^2 + \mu_1^2} \mp i\epsilon, \quad (4.127)$$

$$q_{20} = \pm\sqrt{\mathbf{q}_2^2 + \mu_2^2} \mp i\epsilon, \quad \text{i.e.} \quad q_{10} = p_{30} \pm \sqrt{(\mathbf{q}_1 - \mathbf{p}_3)^2 + \mu_2^2} \mp i\epsilon,$$

$$q_{30} = \pm\sqrt{\mathbf{q}_3^2 + \mu_3^2} \mp i\epsilon, \quad \text{i.e.} \quad q_{10} = -p_{20} \pm \sqrt{(\mathbf{q}_1 + \mathbf{p}_2)^2 + \mu_3^2} \mp i\epsilon.$$

with $\epsilon \to +0$.

Obviously, the integral (4.126) can have a singularity only in the case when the poles in the q_{10} plane of the integrand pinch the integration contour, *e.g.* when the position of a cross above the real axis coincides (up to the accuracy of the sign of the infinitesimal imaginary part) with that of a circle which is situated below the real axis. Hence, the integral (4.126) has a singularity only when at least two of all three denominators in (4.126) are turning into zero at the same time:

$$q_i^2 = \mu_i^2 \quad (i = 1, 2, \quad \text{or} \quad i = 1, 3, \quad \text{or} \quad i = 2, 3), \quad (4.128)$$

or

$$q_i^2 = \mu_i^2 \quad (i = 1, 2, 3). \quad (4.129)$$

The conditions (4.128) are, essentially, conditions to find the singularities of simpler diagrams in which one of the internal lines is tightened into a point. We can check this immediately by writing an expression for such a simplified diagram. We shall consider now only singularities which are related to the condition (4.129). Using the same logics as in the following, it is easy to demonstrate that the conditions (4.128) lead to the simple threshold singularities

$$p_1^2 = (\mu_3 + \mu_2)^2, \quad p_2^2 = (\mu_1 + \mu_3)^2, \quad p_3^2 = (\mu_2 + \mu_1)^2 .$$

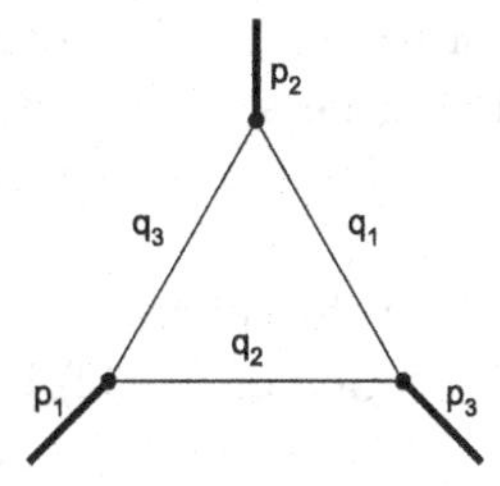

Fig. 4.13 Triangle diagram considered for illustration of the Landau rules.

The conditions (4.129) can, apparently, not be sufficient for the appearance of a singularity. Indeed, as we saw, a part of the coinciding singularities of the integrand has to be above the real axis, the other part – below it. Since the sign of the infinitesimal imaginary addition is the opposite of the sign of q_{i0} (see (4.127)), it is necessary that one of the three quantities q_{10}, q_{20}, q_{30} has a sign opposite to those of the other two. In a more general case, when a larger number of $q_i^2 - \mu_i^2$ turns into zero, we have to require that among them there are quantities q_{i0} with different signs. Note that this requirement is invariant, because the sign of the variable component of the time vector does not depend on the choice of the centre-of-mass system.

Let us consider now three positive numbers $\alpha_1, \alpha_2, \alpha_3$ which will be determined later. We write a three-dimensional vector $\alpha_1 q_1 + \alpha_2 q_2 + \alpha_3 q_3$ and chose a coordinate system in such a way that the momentum turns into zero. (Remark, here the 4-vector $\alpha_1 q_1 + \alpha_2 q_2 + \alpha_3 q_3$ is considered in the beginning as time-like. Later, according to the choice of the corresponding value of α_i, it becomes zero. Such a zero vector can, obviously, always be obtained as a particular case of the time-like vector.) Hence, by definition,

$$\alpha_1 q_1 + \alpha_2 q_2 + \alpha_3 q_3 \; = \; 0. \qquad (4.130)$$

Consider now $\alpha_1 q_{10} + \alpha_2 q_{20} + \alpha_3 q_{30}$. Since one of the quantities q_{i0} has a sign opposite to that of the other two, we can always chose α_1, α_2 and α_3 in such a way that the combination $\alpha_1 q_{10} + \alpha_2 q_{20} + \alpha_3 q_{30}$ becomes zero:

$$\alpha_1 q_{10} + \alpha_2 q_{20} + \alpha_3 q_{30} \; = \; 0 . \qquad (4.131)$$

We can formulate now (4.130) and (4.131) in the following invariant way. For the q_i values corresponding to the appearance of the singularity in the integral (4.126), we can always find such positive numbers α_i, which lead to a zero 4-vector $\alpha_1 q_1 + \alpha_2 q_2 + \alpha_3 q_3$:

$$\sum_{i=1}^{3} \alpha_i \, q_i = 0 . \tag{4.132}$$

The conditions (4.129) (or (4.128)) and (4.132) are the so-called Landau rules; they bear an absolutely general character. Investigating real singularities, it can be shown that the above conditions are sufficient for the singularities to appear.

In the following we consider the concrete case of the diagram Fig. 26. Carrying out a scalar multiplication of (4.132) by q_1, q_2 and q_3, we obtain

$$\left. \begin{aligned} \alpha_1 q_1^2 + \alpha_2 (q_1 q_2) + \alpha_3 (q_1 q_3) &= 0 , \\ \alpha_1 (q_1 q_2) + \alpha_2 q_2^2 + \alpha_3 (q_2 q_3) &= 0 , \\ \alpha_1 (q_1 q_3) + \alpha_2 (q_2 q_3) + \alpha_3 q_3^2 &= 0 . \end{aligned} \right\} \tag{4.133}$$

(4.133) determines α_i different from zero only if there is an additional condition, namely, that the determinant

$$\begin{vmatrix} q_1^2 , & (q_1 q_2) , & (q_1 q_3) \\ (q_1 q_2) , & q_2^2 , & (q_2 q_3) \\ (q_1 q_3) , & (q_2 q_3) , & q_3^2 \end{vmatrix} = 0 . \tag{4.134}$$

equals zero. As a consequence of (4.129) $q_i^2 = \mu_i^2$, or, for example,

$$(q_1 q_2) = -\frac{1}{2}\left[(q_1 - q_2)^2 - q_1^2 - q_2^2\right] = -\frac{1}{2}\left[m_3^2 - \mu_1^2 - \mu_2^2\right] \equiv -\mu_1 \mu_2 z_3 .$$

With the help of the variables z introduced this way, the condition (4.134) can be rewritten in the form:

$$z_1^2 + z_2^2 + z_3^2 - 2 z_1 z_2 z_3 - 1 = 0 . \tag{4.135}$$

Solving (4.135) for one of the invariants m_1^2, m_2^2, m_3^2 or, what is the same, for one of the quantities z_1, z_2, z_3, two solutions are possible. It can be shown that only one of them leads to a positive α_i. The corresponding singularity turns out to be the only one on the physical sheet which is given by the exact definition of the situation of singularities in the integral (4.129), *i.e.* by the formulae (4.127).

4.7 Appendix B. Anomalous thresholds and final state interaction

In the present appendix we investigate the influence of the anomalous singularities of the three particle production amplitude on the analytic properties of the amplitude if the energy of two produced particles is small. As an example, let us consider a diagram for the production of a meson in meson-nucleon scattering processes shown in Fig. 4.14. We will see that the presence of anomalous terms in the dispersion relations does not influence the expansion of the amplitude over the degrees of the threshold momenta. In fact we shall demonstrate here that the anomalous singularity does not violate the threshold Landau singularity. This is not quite trivial, there were pronounced opposite statements [24].

We shall show now that the diagram Fig. 4.14 has a standard square-root singularity at $s_{23} = (k_2 + k_3)^2 = 4\mu^2$ (where μ is the meson mass). The logarithm-type singularity which is related to the anomaly in a definite way is, generally speaking, far from the physical threshold.

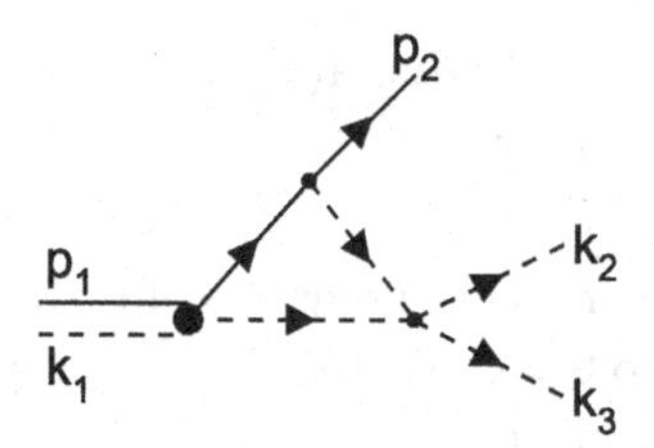

Fig. 4.14 Triangle diagram in the process $\pi N \to \pi\pi N$ which is used for analysis of the interplay of anomalous and threshold singularities.

At $(k_1 + p_1)^2 \equiv s \sim M^2$ (M is the nucleon mass) the diagram Fig. D1(4.14) has logarithmic singularities in s_{23} at

$$s_{23\pm} = \frac{\mu^2}{2M^2}\left[s + 3M^2 - \mu^2 \pm \sqrt{\left(\frac{4M^2}{\mu^2} - 1\right)\left((M+\mu)^2 - s\right)\left(s - (M-\mu)^2\right)}\right], \quad (4.136)$$

which are on the second sheet due to the square-root singularity at $s_{23} = 4\mu^2$. One of them, s_{23+}, appears at $s > M^2 + 2\mu^2$ and occurs on the lower half of the first sheet at $s > (M + \mu)^2$. This happens in the following way.

At $s \sim M^2$ we have for the amplitude of the considered triangle diagram a standard dispersion relation representation:

$$A(s_{23}, s \sim M^2) = \int_{4\mu^2}^{\infty} \frac{ds'_{23}}{\pi} \frac{discA(s'_{23}, s \sim M^2)}{s'_{23} - s_{23}} \qquad (4.137)$$

The integration contour and the location of the logarithmic singularities of the integrand is shown in Fig. 4.15a.

We have to arrive at energies above the threshold of the reaction $s > (M + 2\mu)^2$ (*i.e.* we have to consider the real energies of the process). Increasing $s + i\epsilon$ we reach $s = M^2 + 2\mu^2$ – this is the s value at which the logarithmic singularity of the second sheet, s_{23+}, is $s_{23} = 4\mu^2$. As $s + i\epsilon$ continues to grow, in the region $s \sim M^2 + 2\mu^2$ the singularity s_{23+} appears on the first sheet of the complex plane s_{23}.

In this case the dispersion representation of the triangle diagram in Fig. 4.14 over s_{23} will be transformed since the logarithmic singularity deforms the integration contour:

$$A(s_{23}, s > M^2 + 2\mu^2) = \int_C \frac{ds'}{\pi} \frac{discA(s'_{23}, s > M^2 + 2\mu^2)}{s' - s_{23}} . \qquad (4.138)$$

Indeed, the contour C starts from the point $s' = 4\mu^2$, reaches the cut which is connected to the anomalous singularity $s' = s_{23+}$, and goes to infinity along the real axis (see Fig. 4.15b).

The integral (4.138) can be represented as the sum of two integrals – one of them is going from s_{23+} to $4\mu^2$, from the jump of the absorption part along the cut, the second one goes from $4\mu^2$ to infinity:

$$\begin{aligned} A(s_{23}, s > M^2 + 2\mu^2) &= \int_{s_{23+}}^{4\mu^2} \frac{ds'_{23}}{\pi} \frac{\epsilon(s'_{23}, s)}{s'_{23} - s_{23}} \\ &+ \int_{4\mu^2}^{\infty} \frac{ds'_{23}}{\pi} \frac{discA(s'_{23}, s > M^2 + 2\mu^2)}{s'_{23} - s_{23}} , \\ \Delta(s'_{23}, s) &= discA(s'_{23} + i\epsilon, s > M^2 + 2\mu^2) \\ &- discA(s'_{23} - i\epsilon, s > M^2 + 2\mu^2) . \end{aligned} \qquad (4.139)$$

At $M^2 + 2\mu^2 < s < (M + \mu)^2$ both integrals are taken along the real axis.

If $s > (M+\mu)^2$, the singularity s_{23+} goes to the lower half-plane:

$$s_{23+} = \frac{\mu^2}{2M^2}\Big[s + 3M^2 - \mu^2 - i\sqrt{\Big(\frac{4M^2}{\mu^2}-1\Big)\Big(s-(M+\mu)^2\Big)\Big(s-(M-\mu)^2\Big)}\Big], \quad (4.140)$$

and draws the integration contour with (see Fig. 4.15c). This means that the amplitude is determined by (4.139) also in the physical region of the reaction, at $s \geq (M+2\mu)^2$, but in this case the contour of the integration will be in the lower half-plane s_{23}.

The first of the integrals (4.139) has a logarithmic singularity at $s_{23} = 4\mu^2$, since the integrand does approach zero at $s'_{23} \to 4\mu^2$. If $discA(s'_{23}, s) \to 0$ at $s'_{23} \to 4\mu^2$, it would be a singularity of the whole amplitude $A(s_{23}, s)$. This would happen, if $discA(s'_{23}, s)$ could be determined by the usual unitarity condition in the interval from $4\mu^2$ to infinity.

The existence of a logarithmic singularity at $s_{23} = 4\mu^2$ would lead to very serious consequences. First, it would be impossible to apply the usual theory for calculating the resonance interaction of two particles in the final state (*i.e.* the isobar models) at sufficiently small momenta of their relative motion in three-particle production reactions. Second, the results obtained for these reactions, which we have considered in the present section, would not be true.

In fact the amplitude $A(s_{23}, s)$ has no logarithmic singularity at $s_{23} = 4\mu^2$. This is due to the fact that $discA(s'_{23}, s)$ is not approaching zero if $s'_{23} \to 4\mu^2$, moreover, $discA(s'_{23}, s) = \Delta(s'_{23}, s)$ at $s'_{23} = 4\mu^2$ [25]. Because of that, the logarithmic singularity of the first integral is compensated exactly by the second integral. The reason is that $discA(s_{23}, s)$ is determined by the usual unitarity condition only if $s_{23} > (\sqrt{s}+M)^2$; in the region of smaller s_{23} the $discA(s_{23}, s)$ amplitude can be obtained by the analytic continuation of that in the region $s_{23} > (\sqrt{M}+M)^2$. At $s > M^2 + 2\mu^2$, *i.e.* just when the integration contour is touched and the anomalous singularity is appearing, the absorption part obtains an addition not going to zero at $s_{23} \to 4\mu^2$. Specifically, for the diagram Fig. 4.14 in which all the vertices are considered as point-like, the absorption part $s_{23} > (\sqrt{s}+M^2)$ is

$$discA(s_{23}, s) = \sqrt{\frac{s_{23}}{[s_{23}-(\sqrt{s}-M)^2][s_{23}-(\sqrt{s}+M)^2]}} \times \ln\frac{s_{23}-s+M^2-2\mu^2-D}{s_{23}-s+M^2-2\mu^2+D}, \quad (4.141)$$

$$D = \sqrt{(s_{23}-4\mu^2)/s_{23}}\sqrt{[s_{23}-(\sqrt{s}-M)^2][s_{23}-(\sqrt{s}+M)^2]}.$$

We have to continue this expression in s_{23} up to $s_{23} = 4\mu^2$. At $s_{23} = (\sqrt{s}+M)^2$ the amplitude $discA(s_{23}, s)$ has no singularities. At $s_{23} \to 4\mu^2$ the value of the logarithm goes to zero, if only in this point $s_{23} - s + M^2 - 2\mu^2 > 0$ (*i.e.* $(s < M^2 + 2\mu^2)$; if, however, $s_{23} - s + M^2 - 2\mu^2 < 0$, $discA(s_{23}, s)$ obtains an addition equal to

$$-2\pi\sqrt{\frac{s_{23}}{[s_{23}-(\sqrt{s}-M)^2][(\sqrt{s}+M)^2-s_{23}]}}. \tag{4.142}$$

The value of this quantity at $s_{23} = 4\mu^2$ is in fact the limit of $discA(s_{23})$ at $s_{23} \to 4\mu^2$ if $s > M^2 + 2\mu^2$, *i.e.* in the case of an anomalous singularity.

The point s_{23b} in Fig. 4.15 corresponds to the zero value of the numerator of the expression under the logarithm (or its denominator, depending on the choice of $\sqrt{s_{23} - (\sqrt{s}+M)^2}$). Turning around it, a jump of the absorption part $\Delta(s_{23}, s)$ appears which equals (4.142) (this can be easily seen). From here it follows, as it was already mentioned, that the amplitude $A(s_{23}, s)$ has no logarithmic singularity at $s_{23} = 4\mu^2$.

We can look at this fact also from a different point of view. The calculation of the jump of the amplitude which appears as the result of turning around the point $s_{23} = 4\mu^2$, can be carried out, for example, with the help of the Landau rule [26]; we can see that this jump approaches zero as $\sqrt{s_{23} - 4\mu^2}$.

The question arises [27] whether the non-Landau type square-root singularity $s_{23a} = (\sqrt{s} - M)^2$ (which is due to $\sqrt{s_{23} - (\sqrt{s} - M)^2} = 0$ in (4.141)) may not be essential in the absorption part being near $s_{23} = 4\mu^2$ when $\sqrt{s} \approx M + 2\mu$. At $\sqrt{s} < M + 2\mu$ we have $s_{23a} < 4\mu^2$, while at $\sqrt{s} > M + 2\mu$ this singularity is situated above the integration contour, since the amplitude has to be continued over the external masses (in the present case this means a continuation over s from the region of small $s + i\epsilon$ values) with a positive imaginary addition $\epsilon > 0$. If this singularity was placed on that edge of the cut of the absorption part along which the integration is carried out near the point $4\mu^2$, this would lead to the singularity $s_{23} = s_{23a}$ of the amplitude $A(s_{23}, s)$ on the non-physical sheet.

Indeed, the physical values of the amplitude correspond to values above the integration contour. Since the singularity s_{23a} of the absorption part is also above the contour of integration, it can be switched into the lower half-plane. After that it becomes obvious that the point $s_{23} = s_{23a}$ is not a singular one. If, however, s_{23} approaches the real axis from the

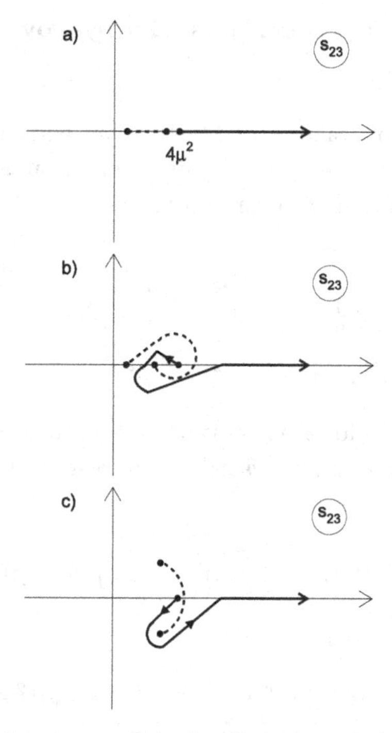

Fig. 4.15 Location of the contour C in the dispersion relation integrals at: (a) $s < M^2+2\mu^2$, Eq. (4.137); (b) $M^2+2\mu^2 < s < (M+\mu)^2$ Eq. (4.138); and (c) $(M+\mu)^2 < s$, Eq. (4.139).

lower half-plane, the integration contour in (4.138) and (4.137) becomes squeezed between the singularities of the integrand $s'_{23} = s_{23}$ and $s'_{23} = s_{23a}$. Consequently, at $s_{23} \to s_{23a}$ from the lower half-plane (i.e. on the non-physical sheet) the point $s_{23} = s_{23a}$ turns out to be singular.

This point, since it is close to $s_{23} = 4\mu^2$, would be essential if considering the reaction near the threshold. In fact the singularity s_{23a} in $discA(s_{23}, s)$ is separated from the integration contour near the point $s_{23} = 4\mu^2$ by the cut of the absorption part, see Fig. 4.15. The singularity s_{23a} appears on those logarithmic sheets where its phase is different from zero and because of that $discA(s_{23}, s)$ obtains an addition of the type of (4.142). However, in the region of integration ($s'_{23} \sim 4\mu^2$) the function $discA(s'_{23}, s)$ does not have any additions, and therefore s_{23+} is placed just in the way as shown in Fig. 4.15, while the whole amplitude does not have any additional singularities close to the square-root singularity $s_{23} = 4\mu^2$.

4.8 Appendix C. Homogeneous Skornyakov–Ter-Martirosyan equation

The homogeneous Skornyakov–Ter-Martirosyan equation for a three-body system determined by short-range pair interaction and large scattering length ($r_0/a << 1$) can be written in a form:

$$A(k^2) = \frac{1}{1/a - i\,k} \int\limits_0^\infty \frac{dp'^2}{\pi} A(\kappa^2) \frac{1}{p} \ln \frac{p'^2 + p^2 + pp' - mE - i0}{p'^2 + p^2 - pp' - mE - i0},$$

$$m\,E - \kappa^2 = \frac{3}{4}p'^2, \quad m\,E - k^2 = \frac{3}{4}p^2. \tag{4.143}$$

The equation is written for a boson-type system, but can be applied to nucleons as well if the spin-isospin variables are responsible for antisymmetry.

Introducing

$$pA(p)\left(1/a - i\sqrt{mE - \frac{3}{4}p^2}\right) = F(p), \tag{4.144}$$

we rewrite the equation as

$$F(p^2) = \int\limits_0^\infty \frac{dp'^2}{\pi} \frac{F(p'^2)}{p'(1/a - i\sqrt{mE - \frac{3}{4}p'^2})} \ln \frac{p'^2 + p^2 + pp' - mE - i0}{p'^2 + p^2 - pp' - mE - i0}. \tag{4.145}$$

The equation (4.145) can be solved by the substitution [19]:

$$p = \sqrt{-\frac{mE}{3}\frac{1-t^2}{t}}, \quad F(p(t)) \Rightarrow F(t) \tag{4.146}$$

and subsequent Mellin transformation:

$$\widetilde{F}(s) = \int\limits_0^\infty dt\, t^{s-1} F(t), \quad F(t) = \frac{1}{2\pi i} \int\limits_{\sigma - i\infty}^{\sigma + i\infty} ds\, t^{-s} \widetilde{F}(s). \tag{4.147}$$

In the limit $1/a \to 0$ we have in the s-representation

$$\left(1 - L(s)\right)\widetilde{F}(s) = 0, \quad L(s) = \frac{8}{\sqrt{3}} \frac{\sin(\pi s/6)}{s\cos(\pi s/2)} \tag{4.148}$$

what leads to $s = \pm i s_0$ with $s_0 = 1.0062...$. The wave function (ψ) and the excitation levels ($-E_n$) are determined as follows [19]:

$$\psi \sim \sin(s_0 \ln t), \quad \sqrt{E_n} = \sqrt{E_0} \exp\left(-\frac{\pi n}{s_0}\right), \; n = 0, \pm 1, \pm 2, \pm 3, ... \tag{4.149}$$

The value of the basic level, $n = 0$, is not determined by the equation, it is a parameter of the model. The set of poles with $n < 0$ corresponds to the three-body collapse [28], while $n > 0$ leads to the concentration of poles towards $E = 0$.

There are several approaches for the zero-range model ($r_0/a << 1$) beyond the scope of the Skornyakov–Ter-Martirosyan equation, (see, *e.g.*, [29, 30, 31] and references therein). Just in the model of this type (with the adiabatic expansion in the configuration space) the effect of three-body spectra concentration on zero total energy at $1/a \to 0$ was re-discovered in [29]. The three-body collapse branch of poles was eliminated in Efimov solution [29] by cutting procedure at large p'^2 – this method of regularization implicitly introduces short range three-body forces.

Discussing helium/tritium states, we should emphasize that the ($n = 0$)-level is the only state in the range of application of the limit $1/a \to 0$ for the Skornyakov–Ter-Martirosyan equation. Indeed, the region of application is determined by conditions for the three-nucleon energy

$$\sqrt{\frac{E}{\epsilon_{deut}}} > 1\,, \quad r_0\sqrt{mE} < 1\,, \tag{4.150}$$

where $\epsilon_{deut} \simeq 1/ma_t^2$ is the deuteron binding energy, and r_0 is the range of nucleon-nucleon forces $r_0 \sim 1/\mu_\pi$. Therefore even the levels $E = E_{n=\pm 1}$ are beyond the constraints of Eq. (4.150).

At present the short range model, and the Skornyakov–Ter-Martirosyan equation, is used extensively in molecular physics, see *e.g.* [32, 33, 34] and references therein.

4.9 Appendix D. Coordinates and observables in the three-body problem

In composite systems a choice of coordinates and related group theory constraints is determined by the interaction of the constituents. Sometimes, however, it is useful to look at the problem of group constraints neglecting interactions. It means to consider only the kinematic part of the Hamiltonian. Below we present some approaches of that type, following to [35, 36].

4.9.1 *Choice of coordinates and group theory properties*

Let $x_i (i = 1, 2, 3)$ be the radius vectors of the three particles, and fix:

$$x_1 + x_2 + x_3 = 0\,. \tag{4.151}$$

The Jacobi coordinates for equal masses are defined as

$$\xi = -\sqrt{\frac{3}{2}}\,(x_1 + x_2), \quad \eta = \sqrt{\frac{1}{2}}\,(x_1 - x_2),$$
$$\xi^2 + \eta^2 = 2x_1^2 + 2x_1x_2 + 2x_2^2 = x_1^2 + x_2^2 + x_3^2 = \rho^2. \tag{4.152}$$

We may define similar coordinates in the momentum space as well. In that case condition (4.152) means that we are in the centre-of-mass frame, and ρ^2 is a quantity proportional to the energy.

The quadratic form (4.152) can be understood as an invariant of the 0(6) group. In fact, we are interested in the direct product $O(3) \times O(2)$, as we have to introduce the total angular momentum observables L and M (group $O(3)$), and quantum numbers of the three-particle permutation group $O(2)$.

To characterize our three-particle system we need five quantum numbers. Thus the $O(6)$ group is too large for our purposes and it is convenient to deal with $SU(3)$ symmetry, in case of which we dispose exactly of the necessary 4 quantum numbers.

Let us introduce the complex vector

$$z = \xi + i\eta\,, \quad z* = \xi - i\eta\,. \tag{4.153}$$

The permutation of two particles leads in terms of these coordinates to rotations in the complex z-plane:

$$P_{12}\begin{pmatrix} z \\ z^* \end{pmatrix} = \begin{pmatrix} z^* \\ z \end{pmatrix}, \quad P_{13}\begin{pmatrix} z \\ z^* \end{pmatrix} = \begin{pmatrix} e^{2i\pi/3}z^* \\ e^{-2\pi/3}z \end{pmatrix}, \quad P_{23}\begin{pmatrix} z \\ z^* \end{pmatrix} = \begin{pmatrix} e^{-2i\pi/3}z^* \\ e^{2i\pi/3}z \end{pmatrix},$$

The condition $\xi^2 + \eta^2 = |z|^2 = \rho^2$ gives the invariant of the group $SU(3) \subset O(6)$. In the following we will take $\rho = 1$.

The generators of $SU(3)$ are defined, as usual:

$$A_{ik} = iz_i \frac{\partial}{\partial z_k} - iz_k^* \frac{\partial}{\partial z_i^*}, \quad (i, k = 1, 2, 3)\,. \tag{4.154}$$

The chain $SU(3) \supset SU(2) \supset U(1)$ familiar from the theory of unitary symmetry of hadrons is of no use for us, because it doesn't contain $O(3)$, *i.e.* going this way we cannot introduce the angular momentum quantum numbers. Instead of that, we consider two subgroups $O(6) \supset O(4) \sim SU(2) \times O(3)$ and $O(6) \supset SU(3)$. In other words, we have to separate from (4.154) the antisymmetric tensor-generator of the rotation group $O(3)$

$$L_{ik} = \frac{1}{2}(A_{ik} - A_{ki}) = \frac{1}{2}\left(iz_i\frac{\partial}{\partial z_k} - iz_k\frac{\partial}{\partial z_i} + iz_i^*\frac{\partial}{\partial z_k^*} - iz_k^*\frac{\partial}{\partial z_i^*}\right). \quad (4.155)$$

The remaining symmetric part

$$B_{ik} = \frac{1}{2}(A_{ik} + A_{ki}) = \frac{1}{2}\left(iz_i\frac{\partial}{\partial z_k} + iz_k\frac{\partial}{\partial z_1} - iz_i^*\frac{\partial}{\partial z_k^*} - iz_k^*\frac{\partial}{\partial z_i^*}\right) \quad (4.156)$$

is the generator of the group of deformations of the triangle which turns out to be locally isomorphic with the rotation group. Finally, we introduce a scalar operator

$$N = \frac{1}{2i}\,\mathrm{Sp}\,A = \frac{1}{2}\sum_{k=1,2,3}\left(z_k\frac{\partial}{\partial z_k} - z_k^*\frac{\partial}{\partial z_k^*}\right). \quad (4.157)$$

For characterizing our system, we choose the following quantum numbers:

$K(K+4)$ – eigenvalue of the Laplace operator (quadratic Casimir operator for $SU(3)$),
$L(L+1)$ – eigenvalue of the square of the angular momentum operator,
$L^2 = 4\sum_{i>k} L_{ik}^2$,
M – eigenvalue of $L_3 = 2L_{12}$,
ν – eigenvalue of N.

(4.158)

Although the generator (4.157) is not a Casimir operator of $SU(3)$, the representation might be characterized by means of its eigenvalue, because, as it can be seen, the eigenvalue of the Casimir operator of third order can be written as a combination of K and ν. (If the harmonic function belongs to the representation (p, q) of $SU(3)$, then it is the eigenfunction of Δ and N with eigenvalues $K(K+4)$ and ν respectively, where $K = p+q$ and $\nu = p-q$).

The fifth quantum number is not included in any of the considered subgroups, we have to take it from $O(6)$. We define it as the eigenvalue of

$$\Omega = \sum_{i,k,l} L_{ik}B_{kl}L_{li} = \mathrm{Sp}\,LBL. \quad (4.159)$$

This cubic generator was first introduced by Racah [37].

4.9.2 *Parametrization of a complex sphere*

Dealing with a three-particle system, we have to introduce coordinates which refer explicitly to the moving axes. One of the possible parametrizations of the vectors z and z^* is the following:

$$z = \frac{1}{\sqrt{2}} e^{-i\lambda/2}\Big(e^{ia/2}l_1 + ie^{-ia/2}l_2\Big),$$
$$z^* = \frac{1}{\sqrt{2}} e^{i\lambda/2}\Big(e^{-ia/2}l_1 - ie^{ia/2}l_2\Big),$$
$$|z|^2 = 1, \quad l_1^2 = l_2^2 = 1, \quad l_1 l_2 = 0. \tag{4.160}$$

In terms of these variables the (diagonal) moment of inertia has the following components:

$$\sin^2\Big(\frac{a}{2} - \frac{\pi}{4}\Big), \quad \cos^2\Big(\frac{a}{2} - \frac{\pi}{4}\Big), \quad l.$$

The three orthogonal unit vectors l_1, l_2 and $l = l_1 \times l_2$ form the moving system of coordinates. Their orientation to the fixed coordinate system can be described with the help of the Euler angles $\varphi_1, \theta, \varphi_2$

$$I_1 = \Big\{ -\sin\varphi_1 \sin\varphi_2 + \cos\varphi_1 \cos\varphi_2 \cos\theta$$
$$- \sin\varphi_1 \cos\varphi_2 - \cos\varphi_1 \sin\varphi_2 \cos\theta; -\cos\varphi_1 \sin\theta\Big\},$$
$$I_2 = \Big\{ -\cos\varphi_1 \sin\varphi_2 - \sin\varphi_1 \cos\varphi_2 \cos\theta;$$
$$-\cos\varphi_1 \cos\varphi_2 + \sin\varphi_1 \sin\varphi_2 \cos\theta; \sin\varphi_1 \sin\theta\Big\},$$
$$I = \Big\{ -\cos\varphi_2 \sin\theta; \sin\varphi_2 \sin\theta; -\cos\theta\Big\}. \tag{4.161}$$

In the following it will be simpler to introduce a new angle

$$a = \alpha - \frac{\pi}{2} \tag{4.162}$$

and work with the vectors

$$z = e^{-i\lambda/2}\Big(\cos\frac{a}{2} l_+ + i\sin\frac{a}{2} l_-\Big)$$
$$l_+ = \frac{1}{\sqrt{2}}(l_1 + il_2), \quad l_- = \frac{1}{\sqrt{2}}(l_1 - l_2). \tag{4.163}$$

Vectors l_+ and l_- have the obvious properties

$$l_+^2 = l_-^2 = 0, \quad l_0 = (l_+ \times l_-) = -il, \quad l_+ l_- = 1, l_+^* = l_-. \tag{4.164}$$

Let us turn our attention to the fact that the components of l_+ and l_- may be expressed in terms of the Wigner D-functions, defined as

$$D^l_{mn}(\varphi_1\theta\varphi_2) = e^{-i(m\varphi_1+n\varphi_2)}P^l_{mn}(\cos\theta) \tag{4.165}$$

in the following way:

$$\begin{aligned} l_+ &= \left\{D^1_{1-1}(\varphi_1\theta\varphi_2);\ D^1_{10}(\varphi_1\theta\varphi_2);\ D^1_{11}(\varphi_1\theta\varphi_2)\right\}, \\ l_0 &= \left\{D^1_{0-1}(\varphi_1\theta\varphi_2);\ D^1_{00}(\varphi_1\theta\varphi_2),\ D^1_{01}(\varphi_1\theta\varphi_2)\right\}, \\ l_- &= \left\{D^1_{-1-1}(\varphi_1\theta\varphi_2);\ D^1_{-10}(\varphi_1\theta\varphi_2);\ D^1_{-11}(\varphi_1\theta\varphi_2)\right\}, \end{aligned} \tag{4.166}$$

These equations demonstrate the possibility to construct the Wigner functions from the unit vectors corresponding to the moving coordinate system, in a way similar to the construction of spherical harmonics from the unit vectors of the fixed coordinate system.

4.9.3 *The Laplace operator*

We have now to write down the operators, the eigenvalues of which we are looking for. First let us construct the Laplace operator. We could do that by a straightforward calculation of $\Delta = |A_{ik}|^2$, but we choose a simpler way. We calculate

$$dz = -\frac{i}{2}\,zd\lambda + \frac{1}{2}\,e^{-i\lambda}(l \times z^*)da - (d\omega \times z)\,. \tag{4.167}$$

This rather simple expression is obtained by introducing the infinitesimal rotation $d\omega$. Its projections onto the fixed coordinate $k_1 = (1,0,0)$, $k_2 = (0,1,0)$, $k_3 = (0,0,1)$ given in terms of the Euler angles are well known:

$$\begin{aligned} d\omega_1 &= \cos\varphi_2 \sin\theta d\varphi_1 - \sin\varphi_2 d\theta\,, \\ d\omega_2 &= -\sin\varphi_2 \sin\theta d\varphi_1 - \cos\varphi_2 d\theta\,, \\ d\omega_3 &= \cos\theta d\varphi_1 + d\varphi_2\,. \end{aligned} \tag{4.168}$$

This provides

$$\begin{aligned} \frac{\partial}{\partial\omega_1} &= \cos\varphi_2\frac{1}{\sin\theta}\frac{\partial}{\partial\varphi_1} - \cos\varphi_2\cot\theta\frac{\partial}{\partial\varphi_2} - \sin\varphi_2\frac{\partial}{\partial\theta}\,, \\ \frac{\partial}{\partial\omega_2} &= -\sin\varphi_2\frac{1}{\sin\theta}\frac{\partial}{\partial\varphi_1} + \sin\varphi_2\cot\theta\frac{\partial}{\partial\varphi_2} - \cos\varphi_2\frac{\partial}{\partial\theta}\,, \\ \frac{\partial}{\partial\omega_3} &= \frac{\partial}{\partial\varphi_2}\,, \end{aligned} \tag{4.169}$$

and the permutation relations

$$\left[\frac{\partial}{\partial\omega_1},\frac{\partial}{\partial\omega_2}\right] = \frac{\partial}{\partial\omega_3}\,,\quad \left[\frac{\partial}{\partial\omega_2},\frac{\partial}{\partial\omega_3}\right] = \frac{\partial}{\partial\omega_1}\,,\quad \left[\frac{\partial}{\partial\omega_3},\frac{\partial}{\partial\omega_1}\right] = \frac{\partial}{\partial\omega_2}\,. \tag{4.170}$$

The effect of this operator on an arbitrary vector A is

$$\frac{\partial}{\partial \omega_i} A = \omega_i \times A ; \tag{4.171}$$

Here ω_i a vector of the length ω_i, directed along the i axis. The expression can be checked using the perturbation relation.

Let us determine now the rotation around the moving axes:

$$d\Omega_i = l_i d\omega . \tag{4.172}$$

In an explicit form $\partial/\partial\Omega_i$ can be written

$$\begin{aligned}
\frac{\partial}{\partial\Omega_1} &= \cos\varphi_1 \cot\theta \frac{\partial}{\partial\varphi_1} - \cos\varphi_1 \frac{1}{\sin\theta}\frac{\partial}{\partial\varphi_2} + \sin\varphi_1 \frac{\partial}{\partial\theta} , \\
\frac{\partial}{\partial\Omega_2} &= -\sin\varphi_1 \cot\theta \frac{\partial}{\partial\varphi_1} + \sin\varphi_1 \frac{1}{\sin\theta}\frac{\partial}{\partial\varphi_2} + \cos\varphi_1 \frac{\partial}{\partial\theta} , \\
\frac{\partial}{\partial\Omega_3} &= -\frac{\partial}{\partial\varphi_1} .
\end{aligned} \tag{4.173}$$

The minus sign in the third component reflects our choice of normalization of the D-function with a minus in the exponent (4.165).

The permutation relations for the operators $\partial/\partial\Omega_i$ are

$$\left[\frac{\partial}{\partial\Omega_1}, \frac{\partial}{\partial\Omega_2}\right] = -\frac{\partial}{\partial\Omega_3}, \quad \left[\frac{\partial}{\partial\Omega_2}, \frac{\partial}{\partial\Omega_3}\right] = -\frac{\partial}{\partial\Omega_1}, \quad \left[\frac{\partial}{\partial\Omega_3}, \frac{\partial}{\partial\Omega_1}\right] = -\frac{\partial}{\partial\Omega_2} . \tag{4.174}$$

The effect on A is defined, correspondingly, as

$$\frac{\partial}{\partial\Omega} A = -\Omega_i \times a , \tag{4.175}$$

which differs from (4.171) by the sign, as a consequence of the different signs in the permutation relations (4.170) and (4.174).

From (4.167) we obtain

$$\begin{aligned}
ds^2 &= dz\, dz^* = g_{ik} x^i x^k \\
&= \frac{1}{4} da^2 + \frac{1}{4} d\lambda^2 + \frac{1}{2} d\Omega_1^2 + \frac{1}{2} d\Omega_2^2 + d\Omega_3^2 - \sin a\, d\Omega_1 d\Omega_2 - \cos a\, d\Omega_3 d\lambda .
\end{aligned} \tag{4.176}$$

This expression determines the components of the metric tensor q_{ik}, and it becomes easy to calculate the Laplace operator

$$\begin{aligned}
\Delta' &= \frac{1}{4}\Delta = \frac{1}{4}\frac{1}{\sqrt{g}}\frac{\partial}{\partial x^i} g^{ik} \sqrt{g} \frac{\partial}{\partial x^k} \\
&= \frac{\partial^2}{\partial a^2} + 2\cot 2a \frac{\partial}{\partial a} + \frac{1}{\sin^2}\left(\frac{\partial^2}{\partial\lambda^2} + \cos a \frac{\partial^2}{\partial\lambda\,\partial\Omega_3} + \frac{1}{4}\frac{\partial^2}{\partial\Omega_3^2}\right) \\
&\quad + \frac{1}{2\cos^2 a}\left[\frac{\partial^2}{\partial\Omega_1^2} + \sin a \left(\frac{\partial^2}{\partial\Omega_1\partial\Omega_2} + \frac{\partial^2}{\partial\Omega_2\partial\Omega_1}\right) + \frac{\partial^2}{\partial\Omega_2^2}\right] .
\end{aligned} \tag{4.177}$$

If Φ is the eigenfunction of Δ', corresponding to a definite representation of $SU(3)$, then

$$\Delta'\Phi = -\frac{1}{4}K(K+4)\Phi = -\frac{K}{2}\left(\frac{K}{2}+2\right),$$
$$\Phi N\Phi = \nu\Phi, \qquad N = i\frac{\partial}{\partial\lambda} \tag{4.178}$$

has to be fulfilled. Expressing (4.177) in terms of the Euler angles, we get the Laplace operator in the form:

$$\Delta' = \Delta_a - \tan a\frac{\partial}{\partial a} + \frac{1}{2\cos^2 a}\left(\Delta_\theta - \frac{\partial^2}{\partial\varphi_1^2}\right)$$
$$-\frac{\sin a}{2\cos^2 a}\left[\cos 2\varphi_1\left(\frac{1+\cos^2\theta}{\sin^2\theta}\frac{\partial}{\partial\varphi_1} - 2\frac{\cos\theta}{\sin^2\theta}\frac{\partial}{\partial\varphi_2} - 2\cot\theta\frac{\partial^2}{\partial\varphi_1\partial\theta}\right.\right.$$
$$\left.\left.+2\frac{1}{2\sin\theta}\frac{\partial^2}{\partial\varphi_2\partial\theta}\right) + \sin 2\varphi_1\left(\Delta_\theta - \frac{\partial^2}{\partial\varphi_1^2} - 2\frac{\partial^2}{\partial\theta^2}\right)\right], \tag{4.179}$$

where Δ_a and Δ_θ are the Laplace operators

$$\Delta_a = \frac{\partial^2}{\partial a^2} + \cot a\frac{\partial}{\partial a} + \frac{1}{\sin^2 a}\left(\frac{\partial^2}{\partial\lambda^2} + \cos a\frac{\partial^2}{\partial\lambda\,\partial\Omega_3} + \frac{1}{4}\frac{\partial^2}{\partial\Omega_3^2}\right),$$
$$\Delta_\theta = \frac{\partial^2}{\partial\theta^2} + \cot\theta\frac{\partial}{\partial\theta} + \frac{1}{\sin^2\theta}\left(\frac{\partial^2}{\partial\varphi_1^2} - 2\cos\theta\frac{\varphi^2}{\partial\varphi_1\partial\varphi_2} + \frac{\partial^2}{\partial\varphi_2^2}\right) \tag{4.180}$$

of the $O(3)$ group. The Laplace operator differs from that calculated in [38] by the parametrization. They are connected, however, by a unitary transformation.

4.9.4 *Calculation of the generators L_{ik} and B_{ik}*

To obtain the generators directly from dz, we have to invert a 5×5 matrix in the case of three particles. That requires rather a long calculation, which is getting hopeless for a larger number of particles. Instead of performing the straightforward calculation, we get the wanted expressions in the following way. Let us first consider L_{ik}, or rather its special case L_{12}. We introduce a parameter σ_{ik} which define the displacement along the particular trajectory which corresponds to the action of the operator L_{ik}. Thus, formally we can write

$$L_{12} = \frac{1}{2}\left(iz_1\frac{\partial}{\partial z_2} - iz_2\frac{\partial}{\partial z_1} + iz_1^*\frac{\partial}{\partial z_2^*} - iz_2^*\frac{\partial}{\partial z_1^*}\right) \equiv \frac{\partial}{\partial\sigma_{12}}. \tag{4.181}$$

Acting with L_{12} on the vectors z and z^*

$$L_{12}\begin{pmatrix} z_1 \\ z_2 \\ z_3 \end{pmatrix} = \frac{1}{2}\begin{pmatrix} -iz_2 \\ iz_1 \\ 0 \end{pmatrix}, \quad L_{12}\begin{pmatrix} z_1^* \\ z_2^* \\ z_3^* \end{pmatrix} = \frac{1}{2}\begin{pmatrix} -iz_2^* \\ iz_1^* \\ 0 \end{pmatrix} \tag{4.182}$$

we see that σ_{12} has to be imaginary. We have from (4.182)

$$\begin{aligned} zL_{12}z &= 0, \quad z^*L_{12}z^* = 0, \\ z^*L_{12}z &= \frac{i}{2}(z\times z^*)_3, \quad lL_{12}z = -\frac{i}{2}(l\times z)_3. \end{aligned} \tag{4.183}$$

Using the expression (4.167) for dz, we can write

$$\begin{aligned} L_{12}z = \frac{\partial z}{\partial\sigma_{12}} &= -\frac{i}{2}z\frac{d\lambda}{d\sigma_{12}} + \frac{1}{2}e^{-i\lambda}(l\times z^*)\frac{\partial a}{\partial\sigma_{12}} - \left(\frac{d\omega}{d\sigma_{12}}\right), \\ L_{12}z^* = \frac{\partial z^*}{\partial\sigma_{12}} &= \frac{i}{2}z^*\frac{d\lambda}{d\sigma_{12}} + \frac{1}{2}e^{i\lambda}(l\times z^*)\frac{\partial a}{\partial\sigma_{12}} - \left(\frac{d\omega}{d\sigma_{12}}\times z^*\right). \end{aligned} \tag{4.184}$$

Here $-iz/2$ are $dz/d\lambda$ etc. With use Eqs. (4.183), (4.184) we obtain

$$\begin{aligned} (l\times z^*) &= ie^{i\lambda/2}\left(\cos\frac{a}{2}l_- + \frac{a}{2}l_+\right), \\ (l\times z) &= -ie^{-i\lambda/2}\left(\cos\frac{a}{2}l_+ - i\sin\frac{a}{2}l_-\right); \\ z^2 = ie^{-i\lambda}\sin a, \quad z^{*2} &= -ie^{i\lambda}\sin a, \end{aligned} \tag{4.185}$$

we have from Eq. (4.183):

$$\frac{\partial a}{\partial\sigma_{12}} = \frac{d\lambda}{d\sigma_{12}} = 0, \quad \frac{a\Omega_3}{d\sigma_{12}} = -\frac{i}{2}l^{(3)}, \tag{4.186}$$

and finally, Eq. (4.184) leads to

$$\frac{d\Omega_2}{d\sigma_{12}} = -i\frac{i}{2}l_2^{(3)}, \quad \frac{d\Omega_1}{d\sigma_{12}} = -\frac{i}{2}l_l^{(3)}, \tag{4.187}$$

where $l_i^{(k)}$ stands for the kth component of vector l_i. Thus we obtain

$$\begin{aligned} L_{12} &= -\frac{i}{2}\left[l_1^{(3)}\frac{\partial}{\partial\omega_1} + l_2^{(3)}\frac{\partial}{\partial\Omega_2} + l_3^{(3)}\frac{\partial}{\partial\Omega_3}\right] = -\frac{i}{2}\frac{\partial}{\partial\omega_3}, \\ L_{23} &= -\frac{i}{2}\left[l_1^{(1)}\frac{\partial}{\partial\Omega_1} + l_2^{(1)}\frac{\partial}{\partial\Omega_2} + l^{(1)}\frac{\partial}{\partial\Omega_3}\right] = -\frac{i}{2}\frac{\partial}{\partial\omega_1}, \\ L_{31} &= -\frac{i}{2}\left[l_1^{(2)}\frac{\partial}{\partial\Omega_1} + l_2^{(2)}\frac{\partial}{\partial\Omega_2} + l^{(2)}\frac{\partial}{\partial\Omega_3}\right] = -\frac{i}{2}\frac{\partial}{\partial\omega_2}. \end{aligned} \tag{4.188}$$

Introducing the notations

$$L_1 = 2L_{23}, \quad L_2 = 2L_{31}, \quad L_3 = 2L_{12}, \tag{4.189}$$

we can write the general expression for the angular momentum operator

$$L_k = -i\left[l_1^{(k)}\frac{\partial}{\partial\Omega_1} + l_2^{(k)}\frac{\partial}{\partial\Omega_2} + l^{(k)}\frac{\partial}{\partial\Omega_3}\right]. \tag{4.190}$$

It fulfills the commutation relations

$$[L_1, L_2] = -iL_3, \quad [L_2, L_3] = -iL_2, \quad [L_3L_1] = -iL_2. \tag{4.191}$$

The square of the angular momentum operator is

$$L^2 = \left(\frac{\partial^2}{\partial\Omega_1^2} + \frac{\partial^2}{\partial\Omega_2^2} + \frac{\partial^2}{\partial\Omega_3^2}\right) = \Delta_\theta . \tag{4.192}$$

Let us now turn our attention to the operator B_{ik}. We consider

$$B_{12} = \frac{1}{2}\Big(iz_1\frac{\partial}{\partial z_2} + iz_2\frac{\partial}{\partial z_1} - iz_1^*\frac{\partial}{\partial z_2^*} - iz_2^*\frac{\partial}{\partial z_1^*}\Big) \equiv \frac{\partial}{\partial \beta_{12}} . \tag{4.193}$$

From the action of B_{12} on z and z^*

$$B_{12}\begin{pmatrix} z_1 \\ z_2 \\ z_3 \end{pmatrix} = \frac{1}{2}\begin{pmatrix} iz_2 \\ iz_1 \\ 0 \end{pmatrix}, \quad B_{12}\begin{pmatrix} z_1^* \\ z_2^* \\ z_3^* \end{pmatrix} = \frac{1}{2}\begin{pmatrix} -z_2^* \\ -z_1^* \\ 0 \end{pmatrix}, \tag{4.194}$$

it is obvious, that β_{12} is real. We make use of the conditions

$$\begin{aligned} zB_{12}z &= iz_1z_2 , \qquad z^*B_{12}z^* = -iz_1^*z_2^* , \\ z^*B_{12}z &= \frac{i}{2}(z_1^*z_2 + z_1z_2^*) , \qquad lB_{12}z = \frac{i}{2}(l^{(1)}z_2 + l^{(2)}z_1) \end{aligned} \tag{4.195}$$

and introduce the notation

$$b_{ik}^{(lm)} = \frac{1}{2}\Big(l_i^{(l)}l_k^{(m)} + l_i^{(m)}l_k^{(l)}\Big). \tag{4.196}$$

Then, following a procedure similar to that in the case of L_{ik}, we obtain from Eq. (4.195)

$$\begin{aligned} \frac{da}{d\beta_{12}} &= b_{11}^{(12)} - b_{22}^{(12)} , \\ \frac{d\lambda}{d\beta_{12}} &= \Big(b_{11}^{(12)} + b_{22}^{(12)}\Big) - 2b_{12}^{(12)}\frac{1}{\sin a} . \end{aligned} \tag{4.197}$$

Equations (4.195) lead to

$$\begin{aligned} \frac{d\Omega_1}{d\beta_{12}} &= -b_{23}^{(12)}\tan a - b_{13}^{(12)}\frac{1}{\cos a} , \\ \frac{d\Omega_2}{d\beta_{12}} &= -b_{23}^{(12)}\frac{1}{\cos a} - b_{13}^{(12)}\tan a , \\ \frac{d\Omega_3}{d\beta_{12}} &= -b_{12}^{(12)}\cot a . \end{aligned} \tag{4.198}$$

Thus the expression for B_{12} will be

$$\begin{aligned} B_{12} = {} & \Big(b_{11}^{(12)} - b_{22}^{(12)}\Big)\frac{\partial}{\partial a} - \Big(b_{11}^{(12)} + b_{22}^{(12)}\Big)\frac{\partial}{\partial\lambda} \\ & - 2b_{12}^{(12)}\left(\frac{1}{\sin a}\frac{\partial}{\partial\lambda} + \frac{1}{2}\cot a\frac{\partial}{\partial\Omega_3}\right) - \left(b_{13}^{(12)}\frac{1}{\cos a} + b_{23}^{(12)}\tan a\right)\frac{\partial}{\partial\Omega_1} \\ & - \left(b_{13}^{(12)}\tan a + b_{23}^{(12)}\frac{1}{\cos a}\right)\frac{\partial}{\partial\Omega_2} . \end{aligned} \tag{4.199}$$

$$B_{ik} = \left(b_{11}^{(ik)} - b_{22}^{(ik)}\right)\frac{\partial}{\partial a} - \left(b_{11}^{(ik)} + b_{22}^{(ik)}\right)\frac{\partial}{\partial \lambda}$$
$$- 2b_{12}^{(ik)}\left(\frac{1}{\sin a}\frac{\partial}{\partial \lambda} + \frac{1}{2}\cot a\frac{\partial}{\partial \Omega_3}\right) - b_{13}^{(ik)}\left(\tan a\frac{\partial}{\partial \Omega_2} + \frac{1}{\cos a}\frac{\partial}{\partial \Omega_1}\right)$$
$$- b_{23}^{(ik)}\left(\tan a\frac{\partial}{\partial \Omega_1} + \frac{1}{\cos a}\frac{\partial}{\partial \Omega_2}\right). \tag{4.200}$$

Acting in the space of polynomials of which include only z (and not z^*), there exist the following operatorial identity:

$$ie^{i\alpha}\frac{\partial}{\partial \Omega_1} = \frac{\partial}{\partial \Omega_2}. \tag{4.201}$$

Thus in the space of polynomials of z, operators B_{ik} might be written as

$$B_{ik} = \left(b_{11}^{(ik)} - b_{22}^{(ik)}\right)\frac{\partial}{\partial a} - \left(b_{11}^{(ik)} + b_{22}^{(ik)}\right)\frac{\partial}{\partial \lambda} + 2\frac{\partial}{\partial \lambda}\delta_{ik}$$
$$- 2b_{12}^{(ik)}\left(\frac{1}{\sin a}\frac{\partial}{\partial \lambda} + \frac{1}{2}\cot a\frac{\partial}{\partial \Omega_3}\right) - ib_{23}^{(ik)}\frac{\partial}{\partial \Omega_1} + ib_{13}^{(ik)}\frac{\partial}{\partial \Omega_2}. \tag{4.202}$$

with

$$[B_{ik}, B_{jl}] = \frac{1}{2}(L_{il}\delta_{kj} - L_{jk}\delta_{il}) + \frac{i}{2}(L_{ij}\delta_{kl} - L_{lk}\delta_{ij}),$$
$$[B_{ik}, L_{jl}] = \frac{i}{2}(B_{il}\delta_{kj} - B_{jk}\delta_{il}) - \frac{i}{2}(B_{ij}\delta_{kl} - B_{ik}\delta_{ij}). \tag{4.203}$$

In particular,

$$[B_{12}, B_{11}] = -iL_{12}, \quad [B_{12}, B_{22}] = iL_{12}, \quad [B_{11}, L_{12}] = iB_{12},$$
$$[B_{22}, L_{12}] = -iB_{12}, \qquad [B_{12}, L_{12}] = -\frac{i}{2}(B_{11} - B_{22}).$$

4.9.5 *The cubic operator* Ω

Operators H_+ and H_- are the usual raising and lowering operators in $SU(2)$ taken at the value of the second Euler angle $-2\Omega_3 = 2\varphi_1 = 0$

$$H_+ = \frac{1}{\sqrt{2}}\left[\frac{\partial}{\partial a} + i\frac{1}{\sin a}\frac{\partial}{\partial \lambda} + \frac{i}{2}\cot a\frac{\partial}{\partial \Omega_3}\right],$$
$$H_- = \frac{1}{\sqrt{2}}\left[\frac{\partial}{\partial a} - i\frac{1}{\sin a}\frac{\partial}{\partial \lambda} - \frac{i}{2}\cot a\frac{\partial}{\partial \Omega_3}\right]. \tag{4.204}$$

Ω can be written in the form

$$\Omega = \sum_{i,j,k} L_{ij}L_{jk}B_{ki} = -\frac{1}{4}\Bigg\{\sqrt{2}\Big(-\frac{\partial^2}{\partial\Omega_+^2}H_+ + \frac{\partial^2}{\partial\Omega_-^2}H_-\Big) + \frac{\partial^2}{\partial\Omega_3^2}\frac{\partial}{\partial\lambda}$$

$$+\Delta_\theta\frac{\partial}{\partial\lambda} - \frac{1}{\cos a}\Big(\Delta_\theta - \frac{\partial^2}{\partial\Omega_3^2} + \frac{1}{2}\Big)\frac{\partial}{\partial\Omega_3}$$

$$+\tan a\Big[i\Big(\frac{\partial^2}{\partial\Omega_+^2} - \frac{\partial^2}{\partial\Omega_-^2}\Big)\frac{\partial}{\partial\Omega^2} - \frac{3}{2}\Big(\frac{\partial^2}{\partial\Omega_+^2} + \frac{\partial^2}{\partial\Omega_-^2}\Big)\Big]\Bigg\}. \tag{4.205}$$

The operator Ω has a simple meaning in the classical approach. Substituting the derivatives by velocities and denoting $\xi = p$ and $\eta = q$, we obtain

$$\frac{1}{2}\Omega = (\xi L)(qL) - (\eta L)(pL). \tag{4.206}$$

It is obvious that the derivative in time of this operator equals zero. Directing the axis z along L and introducing two two-dimensional vectors in the permutation space

$$x = (\xi_z, \eta_2) \text{ and } y = (p_z, q_z), \tag{4.207}$$

(4.206) can be written as

$$\frac{1}{2}\Omega = (x \times y)_3. \tag{4.208}$$

The operator has the form of the third component of the momentum in the permutation space. Hence the symmetry of the problem becomes clear: it is spherical in the coordinate space and axial in the space of permutations.

At small K and ν, when the degeneracy is negligible, the eigenvalues of this operator are not needed. Indeed, at given K and ν values the number of states is determined by the usual $SU(3)$ expression

$$n(K,\nu) = \frac{1}{8}(K+2)(K+2-2\nu)(K+2+2\nu). \tag{4.209}$$

The summation of this formula over 2ν from $-K$ to K leads to the well-known expression [39]

$$n(K) = \frac{(K+3)(K+2)^2(K+1)}{12}. \tag{4.210}$$

The maximal degenerations correspond to terms with $\nu = 0$ for even K values or $\nu - 1/2$ for odd ones.

$$n(K,0) = \begin{cases} \frac{1}{8}(K+2)^3, & K \; - \; \text{odd}, \\ \frac{1}{8}(K+1)(K+2)(K+3), & K \; - \; \text{even}. \end{cases} \tag{4.211}$$

On the other hand, since $L \in (0,K)$ and $M \in (-l, L)$, at a given K there are $(K+1)^2$ terms with different L and M values. $n(K,0) > (K+1)^2$ for $K \geq 4$, and thus the additional quantum number is necessary only for such K.

4.9.6 Solution of the eigenvalue problem

$$\Phi^L_M = \sum_\lambda \sum_{M'=-\Lambda}^{\lambda} a(\Lambda, M')a(\Lambda, M')\, D^\Lambda_{\nu,M'}(\lambda, a, 0)\, D^L_{2M',M}(\varphi_1, \theta, \varphi_2)\,. \tag{4.212}$$

Let us finally consider a few special cases of the solution. As it is discussed by Dragt [40], in the low-dimensional representations of $SU(3)$ ($L = 0, 1$) the Ω is not needed. Indeed, in the case of $L = 0$ the Laplace operator obtains the form

$$\Delta = \frac{\partial^2}{\partial a^2} + 2\cot 2a \frac{\partial}{\partial a} + \frac{1}{\sin^2 a}\frac{\partial^2}{\partial \lambda^2}\,. \tag{4.213}$$

Obviously, the eigenfunction will be the following

$$\Phi_0 = D^\Lambda_{\nu,0}(\lambda, a, 0)\,, \tag{4.214}$$

which obeys the equation

$$\Delta\Phi_0 = -\Lambda(\Lambda+1)\Phi_0\,, \qquad \lambda = 0, 1, \ldots\,. \tag{4.215}$$

This solution demonstrates clearly the $SU(2)$ nature of a non-rotating triangles.

In the case of $L = 1$ the solutions are

$$z_M = \sum_{M'=\pm 1/2} D^{1/2}_{1/2,M'}(\lambda, a, 0)\, D^1_{2M',M}(\varphi_1, \theta, \varphi_2),$$

$$z^*_M = D^{1/2}_{-1/2,-1/2}(\lambda, a, 0) D^1_{-1,M}(\varphi_1\theta\varphi_2) - D^{1/2}_{-1/2,1/2}(\lambda, a, 0)\, D^1_{1,M}(\varphi_1\theta\varphi_2),$$

fulfilling the Laplace equation with the value $K = 1$. Simultaneously z_M obeys the equations

$$\begin{aligned} &L^2 z_M = -2z_M\,, \quad L_3 z_M = -Mz_M\,, \quad M = -1, 0, 1\,, \\ &\Delta_a z_M = -\frac{3}{4} z_M\,, \quad N z_M = \frac{1}{2} z_M\,, \quad \Omega z_M = -\frac{3}{4} i z_M, \end{aligned} \tag{4.216}$$

and, accordingly, z^* obeys complex conjugated equations.

References

[1] L.D. Landau, Nucl. Phys. **13**, 181 (1959).
[2] V.V. Anisovich, A.A. Anselm, V.N. Gribov, I.T. Dyatlov, ZhETP **43**, 906 (1962).
[3] V.N. Gribov, Nucl. Phys. **5**, 653 (1958).
[4] I.T. Dyatlov, ZhETP **37**, 1330 (1959).

[5] G.V. Skornyakov and K.A. Ter-Martirosyan, ZhETP **31**, 775 (1956), [Sov. Phys. JETP **4**, 648 (1956)].
[6] G.S. Danilov, ZhETP **40**, 498 (1961); **42**, 1449 (1962).
[7] L.D. Faddeev, ZhETF **39** 1459 (1960).
[8] J. Nyiri, ZhETP **46**, 671 (1964).
[9] L.G. Dakhno, Yad. Fiz. **12**, 840 (1970).
[10] V.V. Anisovich, P.E. Volkovitsky, Yad. Fiz. **14**, 1055 (1971); Phys. Lett. **B35**, 443 (1971).
[11] V.N. Gribov, ZhETP **38**, 553 (1960).
[12] V.V. Anisovich, L.G. Dakhno, ZhETP **46**, 1307 (1964).
[13] V.V. Anisovich, ZhETP **44**, 1593 (1963).
[14] V.V. Anisovich and L.G. Dakhno, Jad. Fiz. **2**, 710 (1965) and stimulated discusion with L.B. Okun.
[15] U.G. Meissner, G. Muller and S. Steininger, Phys. Lett. **B406**, 154 (1997), U.G. Meissner **A629**, 72 (1998).
[16] N. Cabibbo, Phys. Rev. Lett. **93**, 121801 (2004).
[17] N. Cabibbo and G. Isidori, JHEP **503**, 21 (2005).
[18] J.R. Batley et al. (The NA48/2 Collab.), Phys. Lett. **B634**, 474 (2006).
[19] R.A. Minlos and L.D. Faddeev, ZhETF **41**, 1850 (1960), [Sov. Phys. JETP **14**, 1315 (1961)].
[20] G.S. Danilov and V.I. Lebedev, ZhETF **44**, 1509 (1963).
[21] A.G. Sitenko, V. F. Kharchenko, Nucl. Phys. **49**, 15 (1963); UFN **103**, 469 (1971).
[22] A.V. Anisovich and V.V. Anisovich, Yad. Fiz. **53**, 1485 (1991); [Sov. J. Nucl. Phys. **53**, 915 (1991)].
[23] J. S. McCarthy, Phys. Rev. C **15**, 1396 (1977).
[24] R.F. Sawyer, Phys. Rev. Lett. **7**, 213 (1961).
[25] R. Blankenbeckler, Y. Nambu, Nuovo Cim. **18**, 595 (1960).
[26] L.D. Landau, Nucl. Phys. **13**, 181 (1959).
[27] V.V. Anisovich, A.A. Anselm, B.N. Gribov, I.T. Dyatlov ZhETF **43**, 906 (1963).
[28] L.H. Thomas, Phys. Rev. **47**, 903 (1935).
[29] V. Efimov, Yad. Fiz. **12**, 1080 (1970); [Sov. J. Nucl. Phys. **12**, 589 (1970)].
[30] D.V. Fedorov, A.S. Jensen, K. Riisager, Phys. Rev. **C 47**, 2372 (1994).
[31] P.F. Bedaque, H-W. Hammer, U. van Klock, Phys. Rev. Lett. **82**, 463 (1999).
[32] F.M. Penkov, ZhETP **97**, 485 (2003).
[33] A.K. Motovilov, W. Sandhas, S.A. Sofianos, E.A. Kolganova, Eur. Phys. J. **D13**, 33 (2001).
[34] F.M. Penkov, W. Sandhas, arXiv:physics/0507031 (2005).
[35] J. Nyiri and Ya.A. Smorodinsky, Yad. Fiz. **9**, 882 (1969) [Sov. J. Nucl. Phys. **9**, 515 (1969)];
Yad. Fiz. **12**, 202 (1970) [Sov. J. Nucl. Phys. **12**, 109 (1971)];
Acta Phys. Hung. **32**, 241 (1972).
Yad. Fiz. **29**, 833 (1979) [Sov. J. Nucl. Phys. **29** 429 (1079].
[36] J. Nyiri, Acta Phys. Slovaca **23**, 82 (1973).

[37] G. Racah, Rev. Mod. Phys. **21** 494 (1949).
[38] J.M. Lévy-Leblond, M. Lévy-Nahas, J. Math. Phys. **6**, 1571 (1965).
[39] W. Zickendraht, Ann. of Phys. **35** 18 (1965).
[40] A.J. Dragt, J. Math. Phys. **6** 533 (1965).

Chapter 5

Propagators of Spin Particles and Relativistic Spectral Integral Equations for Three-Hadron Systems

The spin particles, their propagations and interactions are subjects of attention in the present chapter.

We present propagators of particles with arbitrary spin meaning their appication in the dispersion relations, in particular, in equations for multiparticle systems. First, the bosons with non-zero spin, $J > 0$, are considered in terms of covariant momentum expansion technique (elements of this technique are given in Chapter 2, Appendix A, see also [1, 2]). Then we present the corresponding propagator operators for fermions subsequently with $J = 1/2\,, 3/2$ and $J > 3/2$, see [2, 3] as well.

Using spin operators in a general form we write down the three particle equations.

As an example of the application we give some quite realistic description of reactions the understanding of which is required: $\bar{p}p(J^{PC} = 0^{-+}) \to 3\ mesons$ measured at the Crystal Barrel Collaboration (see [4, 5] and references therein). These reactions were successfully analyzed by extracting the leading amplitude singularities – pole singularities – with the aim to obtain information about two-meson resonances. But these analyses do not take into account three-body final-state interactions in an explicitly correct way (though the logarithmic singularity, due to the triangle diagram, was included in the fit, see [6]). Here, following [7], we demonstrate how the coupled three-particle equations may be written for the $\pi^0\pi^0\pi^0$, $\eta\pi^0\pi^0$, $\eta\eta\pi^0$, $\bar{K}K\pi^0$ channels in the $\bar{p}p(0^{-+})$ annihilation at rest to take into consideration produced hadron rescatterings. For the final-state mesons the $S-$, $P-$, and $D-$ wave interactions are taken into account. The realistic description of meson interactions requires a many-channel approach and the consideration of a large number of resonances with the corresponding spins.

5.1 Boson propagators

First, we recall properties of the momentum operators for a two-particle system. Masses and momenta of particles are denoted as m_1, k_1 and m_2, k_2. We work with $p = k_1 + k_2$ and vectors which are orthogonal to the total momentum of the system p:

$$g^{\perp}_{\mu\nu} = g_{\mu\nu} - \frac{p_\mu p_\nu}{p^2},$$
$$k^{\perp}_{1\mu} = g^{\perp}_{\mu\nu} k_{1\nu}, \quad k^{\perp}_{2\mu} = g^{\perp}_{\mu\nu} k_{2\nu}. \tag{5.1}$$

We denote $k^{\perp}_{1\mu} = -k^{\perp}_{2\mu} \equiv k^{\perp}_{\mu}$.

5.1.1 *Projection operators and denominators of the boson propagators*

For the lowest states the projection operators read:

$$O = 1\,, \qquad O^{\mu}_{\nu} = g^{\perp}_{\mu\nu}\,,$$
$$O^{\mu_1\mu_2}_{\nu_1\nu_2} = \frac{1}{2}\left(g^{\perp}_{\mu_1\nu_1} g^{\perp}_{\mu_2\nu_2} + g^{\perp}_{\mu_1\nu_2} g^{\perp}_{\mu_2\nu_1} - \frac{2}{3} g^{\perp}_{\mu_1\mu_2} g^{\perp}_{\nu_1\nu_2}\right). \tag{5.2}$$

For higher states, the operator can be calculated using the recurrent expression:

$$O^{\mu_1\ldots\mu_J}_{\nu_1\ldots\nu_J} = \frac{1}{J^2}\Bigg(\sum_{i,j=1}^{J} g^{\perp}_{\mu_i\nu_j} O^{\mu_1\ldots\mu_{i-1}\mu_{i+1}\ldots\mu_J}_{\nu_1\ldots\nu_{j-1}\nu_{j+1}\ldots\nu_J}$$
$$-\frac{4}{(2J-1)(2J-3)} \sum_{\substack{i<j\\k<m}}^{J} g^{\perp}_{\mu_i\mu_j} g^{\perp}_{\nu_k\nu_m} O^{\mu_1\ldots\mu_{i-1}\mu_{i+1}\ldots\mu_{j-1}\mu_{j+1}\ldots\mu_J}_{\nu_1\ldots\nu_{k-1}\nu_{k+1}\ldots\nu_{m-1}\nu_{m+1}\ldots\nu_J}\Bigg). \tag{5.3}$$

The projection operator $O^{\mu_1\ldots\mu_J}_{\nu_1\ldots\nu_J}$ is normalized in a such way that:

$$O^{\mu_1\ldots\mu_J}_{\alpha_1\ldots\alpha_J} O^{\alpha_1\ldots\alpha_J}_{\nu_1\ldots\nu_J} = O^{\mu_1\ldots\mu_J}_{\nu_1\ldots\nu_J} \tag{5.4}$$

and

$$O^{\alpha\mu_1\ldots\mu_J}_{\alpha\nu_1\ldots\nu_J} = \frac{2J+3}{2J+1} O^{\mu_1\ldots\mu_J}_{\nu_1\ldots\nu_J}\,. \tag{5.5}$$

The tensor part of the numerator of the boson propagator is defined by the projection operator. Let us write it as follows:

$$F^{\mu_1\ldots\mu_J}_{\nu_1\ldots\nu_J} = (-1)^J\, O^{\mu_1\ldots\mu_J}_{\nu_1\ldots\nu_J}\,, \tag{5.6}$$

with the definition of the propagator

$$\frac{F^{\mu_1\ldots\mu_J}_{\nu_1\ldots\nu_J}(\perp p)}{M^2 - p^2 - i0}. \tag{5.7}$$

Here in the propagator we indicate the vector p, which is orthoganal to the space of the particle polarization vectors, $g^{\perp}_{\mu\nu}p_\nu = 0$.

Taking into account the definition of projection operators (5.3), it is possible to obtain the X-operators which where introduced in Chapter 2:

$$k_{\mu_1}\dots k_{\mu_J} O^{\mu_1\dots\mu_J}_{\nu_1\dots\nu_J} = \frac{1}{\alpha_J} X^{(J)}_{\nu_1\dots\nu_J}(k^\perp), \quad \alpha_L = \prod_{l=1}^{L}\frac{2l-1}{l}. \tag{5.8}$$

The product of two X-operators integrated over a solid angle (that is equivalent to the integration over internal momenta) depends only on the external momenta and the metric tensor. Therefore, it must be proportional to the projection operator. After straightforward calculations we obtain

$$\int\frac{d\Omega}{4\pi} X^{(J)}_{\mu_1\dots\mu_J}(k^\perp) X^{(J)}_{\nu_1\dots\nu_J}(k^\perp) = \frac{\alpha_J\, k^{2J}_\perp}{2J+1} O^{\mu_1\dots\mu_J}_{\nu_1\dots\nu_J}\,. \tag{5.9}$$

5.1.1.1 *The photon projection operator*

The sum over the polarisations of the virtual photon which is described by the polarisation vector $\epsilon^{(\gamma^*)}_\mu$ and momentum q ($q^2 \neq 0$) sets up the metric operator:

$$-\sum_{a=1,2,3} \epsilon^{(\gamma^*)a}_\mu \epsilon^{(\gamma^*)a+}_\nu = O^\mu_\nu = g^{\perp q}_{\mu\nu}\,, \qquad g^{\perp q}_{\mu\nu} = g_{\mu\nu} - \frac{q_\mu q_\nu}{q^2}\,. \tag{5.10}$$

The three independent polarisation vectors are orthogonal to the momentum of the particle,

$$q_\mu \epsilon^{(\gamma^*)a}_\mu = 0 \tag{5.11}$$

and are normalised as

$$\epsilon^{(\gamma^*)a+}_\mu \epsilon^{(\gamma^*)b}_\mu = -\delta_{ab}\,. \tag{5.12}$$

A real photon has, however, only two independent polarisations. The invariant expression for the photon projection operator can be constructed only for the photon interacting with another particle. In this case (here we consider the photon–baryon interaction, $\gamma + N \to$ *baryon state*) the completeness condition reads:

$$-\sum_{a=1,2} \epsilon^{(\gamma)a}_\mu \epsilon^{(\gamma)a+}_\nu = g_{\mu\nu} - \frac{P_\mu P_\nu}{P^2} - \frac{k^\perp_\mu k^\perp_\nu}{k^2_\perp} = g^{\perp\perp}_{\mu\nu}(P, p_N)\,. \tag{5.13}$$

In (5.13) the baryon and photon momenta are p_N and q_γ (remind that $q^2_\gamma = 0$), the total momentum is denoted as $P = p_N + q_\gamma$; we have introduced $k = \frac{1}{2}(p_N - q_\gamma)$ and $k^\perp$:

$$k^\perp_\mu \equiv k^{\perp P}_\mu = \frac{1}{2}(p_N - q_\gamma)_\nu g^{\perp P}_{\mu\nu} = \frac{1}{2}(p_N - q_\gamma)_\nu \left(g_{\mu\nu} - \frac{P_\mu P_\nu}{P^2}\right). \tag{5.14}$$

In the c.m. system ($\mathbf{p}_N + \mathbf{q}_\gamma = 0$ and $P = (\sqrt{s}, 0, 0, 0)$), if the momenta of $\mathbf{p}_N$ and $\mathbf{q}_\gamma$ are directed along the z-axis , the metric tensor $g^{\perp\perp}_{\mu\nu}(P, p_N)$ has only two non-zero elements: $g^{\perp\perp}_{xx} = g^{\perp\perp}_{yy} = -1$ (the four-vector components are defined as $p = (p_0, p_x, p_y, p_z)$). For the photon polarisation vector we can use the linear basis: $\epsilon^{(\gamma)x} = (0, 1, 0, 0)$ and $\epsilon^{(\gamma)y} = (0, 0, 1, 0)$, as well as the circular one with helicities ± 1: $\epsilon^{(\gamma)+1} = -(0, 1, +i, 0)/\sqrt{2}$ and $\epsilon^{(\gamma)-1} = (0, 1, -i, 0)/\sqrt{2}$.

The tensor $g^{\perp\perp}_{\mu\nu}(P, p_N)$ acts in the space which is orthogonal to the momenta of both particles, p_N and q_γ, and extracts the gauge invariant part of the amplitude: $A = A_\mu \epsilon^{(\gamma)}_\mu = A_\nu g^{\perp\perp}_{\nu\mu}(P, p_N)\epsilon^{(\gamma)}_\mu$. Indeed, $A_\nu g^{\perp\perp}_{\nu\mu}(P, p_N)$ is gauge invariant: $A_\nu g^{\perp\perp}_{\nu\mu}(P, p_N) q_{\gamma\mu} = 0$.

5.2 Propagators of fermions

Here, using the operator expansion method, we construct the partial wave amplitudes for the production and the decay of baryon resonances.

5.2.1 *The classification of the baryon states*

The baryon states are classified by isospin, total spin and P-parity. The states with isospin $I = 1/2$ are called nucleon states and states with $I = 3/2$ are delta-states. In the literature baryon states are often classified by their decay properties into a nucleon and a pseudoscalar meson: for the sake of simplicity let us consider a πN system. Thus a state called $L_{2I\,2J}$ decays into a nucleon and a pion with the orbital momentum $L = 0, 1, 2, 3, 4, \ldots$, it has an isospin I and a total spin J.

A system of a pseudoscalar meson and a nucleon with orbital momentum L can form a baryon state with total spin either equal to $J = L - 1/2$ or to $J = L + 1/2$ and parity $P = (-1)^{L+1}$. The first set of states is called '−' states and the second set '+' states. For each set, the vertex for the decay of a baryon into a pion-nucleon system is formed by the same convolution of the spin operators (Dirac matrices) and orbital momentum operators. In the nucleon sector the '−' states are:

$$I\, J^P(L_{2I\,2J}) = \frac{1}{2}\frac{1}{2}^+ (P_{11}),\ \frac{1}{2}\frac{3}{2}^- (D_{13}),\ \frac{1}{2}\frac{5}{2}^+ (F_{15}),\ \frac{1}{2}\frac{7}{2}^- (G_{17}), \ldots$$

and the '+' states:

$$I\, J^P(L_{2I\,2J}) = \frac{1}{2}\frac{1}{2}^- (S_{11}),\ \frac{1}{2}\frac{3}{2}^+ (P_{13}),\ \frac{1}{2}\frac{5}{2}^- (D_{15}),\ \frac{1}{2}\frac{7}{2}^+ (F_{17}), \ldots$$

5.2.2 *Spin-1/2 wave functions*

We work with baryon wave functions $\psi(p)$ and $\bar{\psi}(p) = \psi^+(p)\gamma_0$ which obey the Dirac equation

$$(\hat{p} - m)\psi(p) = 0, \qquad \bar{\psi}(p)(\hat{p} - m) = 0. \tag{5.15}$$

The following γ-matrices are used:

$$\begin{aligned}
\gamma_0 &= \begin{pmatrix} I & 0 \\ 0 & -I \end{pmatrix}, \quad \boldsymbol{\gamma} = \begin{pmatrix} 0 & \boldsymbol{\sigma} \\ -\boldsymbol{\sigma} & 0 \end{pmatrix}, \\
\gamma_5 &= i\gamma_1\gamma_2\gamma_3\gamma_0 = -\begin{pmatrix} 0 & I \\ I & 0 \end{pmatrix}, \\
\gamma_0^+ &= \gamma_0\,, \quad \boldsymbol{\gamma}^+ = -\boldsymbol{\gamma}\,,
\end{aligned} \tag{5.16}$$

and the standard Pauli matrices:

$$\sigma_1 = \begin{pmatrix} 0 & 1 \\ 1 & 0 \end{pmatrix}, \quad \sigma_2 = \begin{pmatrix} 0 & -i \\ i & 0 \end{pmatrix}, \quad \sigma_3 = \begin{pmatrix} 1 & 0 \\ 0 & -1 \end{pmatrix}, \tag{5.17}$$

$$\sigma_a\sigma_b = I_{ab} + i\varepsilon_{abc}\sigma_c\,.$$

The solution of the Dirac equation gives us four wave functions:

$$\begin{aligned}
j = 1,2: \quad & \psi_j(p) = \sqrt{p_0 + m}\begin{pmatrix} \varphi_j \\ \frac{(\boldsymbol{\sigma}\mathbf{p})}{p_0+m}\,\varphi_j \end{pmatrix}, \\
& \bar{\psi}_j(p) = \sqrt{p_0 + m}\left(\varphi_j^+, -\varphi_j^+\frac{(\boldsymbol{\sigma}\mathbf{p})}{p_0+m}\right), \\
j = 3,4: \quad & \psi_j(-p) = i\sqrt{p_0 + m}\begin{pmatrix} \frac{(\boldsymbol{\sigma}\mathbf{p})}{p_0+m}\,\chi_j \\ \chi_j \end{pmatrix}, \\
& \bar{\psi}_j(-p) = -i\sqrt{p_0 + m}\left(\chi_j^+\frac{(\boldsymbol{\sigma}\mathbf{p})}{p_0+m}, -\chi_j^+\right),
\end{aligned} \tag{5.18}$$

where φ_j and χ_j are two-component spinors,

$$\varphi_j = \begin{pmatrix} \varphi_{j1} \\ \varphi_{j2} \end{pmatrix}, \qquad \chi_j = \begin{pmatrix} \chi_{j1} \\ \chi_{j2} \end{pmatrix}, \tag{5.19}$$

normalised as

$$\varphi_j^+\varphi_\ell = \delta_{j\ell}, \quad \chi_j^+\chi_\ell = \delta_{j\ell}\,. \tag{5.20}$$

Solutions with $j = 3,4$ refer to antibaryons. The corresponding wave function is defined as

$$j = 3,4: \quad \psi_j^c(p) = C\bar{\psi}_j^T(-p)\,, \tag{5.21}$$

where the matrix C obeys the requirement

$$C^{-1}\gamma_\mu C = -\gamma_\mu^T \ . \tag{5.22}$$

One can use

$$C = \gamma_2\gamma_0 = \begin{pmatrix} 0 & -\sigma_2 \\ -\sigma_2 & 0 \end{pmatrix} . \tag{5.23}$$

We see that

$$C^{-1} = C = C^+ , \tag{5.24}$$

and $\psi_j^c(p)$ satisfies the equation:

$$(\hat{p} - m)\psi_j^c(p) = 0 \ . \tag{5.25}$$

Let us present $\psi_j^c(p)$ defined by (5.21) in more detail:

$$j = 3,4 : \psi_j^c(p) = \begin{pmatrix} 0 & -\sigma_2 \\ -\sigma_2 & 0 \end{pmatrix} (-i)\sqrt{p_0 + m} \begin{pmatrix} \frac{(\boldsymbol{\sigma}^T \boldsymbol{p})}{p_0+m}\ \chi_j^* \\ -\chi_j^* \end{pmatrix} \tag{5.26}$$

$$= -\sqrt{p_0 + m} \begin{pmatrix} \sigma_2\chi_j^* \\ \frac{(\boldsymbol{\sigma}\boldsymbol{p})}{p_0+m}\ \sigma_2\chi_j^* \end{pmatrix} = \sqrt{p_0 + m} \begin{pmatrix} \varphi_j^c \\ \frac{(\boldsymbol{\sigma}\boldsymbol{p})}{p_0+m}\ \varphi_j^c \end{pmatrix} .$$

In (5.26) we have used the commutator $-\sigma_2(\boldsymbol{\sigma}^T\boldsymbol{p}) = \sigma_1 p_1 \sigma_2 + \sigma_2 p_2 \sigma_2 + \sigma_3 p_3 \sigma_2 = (\boldsymbol{\sigma}\boldsymbol{p})\sigma_2$. Also, we defined the spinor for the antibaryon as

$$\varphi_j^c = -i\sigma_2\chi_j^* = \begin{pmatrix} 0 & -1 \\ 1 & 0 \end{pmatrix} \chi_j^* = \begin{pmatrix} -\chi_{j2}^* \\ \chi_{j1}^* \end{pmatrix} . \tag{5.27}$$

Wave functions defined by (5.18) are normalised as follows:

$$\begin{aligned} j,\ell = 1,2 : \quad & (\bar{\psi}_j(p)\psi_\ell(p)) = 2m\ \delta_{j\ell}, \\ j,\ell = 3,4 : \quad & (\bar{\psi}_j(p)\psi_\ell(p)) = -2m\ \delta_{j\ell}, \end{aligned} \tag{5.28}$$

and, after summing over polarisations, they obey the completeness conditions:

$$\begin{aligned} \sum_{j=1,2} \psi_{j\alpha}(p)\,\bar{\psi}_{j\beta}(p) \ &= \ (\hat{p} + m)_{\alpha\beta} \ , \\ \sum_{j=3,4} \psi_{j\alpha}(p)\,\bar{\psi}_{j\beta}(p) \ &= \ -(\hat{p} + m)_{\alpha\beta} \ . \end{aligned} \tag{5.29}$$

As an example, let us present the calculation of the normalisation conditions (5.28) in more detail:

$$\begin{aligned} j,\ell = 1,2 : \quad & (\bar{\psi}_j(p)\psi_\ell(p)) = (p_0 + m)\left(\varphi_j^+\varphi_\ell - \varphi_j^+ \frac{(\boldsymbol{\sigma}\boldsymbol{p})}{p_0+m}\frac{(\boldsymbol{\sigma}\boldsymbol{p})}{p_0+m}\,\varphi_\ell\right) \\ &= \frac{p_0^2 + 2p_0 m + m^2 - \boldsymbol{p}^2}{p_0 + m} = \frac{2m^2 + 2p_0 m}{p_0 + m} = 2m\ \delta_{j\ell} \ , \\ j,\ell = 3,4 : \quad & (\bar{\psi}_j(p)\psi_\ell(p)) = (p_0 + m)\left(\chi_j^+ \frac{(\boldsymbol{\sigma}\boldsymbol{p})}{p_0+m}\frac{(\boldsymbol{\sigma}\boldsymbol{p})}{p_0+m}\,\chi_\ell - \chi_j^+\chi_\ell\right) \\ &= -2m\delta_{j\ell} \ . \end{aligned} \tag{5.30}$$

Sometimes it is more convenient to use the four-component spinors with a different normalisation, substituting $\psi(p) \to u(p)$:

$$(\bar{u}(p)_j u_\ell(p)) = -(\bar{u}_j(-p)u_\ell(-p)) = \delta_{j\ell} ,$$
$$\sum_{j=1,2} u_j(p)\bar{u}_j(p) = \frac{m+\hat{p}}{2m} ,$$
$$\sum_{j=3,4} u_j(-p)\bar{u}_j(-p) = \frac{-m+\hat{p}}{2m} . \tag{5.31}$$

Below we use both types of four-component spinors, $\psi(p)$ and $u(p)$.

5.2.3 *Spin-3/2 wave functions*

To describe Δ and $\bar{\Delta}$, we use the wave functions $\psi_\mu(p)$ and $\bar{\psi}_\mu(p) = \psi^+_\mu(p)\gamma_0$ which satisfy the following constraints:

$$(\hat{p}-m)\psi_\mu(p) = 0, \qquad \bar{\psi}_\mu(p)(\hat{p}-m) = 0,$$
$$p_\mu\psi_\mu(p) = 0, \qquad \gamma_\mu\psi_\mu(p) = 0 . \tag{5.32}$$

Here $\psi_\mu(p)$ is a four-component spinor and μ is a four-vector index. Sometimes, to underline spin variables, we use the notation $\psi_\mu(p;a)$ for the spin-$\frac{3}{2}$ wave functions.

5.2.3.1 *Wave function for* Δ

The equation (5.32) gives four wave functions for the Δ:

$$a=1,2: \quad \psi_\mu(p;a) = \sqrt{p_0+m}\begin{pmatrix} \varphi_{\mu\perp}(a) \\ \frac{(\boldsymbol{\sigma p})}{p_0+m}\varphi_{\mu\perp}(a) \end{pmatrix} ,$$
$$\bar{\psi}_\mu(p;a) = \sqrt{p_0+m}\left(\varphi^+_{\mu\perp}(a), -\varphi^+_{\mu\perp}(a)\frac{(\boldsymbol{\sigma p})}{p_0+m}\right) , \tag{5.33}$$

where the spinors $\varphi_{\mu\perp}(a)$ are determined to be perpendicular to p_μ:

$$\varphi_{\mu\perp}(a) = g^{\perp p}_{\mu\mu'}\varphi_{\mu'}(a), \quad g^{\perp p}_{\mu\mu'} = g_{\mu\mu'} - p_\mu p_{\mu'}/p^2 . \tag{5.34}$$

The requirement $\gamma_\mu\psi_\mu(p;a)$ results in the following constraints for $\varphi_{\mu\perp}$:

$$m\varphi_{0\perp}(a) = (\boldsymbol{p}\boldsymbol{\varphi}_\perp(a)) ,$$
$$m(p_0+m)(\boldsymbol{\sigma}\boldsymbol{\varphi}_\perp(a)) + (\boldsymbol{p\sigma})(\boldsymbol{p}\boldsymbol{\varphi}_\perp(a)) = 0. \tag{5.35}$$

In the limit $p \to 0$ (the Δ at rest), we have:

$$m\varphi_{0\perp}(a) = 0 ,$$
$$(\boldsymbol{\sigma}\boldsymbol{\varphi}_\perp(a)) = 0, \tag{5.36}$$

thus keeping for Δ four independent spin components $\mu_z = 3/2, 1/2, -1/2, -3/2$ related to the spin $S = 3/2$ and removing the components with $S = 1/2$.

The completeness conditions for the spin-$\frac{3}{2}$ wave functions can be written as follows:

$$\sum_{a=1,2} \psi_\mu(p;a)\,\bar{\psi}_\nu(p;a) = (\hat{p}+m)\left(-g^\perp_{\mu\nu} + \frac{1}{3}\gamma^\perp_\mu\gamma^\perp_\nu\right)$$
$$= (\hat{p}+m)\frac{2}{3}\left(-g^\perp_{\mu\nu} + \frac{1}{2}\sigma^\perp_{\mu\nu}\right), \qquad (5.37)$$

where $g^\perp_{\mu\nu} \equiv g^{\perp p}_{\mu\nu}$ and $\gamma^\perp_\mu = g^{\perp p}_{\mu\mu'}\gamma_{\mu'}$. The factor $(\hat{p}+m)$ commutates with $(g^\perp_{\mu\nu} - \frac{1}{3}\gamma^\perp_\mu\gamma^\perp_\nu)$ in (5.37) because $\hat{p}\gamma^\perp_\mu\gamma^\perp_\nu = \gamma^\perp_\mu\gamma^\perp_\nu\hat{p}$. The matrix $\sigma^\perp_{\mu\nu}$ is determined in a standard way, $\sigma^\perp_{\mu\nu} = \frac{1}{2}(\gamma^\perp_\mu\gamma^\perp_\nu - \gamma^\perp_\nu\gamma^\perp_\mu)$.

5.2.3.2 *Wave function for* $\bar{\Delta}$

The anti-delta, $\bar{\Delta}$, is determined by the following four wave functions:

$$b = 3,4: \ \psi_\mu(-p;b) = i\sqrt{p_0+m}\begin{pmatrix} \frac{(\boldsymbol{\sigma p})}{p_0+m}\,\chi_{\mu\perp}(b) \\ \chi_{\mu\perp}(b)\end{pmatrix},$$
$$\bar{\psi}_\mu(-p;b) = -i\sqrt{p_0+m}\left(\chi^+_{\mu\perp}(b)\frac{(\boldsymbol{\sigma p})}{p_0+m}, -\chi^+_{\mu\perp}(b)\right), \qquad (5.38)$$

where in the system at rest $(p \to 0)$ the spinors $\chi_{\mu\perp}(b)$ obey the relations:

$$m\chi_{0\perp}(b) = 0\,, \quad (\boldsymbol{\sigma\chi}_\perp(b)) = 0, \qquad (5.39)$$

that take away the spin-$\frac{1}{2}$ components.

The completeness conditions for spin-$\frac{3}{2}$ wave functions with $b = 3,4$ are

$$\sum_{b=3,4} \psi_\mu(-p;b)\,\bar{\psi}_\nu(-p) = -(\hat{p}+m)\left(-g^\perp_{\mu\nu} + \frac{1}{3}\gamma^\perp_\mu\gamma^\perp_\nu\right)$$
$$= -(\hat{p}+m)\frac{2}{3}\left(-g^\perp_{\mu\nu} + \frac{1}{2}\sigma^\perp_{\mu\nu}\right). \qquad (5.40)$$

The equation (5.38) can be rewritten in the form of (5.33) using the charge conjugation matrix C which was introduced for spin-$\frac{1}{2}$ particles, $C = \gamma_2\gamma_0$. It satisfies the relations $C^{-1}\gamma_\mu C = -\gamma^T_\mu$ and $C^{-1} = C = C^+$. We write:

$$b = 3,4: \psi^c_\mu(p;b) \ = \ C\bar{\psi}^T_\mu(-p;b). \qquad (5.41)$$

The wave functions $\psi^c_\mu(p;b)$ with $b = 3,4$ obey the equation:

$$(\hat{p}-m)\psi^c_\mu(p;b) = 0\,. \qquad (5.42)$$

In the explicit form the charge conjugated wave functions read:

$$b = 3,4 : \psi^c_\mu(p;b) = -\sqrt{p_0+m}\begin{pmatrix} \sigma_2\chi^*_{\mu\perp}(b) \\ \frac{(\boldsymbol{\sigma p})}{p_0+m}\,\sigma_2\chi^*_{\mu\perp}(b)\end{pmatrix}$$
$$= \sqrt{p_0+m}\begin{pmatrix} \varphi^c_{\mu\perp}(b) \\ \frac{(\boldsymbol{\sigma p})}{p_0+m}\,\varphi^c_{\mu\perp}(b)\end{pmatrix}, \qquad (5.43)$$

with $\varphi^c_{\mu\perp}(b) = -\sigma_2\chi^*_{\mu\perp}(b)$.

5.2.3.3 *Baryon projection operators*

In this chapter it is convenient to use the baryon wave functions introduced in Chapter 4, $u_j(p)$ and $\bar{u}_j(p)$, which are normalised as $\bar{u}_j(p)u_\ell(p) = \delta_{j\ell}$ and obey the completeness condition $\sum_{j=1,2} u_j(p)\bar{u}_j(p) = (m+\hat{p})/2m$. For a baryon with fixed polarisation one has:

$$u_S(p)\bar{u}_S(p) = \frac{m+\hat{p}}{2m}\left(\frac{1}{2} - \frac{1}{2}\gamma_5\hat{S}\right), \qquad (5.44)$$

with the following constraints for the polarisation vector S_μ:

$$S^2 = -1, \quad (pS) = 0. \qquad (5.45)$$

5.2.3.4 *Projection operators for particles with $J > 1/2$.*

The wave function of a particle with spin $J = n + 1/2$, momentum p and mass m is given by a tensor four-spinor $\Psi_{\mu_1\ldots\mu_n}$. It satisfies the constraints

$$(\hat{p} - m)\Psi_{\mu_1\ldots\mu_n} = 0, \quad p_{\mu_i}\Psi_{\mu_1\ldots\mu_n} = 0, \quad \gamma_{\mu_i}\Psi_{\mu_1\ldots\mu_n} = 0, \qquad (5.46)$$

and the symmetry properties

$$\Psi_{\mu_1\ldots\mu_i\ldots\mu_j\ldots\mu_n} = \Psi_{\mu_1\ldots\mu_j\ldots\mu_i\ldots\mu_n},$$
$$g_{\mu_i\mu_j}\Psi_{\mu_1\ldots\mu_i\ldots\mu_j\ldots\mu_n} = g^{\perp p}_{\mu_i\mu_j}\Psi_{\mu_1\ldots\mu_i\ldots\mu_j\ldots\mu_n} = 0. \qquad (5.47)$$

Conditions (5.46), (5.47) define the structure of the denominator of the fermion propagator (the projection operator) which can be written in the following form:

$$F^{\mu_1\ldots\mu_n}_{\nu_1\ldots\nu_n}(p) = (-1)^n\frac{m+\hat{p}}{2m}R^{\mu_1\ldots\mu_n}_{\nu_1\ldots\nu_n}(\perp p)\,. \qquad (5.48)$$

The operator $R^{\mu_1\ldots\mu_n}_{\nu_1\ldots\nu_n}(\perp p)$ describes the tensor structure of the propagator. It is equal to 1 for a ($J = 1/2$)-particle and is proportional to $g^{\perp p}_{\mu\nu} - \gamma^\perp_\mu\gamma^\perp_\nu/3$ for a particle with spin $J = 3/2$ (remind that $\gamma^\perp_\mu = g^{\perp p}_{\mu\nu}\gamma_\nu$, see Chapter 4, Subsection 4.3.1).

The conditions (5.47) are identical for fermion and boson projection operators and therefore the fermion projection operator can be written as:

$$R^{\mu_1\ldots\mu_n}_{\nu_1\ldots\nu_n}(\perp p) = O^{\mu_1\ldots\mu_n}_{\alpha_1\ldots\alpha_n}(\perp p) T^{\alpha_1\ldots\alpha_n}_{\beta_1\ldots\beta_n}(\perp p) O^{\beta_1\ldots\beta_n}_{\nu_1\ldots\nu_n}(\perp p)\,. \tag{5.49}$$

The operator $T^{\alpha_1\ldots\alpha_n}_{\beta_1\ldots\beta_n}(\perp p)$ can be expressed in a rather simple form since all symmetry and orthogonality conditions are imposed by O-operators. First, the T-operator is constructed of metric tensors only, which act in the space of $\perp p$ and $\gamma^\perp$-matrices. Second, a construction like $\gamma^\perp_{\alpha_i}\gamma^\perp_{\alpha_j} = \frac{1}{2}g^\perp_{\alpha_i\alpha_j} + \sigma^\perp_{\alpha_i\alpha_j}$ (remind that here $\sigma^\perp_{\alpha_i\alpha_j} = \frac{1}{2}(\gamma^\perp_{\alpha_i}\gamma^\perp_{\alpha_j} - \gamma^\perp_{\alpha_j}\gamma^\perp_{\alpha_i})$ gives zero if multiplied by an $O^{\mu_1\ldots\mu_n}_{\alpha_1\ldots\alpha_n}$-operator: the first term is due to the tracelessness conditions and the second one to symmetry properties. The only structures which can then be constructed are $g^\perp_{\alpha_i\beta_j}$ and $\sigma^\perp_{\alpha_i\beta_j}$. Moreover, taking into account the symmetry properties of the O-operators, one can use any pair of indices from sets $\alpha_1\ldots\alpha_n$ and $\beta_1\ldots\beta_n$, for example, $\alpha_i \to \alpha_1$ and $\beta_j \to \beta_1$. Then

$$T^{\alpha_1\ldots\alpha_n}_{\beta_1\ldots\beta_n}(\perp p) = \frac{n+1}{2n+1}\left(g^\perp_{\alpha_1\beta_1} - \frac{n}{n+1}\sigma^\perp_{\alpha_1\beta_1}\right)\prod_{i=2}^{n} g^\perp_{\alpha_i\beta_i}\,. \tag{5.50}$$

Since $R^{\mu_1\ldots\mu_n}_{\nu_1\ldots\nu_n}(\perp p)$ is determined by convolutions of O-operators, see Eq. (5.49), we can replace in (5.49)

$$T^{\alpha_1\ldots\alpha_n}_{\beta_1\ldots\beta_n}(\perp p) \to T^{\alpha_1\ldots\alpha_n}_{\beta_1\ldots\beta_n}(p) = \frac{n+1}{2n+1}\left(g_{\alpha_1\beta_1} - \frac{n}{n+1}\sigma_{\alpha_1\beta_1}\right)\prod_{i=2}^{n} g_{\alpha_i\beta_i}\,. \tag{5.51}$$

The coefficients in (5.51) are chosen to satisfy the constraints (5.46) and the convolution condition:

$$F^{\mu_1\ldots\mu_n}_{\alpha_1\ldots\alpha_n}(p) F^{\alpha_1\ldots\alpha_n}_{\nu_1\ldots\nu_n}(p) = (-1)^n F^{\mu_1\ldots\mu_n}_{\nu_1\ldots\nu_n}(p)\,. \tag{5.52}$$

5.3 Spectral integral equations for the coupled three-meson decay channels in $\bar{p}p$ ($J^{PC} = 0^{-+}$) annihilation at rest

The lowest positronium-like states $\bar{p}p(0^{-+})$ have isospins $I = 0$ and $I = 1$. Due to isospin conservation in strong interactions there exist two sets of the coupled decay channels:

$$\begin{aligned} \bar{p}p(IJ^{PC} = 10^{-+}) &\to \pi^0\pi^0\pi^0,\ \eta\eta\pi^0,\ \bar{K}K\pi^0 \\ \bar{p}p(IJ^{PC} = 00^{-+}) &\to \eta\pi^0\pi^0,\ \bar{K}K\pi^0\,. \end{aligned} \tag{5.53}$$

We present spectral integral equations for the decay amplitude of an $I = 1$ state. The set of equations with $I = 0$ can be considered quite similarly.

The high statistics data on three-meson production from the $\bar{p}p$ annihilation at rest (from the lowest 0^{-+}-level) were presented by Crystal Barrel Collaboration (CERN). The data were successfully analyzed (see reviews [4, 5] and, for example, publications [8, 9, 10]) with searching for new meson resonances in the region 1000—2300 MeV. The K-matrix formalism as well as the isobar model with elements of the dispersion relation N/D-method were used. There were expectations [11, 12, 13] that the lowest scalar and tensor glueballs are located in this region. Thus, the identification of scalar resonances in the mass region 800—2300 MeV and their classification as quark-antiquark states ($\bar{q}q$) or gluoniums (gg) is the decisive step in low energy physics. The quark/qluon structure of these resonances can be determined from the analysis of coupling constants of these states to pseudoscalar and vector mesons based on quark combinatorics [2, 14]. Such analyses were performed for scalar mesons [15] and tensor ones [16] and indicated on the glueball nature of the broad state $f_0(1200-1500)$ and resonance $f_2(\sim 2000)$. The question is whether the K-matrix approximation – or a simplified N/D-method – is sufficient for this purpose. The way to give an answer to these justified doubts lays in applying spectral integral technique and three-particle equations to analysis of data.

5.3.1 *The S-P-D-wave meson rescatterings*

Three-particle forces being neglected, the amplitude $A_{p\bar{p}\to 123}$ for the three-particle production is a sum of four terms:
(i) the direct production amplitude $\lambda(s_{12}, s_{13}, s_{23})$, which is free of singularities;
(ii) three amplitudes A_{ij}, where the last interaction is that of particles i and j.

We take into account final-state interactions with different orbital momenta L, hence the production amplitude can be expressed through the L-wave amplitudes $A^{(L)}(s_{12}, s_{13}, s_{23},)$ and can written as follows:

$$A_{p\bar{p}\to 123} = \lambda(s_{12}, s_{13}, s_{23}) + \sum_{L=0,1,2}\sum_{\ell\neq i\neq j} F_L(k^\perp_\ell, k^\perp_{ij}) A^{(L)}_{ij}(s_{ij}),$$
$$F_0(k^\perp_\ell, k^\perp_{ij}) = 1\,, \quad F_1(k^\perp_\ell, k^\perp_{ij}) = k^\perp_{\ell\mu} k^\perp_{ij\mu}\,,$$
$$F_2(k^\perp_\ell, k^\perp_{ij}) = k^\perp_{\ell\mu} k^\perp_{\ell\nu}\left(k^\perp_{ij\mu} k^\perp_{ij\nu} - \frac{1}{3} g^\perp_{\mu\nu} k^{\perp 2}_{ij}\right). \tag{5.54}$$

The function $F_L(k^\perp_\ell, k^\perp_{ij})$ defines the angular distribution of particles in the final state. We mean here that $k^\perp_\ell \perp (k_i + k_j)$, $k^\perp_{ij} \perp (k_i + k_j)$ and

$g^{\perp}_{\mu\nu}(k_i+k_j)_\nu = 0$. For the left-hand side angular momentum operator we use $k_\ell^\perp \perp (k_i+k_j)$ – such a possibility was discussed in Chapter 3.

In order to avoid a double account, it is safer not to include here the analytic part of the amplitude, $\lambda(s_{12}, s_{13}, s_{23})$, into the expansion over L. The singular terms in the right-hand side of (5.54) describe rescatterings of outgoing particles.

Let us consider in (5.54) the term $A^{(L)}_{12}(s_{12})$ with the last S-wave binary interaction in the 12 channel. We have $F_0 = 1$ and the following equation for $A^{(0)}_{12}(s_{12})$:

$$A^{(0)}_{12}(s_{12}) = \lambda^{(0)}_{12}(s_{12})\frac{B_0(s_{12})}{1-B_0(s_{12})} + \frac{1}{1-B_0(s_{12})} \times$$
$$\times \int\limits_{(m_1+m_2)^2}^{\infty} \frac{ds'_{12}}{\pi}\frac{\rho_{12}(s'_{12})G^L_0(s'_{12})}{s'_{12}-s_{12}}\left(\left\langle A^{(0)}_{13}(s'_{13})\right\rangle^{(0)}_{12} + \left\langle A^{(0)}_{23}(s'_{23})\right\rangle^{(0)}_{12}\right). \quad (5.55)$$

As previously, the integration over the angular variables reads:

$$\left\langle \dots \right\rangle^{(0)}_{12} = \int\limits_{-1}^{1}\frac{dz_{13}}{2} \to \int\limits_{C(s_{13})}\frac{dz_{13}}{2} = \int\limits_{s_{13-}}^{s_{13+}}\frac{ds_{13}}{4\mid \mathbf{k}_1 \mid\mid \mathbf{k}_3 \mid}$$

$$s_{13\pm} = m_1^2 + m_3^2 - 2k_{10}k_{30} \pm 2\mid \mathbf{k}_1 \mid\mid \mathbf{k}_3 \mid ,$$
$$k_{10} = \frac{s_{12}+m_1^2-m_2^2}{2\sqrt{s_{12}}}, \qquad \mid \mathbf{k}_1 \mid = \sqrt{k_{10}^2 - m_1^2},$$
$$k_{30} = \frac{s_{12}+m_3^2-s}{2\sqrt{s_{12}}}, \qquad \mid \mathbf{k}_3 \mid = \sqrt{k_{30}^2 - m_3^2}. \quad (5.56)$$

The replacement $1 \leftrightarrow 2$ gives us the integration over $A^{(0)}_{23}(s_{23})$ (second term in the integral (5.55)).

Here we would like to repeat that at small s_{12}, the integral over the angle variables in (5.55) coincides completely with the phase-space integration contour $-1 \le z_{13} \le 1$, while it contains an additional piece at $s_{12} \ge s\frac{m_1}{m_1+m_3} + \frac{m_3}{m_1+m_3}m_2^2 - m_1m_3$, – this point was considered in Chapter 3 (see Fig. 3.3 and correspnding discussion).

Dispersion relation equations for interactions with $L > 0$ are written analogously. Nevertheless, to be specified we present below the cases with P-wave and D-wave interactions.

5.3.1.1 *P-wave interaction in the final state*

We turn now to the amplitude where particles 1 and 2 interact in the P-wave while the particle 3 is a spectator. The assumed form of the amplitude reads:

$$X^{(1)}_\mu(k_3^\perp)X^{(1)}_\mu(k_{12}^\perp)A^{(1)}_{12}(s_{12}), \tag{5.57}$$

where the the angular momentum operators describe the P-wave angular distribution of particles 1 and 2 in the final state, it is defined as the relative momentum of these particles

$$X^{(1)}_\mu(k_{12}^\perp) = k_{1\mu} - k_{2\mu} - \frac{m_1^2 - m_2^2}{s_{12}}(k_1 + k_2)_\mu \equiv k_{12\mu}^\perp. \tag{5.58}$$

The operator $X^{(1)}_\mu(k_3^\perp)$ should be constructed as a relative momentum of the initial state and the antiparticle 3. Taking into account that $X^{(1)}_\mu(k_{12}^\perp)\ (k_1 + k_2)_\mu = 0$, we define:

$$X^{(1)}_\mu(k_3^\perp) = k_{3\mu}^\perp, \quad \text{and} \quad F_1(k_3^\perp k_{12}^\perp) = X^{(1)}_\mu(k_3^\perp)X^{(1)}_\mu(k_{12}^\perp) \tag{5.59}$$

The product $F_1(k_3^\perp k_{12}^\perp)$ is proportional to z_{13} in the c.m.s. of particles 1 and 2.

First, the P-wave two-particle scattering amplitude of particles i and j is written as

$$A^{(1)}_{2\to 2}(s_{ij}) = X^{(1)}_\mu(k_{ij}^\perp)\frac{G_1^L(s_{ij})}{1 - B_1(s_{ij})}X^{(1)}_\mu(k_{ij}^\perp),$$
$$B_1(s_{ij}) = \int\limits_{(m_i+m_j)^2}^{\infty} \frac{ds'}{\pi}\frac{G_1^L(s')\rho_{ij}(s')k_{ij}'^{\perp 2}}{s' - s_{ij} - i0}, \tag{5.60}$$

As the next step in the presentation of the three-particle production amplitude, one should consider a triangle diagram with one P-wave rescattering of particles i and j (let to be $i = 1$ and $j = 2$). The discontinuity of the triangle diagram is equal to:

$$k_{3\mu}^\perp \mathrm{disc}_{12} B^{(1)}_{13\to 12}(s_{12}) = \int d\Phi_{12}(k_1, k_2)A_{13}(s_{13}, z_{13})X^{(1)}_\mu(k_{12}^\perp)G_1^L(s_{12})$$
$$= G_1^L(s_{12})\rho_{12}(s_{12})\int\frac{d\Omega}{4\pi}k_{12\mu}A_{13}(s_{13}, z_{13}). \tag{5.61}$$

The integration over the space angle is performed in the c.m. frame of particles 1 and 2. In this frame $X^{(1)}_\mu(k_{12}^\perp)$ turns into $\mathbf{k}_{12}$. The z-axis being directed along $\mathbf{k}_3$, the components of $\mathbf{k}_{12}$ are equal to

$$k_{12x} = k_{12}\sin\theta_{13}\cos\phi, \quad k_{12y} = k_{12}\sin\theta_{13}\sin\phi, \quad k_{12z} = k_{12}\cos\theta_{13},$$
$$k_{12} = \sqrt{\frac{1}{s_{12}}[s_{12} - (m_1 + m_2)^2][s_{12} - (m_1 - m_2)^2]}, \tag{5.62}$$

where ϕ is the azimuthal angle. The integration over ϕ in (5.61) keeps the components of $X^{(1)}_\mu$ with $\mu = z$ only:

$$\int \frac{d\Omega}{4\pi} X^{(1)}_x \ldots = \int \frac{d\Omega}{4\pi} X^{(1)}_y \ldots = 0,$$
$$\int \frac{d\Omega}{4\pi} X^{(1)}_z \ldots = k_{12} \int \frac{dz_{13}}{2} z_{13} \ldots \tag{5.63}$$

Let us introduce the vector $k_3^\perp$ with only the z component being different from zero in the c.m. frame:

$$k^\perp_{3\mu} = k_{3\mu} - p_{12\mu}\frac{(p_{12}\, k_3)}{p^2_{12}}, \qquad p_{12} = k_1 + k_2\,,$$
$$-(k_3^\perp)^2 = \frac{[(M-m_3)^2 - s_{12}][(M+m_3)^2 - s_{12}]}{4s_{12}}. \tag{5.64}$$

Then it follows from (5.61):

$$k^\perp_{3\mu}\mathrm{disc}_{12} B^{(1)}_{13\to 12}(s_{12}) = k^\perp_{3\mu}\rho_{12}(s_{12}) G^L_1(s_{12}) \frac{k_{12}}{\sqrt{-(k_3^\perp)^2}}$$
$$\times \int\limits_{C_3(s_{12})} \frac{dz_{13}}{2}\, z_{13}\, A_{13}(s'_{13}, z_{13}), \tag{5.65}$$

The invariant part of the P-wave amplitude can be written as a dispersion integral over the energy squared of particles 1 and 2 in the intermediate state, while $k_3^\perp$ in (5.65) defines the operator structure of the P-wave amplitude. Thus, the P-wave triangle diagram is equal to

$$k^\perp_{3\mu} \int\limits_{(m_1+m_2)^2}^{\infty} \frac{ds'_{12}}{\pi} \frac{\rho_{12}(s'_{12}) G^L_1(s'_{12})}{s'_{12} - s_{12}} \Big\langle A_{13}(s'_{13}, z)\Big\rangle^{(1)}_{12},$$
$$\Big\langle A_{13}(s'_{13}, z)\Big\rangle^{(1)}_{12} = \frac{k'_{12}}{\sqrt{-(k'^\perp_3)^2}} \int\limits_{C_3(s'_{12})} \frac{dz}{2} z\, A_{13}(s'_{13}, z). \tag{5.66}$$

The amplitude (5.66) should be multiplied by the operator $X^{(1)}_\mu = k_{12\mu}$ which describes the angular distribution of particles 1 and 2 in the final state. To take into account binary rescattering in the final state it is necessary to multiply (5.66) by the factor $(1 - B_1(s_{12}))^{-1}$.

The same steps should be done, if A_{13} is replaced by A_{23} and the direct production term $\lambda(s_{12}, s_{13}, s_{23})$. Then the following integral equation for the amplitude $A^{(1)}_{12}$ can be written:

$$A^{(1)}_{12}(s_{12}) = \lambda^{(1)}_{12}(s_{12}) \frac{B_1(s_{12})}{1 - B_1(s_{12})} + \frac{1}{1 - B_1(s_{12})} \tag{5.67}$$
$$\times \int\limits_{(m_1+m_2)^2}^{\infty} \frac{ds'_{12}}{\pi} \frac{\rho_{12}(s'_{12}) G^L_1(s'_{12})}{s'_{12} - s_{12} - i0} \left(\Big\langle A_{13}(s'_{13})\Big\rangle^{(1)}_{12} + \Big\langle A_{23}(s'_{23})\Big\rangle^{(1)}_{12}\right).$$

5.3.1.2 *D-wave interaction in the final state*

Likewise, the amplitude with D-wave interaction in the final state may be presented. The amplitude, where particles 1 and 2 have the last interaction in the D-wave, reads:

$$O^{(2)}_{\mu\nu}(k^{\perp}_3)\, A^{(2)}_{12}(s_{12})\, Q^{(2)}_{\mu\nu}(k^{\perp}_{12}) \tag{5.68}$$

$$O^{(2)}_{\mu\nu}(k^{\perp}_3) \;=\; k^{\perp}_{3\mu}k^{\perp}_{3\nu}, \quad Q^{(2)}_{\mu\nu}(k^{\perp}_{12}) = \frac{2}{3}X^{(2)}_{\mu\nu}(k^{\perp}_{12}) = k^{\perp}_{12\mu}k^{\perp}_{12\nu} - \frac{1}{3}k^{\perp 2}_{12}\, g^{\perp}_{\mu\nu}.$$

Note that the operator part of the D-wave decaying amplitude is proportional to $3z^2_{13} - 1$.

The D-wave scattering amplitude of particles 1 and 2 within the operators (5.68) reads:

$$A^{(2)}_{2\to 2}(s_{12}) = Q_{2\mu\nu}(k^{\perp}_{12})\,\frac{G^L_2(s_{12})}{1 - B_2(s_{12})}\,Q_{2\mu\nu}(k^{\perp}_{12}). \tag{5.69}$$

The discontinuity of the triangle diagram with the D-wave rescattering of particles 1 and 2 is equal to

$$\int d\Phi_{12}(k_1,k_2)A_{13}(s_{13},z_{13})Q_{2\mu\nu}(k^{\perp}_{12})G^L_2(s_{12}) = G^L_2(s_{12})\rho_{12}(s_{12}) \times \int \frac{d\Omega}{4\pi}Q_{2\mu\nu}(k^{\perp}_{12})A_{13}(s_{13},z_{13}). \tag{5.70}$$

The integration over the space angle is performed in the c.m. frame of particles 1 and 2. Equation (5.62) is used, and only the following components of tensor $k_{12\mu}k_{12\nu}$ are not equal to zero:

$$\int \frac{d\Omega}{4\pi}\, k_{12x}k_{12x}\ldots = \int \frac{d\Omega}{4\pi}\, k_{12y}k_{12y}\ldots \;=\; k^2_{12}\int \frac{dz_{13}}{4}\,(1-z^2_{13})\ldots$$

$$\int \frac{d\Omega}{4\pi}\, k_{12z}k_{12z}\ldots = k^2_{12}\int \frac{dz_{13}}{2}\, z^2_{13}\ldots \tag{5.71}$$

Introducing

$$\Big\langle A_{13}(s_{13},z)\Big\rangle^{(2)}_{12} \;=\; \frac{k^2_{12}}{(k^{\perp}_3)^2} \int\limits_{C_3(s_{12})} \frac{dz}{4}(1-3z^2)\, A_{13}(s_{13},z), \tag{5.72}$$

we write the invariant part of the D-wave amplitude as a dispersion integral. The full set of binary rescatterings in the final state is defined by the factor $(1 - B_2(s_{12}))^{-1}$. As a result, we have the following integral equation for the amplitude $A^{(2)}_{12}(s_{12})$:

$$A^{(2)}_{12}(s_{12}) = \lambda^D_{12}(s_{12})\frac{B_2(s_{12})}{1-B_2(s_{12})} + \frac{1}{1-B_2(s_{12})} \times \int\limits_{(m_1+m_2)^2}^{\infty} \frac{ds'_{12}}{\pi}\frac{\rho_{12}(s'_{12})G^L_2(s'_{12})}{s'_{12}-s_{12}-i0}\left(\Big\langle A_{13}(s_{13},z)\Big\rangle^{(2)}_{12} + \Big\langle A_{23}(s_{23},z)\Big\rangle^{(2)}_{12}\right), \tag{5.73}$$

where the constant $\lambda^D_{ij}(s_{12})$ stands for the direct production amplitude of particles i and j in the D-wave.

5.3.2 *Equations with inclusion of resonance production*

Let us consider the case when two-particle interaction is determined by the s-channel resonances only.

5.3.2.1 *Two-particle scattering amplitude*

We write a scattering amplitude as follows:

$$A_L = Q_{L,\mu_1,\ldots}(k^\perp)\frac{\sum\limits_\alpha \frac{g^{(\alpha)2}(s)}{M^2_\alpha - s}}{1 - b_L(s)\sum\limits_\alpha \frac{g^{(\alpha)2}}{M^2_\alpha - s}} Q_{L,\mu_1,\ldots}(k^\perp). \tag{5.74}$$

the subtraction procedure in loop diagrams has occurred due to couplings $g^{(\alpha)2}(s)$. Here we suppose an universal s-dependence of couplings:

$$g^{(\alpha)}(s) = g^{(\alpha)}\phi(s), \tag{5.75}$$

that is equivalent to an universal cutting procedure in the loop diagrams:

$$b_L(s) = \int \frac{ds'}{\pi}\frac{\rho(s')\phi^2(s')}{s' - s - i0}\langle Q_{L,\mu}(k'^\perp)Q_{L,\mu}(k'^\perp)\rangle, \tag{5.76}$$

We should keep in mind that $\phi(s)$ can depend on L as well, so $\phi(s) \to \phi_L(s)$.

The multichannel amplitudes can be treated similarly. The decay coupling of resonance α into particles i and j is given by the function

$$g^{(\alpha)}_{ij}(s) = g^{(\alpha)}_{ij}\phi_{ij}(s). \tag{5.77}$$

Then the multichannel scattering amplitude reads:

$$\begin{aligned}
\hat A_L &= Q_{L,\,\mu_1\ldots\mu_L}(k^\perp)\,\hat B\frac{\hat I}{\hat I - \hat b\hat B}Q_{L,\,\mu_1\ldots\mu_L}(k^\perp) \\
\hat B \to B_{ij;i'j'}(s) &= \sum_\alpha \frac{g^{(\alpha)}_{ij}g^{(\alpha)}_{i'j'}}{M^2_\alpha - s}, \\
\hat b \to b_{ij;i'j'}(s) &= \int\limits_{(m_i+m_j)^2}^{\infty}\frac{ds'}{\pi}\frac{\rho_{ij}(s')}{s' - s - i0}\phi^2_{ij}(s')\langle Q^2_{L,\mu}(k'^\perp)\rangle\delta_{ij;i'j'},
\end{aligned} \tag{5.78}$$

The denominator, which describes the rescattering of particles in multichannel case, has the matrix form $(\hat I - \hat b\hat B)^{-1}$, where $\hat I$ is a unit matrix and $\hat b$ is diagonal one.

Actually Eq. (5.78) presents the scattering amplitude in terms of the dispersion relation D-matrix within hypothesis about universal cuttings in loop diagrams (the cutting is determined only on sort of loop diagram particles).

Terms which describe background contribution, if it exists, can be easily included into Eq. (5.78), for example, by adding into $\hat{B}$ poles located far from the region of consideration. The direct way means

$$\sum_{\alpha} \frac{g_{ij}^{(\alpha)} g_{i'j'}^{(\alpha)}}{M_{\alpha}^2 - s} \quad \to \quad \sum_{\alpha} \frac{g_{ij}^{(\alpha)} g_{i'j'}^{(\alpha)}}{M_{\alpha}^2 - s} + f_{ij\,;\,i'j'}(s), \tag{5.79}$$

where $f_{ij\,;\,i'j'}(s)$ is responsible for the background.

5.3.2.2 *Three-particle production amplitude*

Inclusion of the resonance production into three-particle amplitude is performed in analogous way. With notations of Eq. (5.54) we write the spectral integral as

$$\sum_{(i'j')} \lambda^{0J}_{i'j';\ell} \left\{ \hat{B} \frac{\hat{I}}{\hat{I} - \hat{b}\hat{B}} \right\}^{0J}_{i'j';\,ij} + \sum_{(i'j')} \Delta^{0J}_{i'j';\ell} \left\{ \hat{B} \frac{\hat{I}}{\hat{I} - \hat{b}\hat{B}} \right\}^{0J}_{i'j';\,ij} \tag{5.80}$$

Here $\lambda^{0J}_{ij;\ell}$ presents the vector of couplings for direct resonance production. The second vector, $\Delta^{0J}_{i'j';\ell}$, presents a set of the trangle diagrams. The summation over intermediate states with the production of $i'j'$ particles is performed. The expression for the triangle diagram, $\Delta^{0J}_{ij;\ell}$, can be written in a standard way:

$$\begin{aligned} \Delta^{0J}_{ij;\ell}(s_{ij}) &= \sum_{i'j'} \int \frac{ds'_{i'j'}}{\pi} \frac{\rho_{i'j'}(s'_{i'j'}) N^{0J}_{i'j';ij}(s'_{i'j'}, s_{ij})}{s'_{i'j'} - s_{ij} - i0} \\ &\quad \times \left(\Big\langle A_{j'\ell;\,i'}(s'_{i'\ell}, z) \Big\rangle_J + \Big\langle A_{i'\ell;j'}(s'_{j'\ell}, z) \Big\rangle_J \right), \\ N^{0J}_{i'j';ij}(s'_{i'j'}, s_{ij}) &= \phi_{i'j'}(s'_{i'j'}) \phi_{ij}(s_{ij}). \end{aligned} \tag{5.81}$$

The dispersion relation integration over $ds'_{i'j'}$ in (5.81) is performed from the threshold singularity, $(m_{i'} + m_{j'})^2$, to $+\infty$.

5.3.3 *The coupled decay channels* $\bar{p}p(IJ^{PC} = 10^{-+}) \to \pi^0\pi^0\pi^0,\ \eta\eta\pi^0,\ \bar{K}K\pi^0$

Let us write down the integral equations for the decay from the $I = 1$ state.

5.3.3.1 *Reaction* $\bar{p}p\ (10^{-+}) \to \pi^0\pi^0\pi^0$.

In these reactions we should take into account the isospin structure of the amplitude. The annihilation amplitude into $3\pi^0$ reads:

$$M_{p\bar{p}\to\pi^0\pi^0\pi^0} = a_{\pi^0\pi^0;\pi^0}(s_{12},z) + a_{\pi^0\pi^0;\pi^0}(s_{13},z) + a_{\pi^0\pi^0;\pi^0}(s_{23},z),$$
$$a_{\pi^0\pi^0;\pi^0} = a^{(0)}_{\pi\pi;\pi^0} + \frac{4}{3}a^{(2)}_{\pi\pi;\pi^0}, \tag{5.82}$$

and $a^{(I)}_{\pi\pi;\pi^0}$ is the amplitude with the last pions interacting in the isospin state I. It should be noted that, due to the C-invariance, there is no term in (5.82) with pion interactions in the state $I = 1$. For the channel with isospin $I = 0$ S- and D-wave interactions are taken into account, that allows one to calculate properly the production of f_0 and f_2 resonances. So

$$a^{(0)}_{\pi\pi;\pi^0} = A^{00}_{\pi\pi;\pi^0} + F_2 A^{02}_{\pi\pi;\pi^0}. \tag{5.83}$$

In the channel with isospin $I = 2$ the S-wave interaction is accounted only, therefore

$$a^{(2)}_{\pi\pi;\pi^0} = A^{20}_{\pi\pi;\pi^0}. \tag{5.84}$$

Integral equations for $A^{0J}_{\pi\pi;\pi^0}$ have the following form:

$$A^{0J}_{\pi\pi;\pi^0}(s_{12}) = \lambda^{0J}_{\pi\pi;\pi^0}\left\{\hat{B}\frac{\hat{I}}{\hat{I}-\hat{b}\hat{B}}\right\}^{0J}_{\pi\pi;\pi\pi}$$
$$+\int\limits_{4\mu_\pi^2}^{\infty}\frac{ds'_{12}}{\pi}\frac{\rho_{\pi\pi}(s'_{12})N^{0J}_{\pi\pi;\pi\pi}(s'_{12},s_{12})}{s'_{12}-s_{12}-i0}$$
$$\times\left(\left\langle\frac{2}{3}a^{(0)}_{\pi\pi;\pi^0}(s'_{13},z)\right\rangle_J + \left\langle\frac{20}{9}a^{(2)}_{\pi\pi;\pi^0}(s'_{23},z)\right\rangle_J\right.$$
$$\left.+\left\langle\frac{4}{3}a^{(1)}_{\pi\pi;\pi^0}(s'_{23},z)\right\rangle_J\right)\left\{\hat{B}\frac{\hat{I}}{\hat{I}-\hat{b}\hat{B}}\right\}^{0J}_{\pi\pi;\pi\pi}$$
$$+\lambda^{0J}_{\eta\eta;\pi^0}\left\{\hat{B}\frac{\hat{I}}{\hat{I}-\hat{b}\hat{B}}\right\}^{0J}_{\eta\eta;\pi\pi}$$
$$+\int\limits_{4\mu_\eta^2}^{\infty}\frac{ds'_{12}}{\pi}\frac{\rho_{\eta\eta}(s'_{12})N^{0J}_{\eta\eta;\pi\pi}(s'_{12},s_{12})}{s'_{12}-s_{12}-i0}\left\langle 2\,a_{\pi^0\eta;\eta}(s'_{13},z)\right\rangle_J$$
$$+\left\{\hat{B}\frac{\hat{I}}{\hat{I}-\hat{b}\hat{B}}\right\}^{0J}_{\eta\eta;\pi\pi}$$

$$+\lambda^{0J}_{\bar{K}K;\pi^0}\left\{\hat{B}\frac{\hat{I}}{\hat{I}-\hat{b}\hat{B}}\right\}^{0J}_{\bar{K}K;\pi\pi}$$
$$+\int\limits_{4\mu_K^2}^{\infty}\frac{ds'_{12}}{\pi}\frac{\rho_{\bar{K}K}(s'_{12})N^{0J}_{\bar{K}K;\pi\pi}(s'_{12},s_{12})}{s'_{12}-s_{12}-i0}$$
$$\times\left(\left\langle a_{\pi^0K;\bar{K}}(s'_{13},z)\right\rangle_J\right.$$
$$\left.+\left\langle a_{\pi^0\bar{K};K}(s'_{23},z)\right\rangle_J\right)\left\{\hat{B}\frac{\hat{I}}{\hat{I}-\hat{b}\hat{B}}\right\}^{0J}_{\bar{K}K;\pi\pi}. \quad (5.85)$$

The amplitude $a^{(1)}_{\pi\pi;\pi^0}$ in (5.85) can be found from the $\bar{p}p$ annihilation into charged pions, for a simple approach one can use isobar model that means to describe it by direct production amplitude of $\rho^+\pi^-$.

In the integral equation for $A^{20}_{\pi\pi;\pi^0}$ we take into account only $\pi\pi$ intermediate states. In this approach one has:

$$A^{20}_{\pi\pi;\pi^0}(s_{12}) = \lambda^{20}_{\pi\pi;\pi}\left\{\hat{B}\frac{\hat{I}}{\hat{I}-\hat{b}\hat{B}}\right\}^{20}_{\pi\pi;\pi\pi}$$
$$+\int\limits_{4\mu_\pi^2}^{\infty}\frac{ds'_{12}}{\pi}\frac{\rho_{\pi\pi}(s'_{12})N^{20}_{\pi\pi;\pi\pi}(s'_{12},s_{12})}{s'_{12}-s_{12}-i0}$$
$$\times\left(\left\langle a^{(0)}_{\pi\pi;\pi^0}(s'_{13},z)\right\rangle_0+\left\langle \frac{1}{3}a^{(2)}_{\pi\pi;\pi^0}(s'_{23},z)\right\rangle_0\right.$$
$$\left.+\left\langle a^{(1)}_{\pi\pi;\pi^0}(s'_{23},z)\right\rangle_0\right)\left\{\hat{B}\frac{\hat{I}}{\hat{I}-\hat{b}\hat{B}}\right\}^{20}_{\pi\pi;\pi\pi}. \quad (5.86)$$

Here in the intermediate state we refer 23 as $\pi^+\pi^-$ and $\pi^0\pi^\pm$ are denoted as 12, 13.

5.3.3.2 *Reaction* $\bar{p}p(0^{-+}) \to \eta\eta\pi^0$

In the $\eta\pi^0$ channel the S and D waves are accounted for and production of a_0 and a_2 resonances are taken into consideration. The annihilation amplitude is as follows:

$$\begin{aligned} M_{\eta\eta\pi^0} &= a_{\eta\eta;\pi^0}(s_{12},z)+a_{\eta\pi^0;\eta}(s_{13},z)+a_{\eta\pi^0;\eta}(s_{23},z),\\ a_{\eta\eta;\pi^0} &= A^{00}_{\eta\eta;\pi^0}+F_2A^{02}_{\eta\eta;\pi^0},\\ a_{\eta\pi^0;\eta} &= A^{10}_{\eta\pi^0;\eta}+F_2A^{12}_{\eta\pi^0;\eta}. \end{aligned} \quad (5.87)$$

The integral equation for the $A^{0J}_{\eta\eta;\pi^0}$ amplitude has the following form:

$$
\begin{aligned}
A^{0J}_{\eta\eta;\pi^0} &= \lambda^{0J}_{\pi\pi;\pi^0}\left\{\hat{B}\frac{\hat{I}}{\hat{I}-\hat{b}\hat{B}}\right\}^{0J}_{\pi\pi;\eta\eta} \\
&+ \int\limits_{4\mu_\pi^2}^{\infty}\frac{ds'_{12}}{\pi}\frac{\rho_{\pi\pi}(s'_{12})N^{0J}_{\pi\pi;\eta\eta}(s'_{12},s_{12})}{s'_{12}-s_{12}-i0} \\
&\quad\times\left(\left\langle\frac{2}{3}a^{(0)}_{\pi\pi;\pi^0}(s'_{13},z)\right\rangle_J+\left\langle\frac{20}{9}a^{(2)}_{\pi\pi;\pi^0}(s'_{23},z)\right\rangle_J \right. \\
&\quad\left. +\left\langle\frac{4}{3}a^{(1)}_{\pi\pi;\pi^0}(s'_{23},z)\right\rangle_J\right)\left\{\hat{B}\frac{\hat{I}}{\hat{I}-\hat{b}\hat{B}}\right\}^{0J}_{\pi\pi;\eta\eta} \\
&+ \lambda^{0J}_{\eta\eta;\pi^0}\left\{\hat{B}\frac{\hat{I}}{\hat{I}-\hat{b}\hat{B})}\right\}^{0J}_{\eta\eta;\eta\eta} \\
&+ \int\limits_{4\mu_\eta^2}^{\infty}\frac{ds'_{12}}{\pi}\frac{\rho_{\eta\eta}(s'_{12})N^{0J}_{\eta\eta;\eta\eta}(s'_{12},s_{12})}{s'_{12}-s_{12}} \\
&\quad\times\left\langle 2\,a_{\pi^0\eta;\eta}(s'_{13},z)\right\rangle_J\left\{\hat{B}\frac{\hat{I}}{\hat{I}-\hat{b}\hat{B}}\right\}^{0J}_{\eta\eta;\eta\eta} \\
&+ \lambda^{0J}_{\bar{K}K;\pi^0}\left\{\hat{B}\frac{\hat{I}}{\hat{I}-\hat{b}\hat{B}}\right\}^{0J}_{\bar{K}K;\eta\eta} \\
&+ \int\limits_{4\mu_K^2}^{\infty}\frac{ds'_{12}}{\pi}\frac{\rho_{\bar{K}K}(s'_{12})N^{0J}_{\bar{K}K;\eta\eta}(s'_{12},s_{12})}{s'_{12}-s_{12}-i0} \\
&\quad\times\left(\left\langle a_{\pi^0K;\bar{K}}(s'_{13},z)\right\rangle_J \right. \\
&\quad\left. +\left\langle a_{\pi^0\bar{K};K}(s'_{23},z)\right\rangle_J\right)\left\{\hat{B}\frac{\hat{I}}{\hat{I}-\hat{b}\hat{B}}\right\}^{0J}_{\bar{K}K;\eta\eta}.
\end{aligned}
\tag{5.88}
$$

The integral equation for the $A^{1J}_{\eta\pi^0;\eta}$ amplitude reads:

$$
\begin{aligned}
A^{1J}_{\eta\pi^0;\eta} &= \lambda^{1J}_{\eta\pi^0;\eta}\left\{\hat{B}\frac{\hat{I}}{\hat{I}-\hat{b}\hat{B}}\right\}^{1J}_{\eta\pi^0;\eta\pi^0} \\
&+ \int\limits_{(\mu_\pi+\mu_\eta)^2}^{\infty}\frac{ds'_{12}}{\pi}\frac{\rho_{\eta\pi^0}(s'_{12})N^{1J}_{\eta\pi^0;\eta\pi^0}(s'_{12},s_{12})}{s'_{12}-s_{12}-i0} \\
&\times\left(\left\langle a_{\eta\eta;\pi^0}(s'_{13},z)\right\rangle_J+\left\langle a_{\eta\pi^0;\eta}(s'_{23},z)\right\rangle_J\right)\left\{\hat{B}\frac{\hat{I}}{\hat{I}-\hat{b}\hat{B}}\right\}^{1J}_{\eta\pi^0;\eta\pi^0}.
\end{aligned}
$$

5.3.3.3 *Reaction* $\bar{p}p(0^{-+}) \to \bar{K}K\pi^0$.

These annihilation amplitudes are:

$$M_{\pi^0\bar{K}K} = a_{\pi^0K;\bar{K}}(s_{12},z) + a_{\pi^0\bar{K};K}(s_{13},z) + a_{\bar{K}K;\pi^0}(s_{23},z)\,. \quad (5.89)$$

As before, in the $\bar{K}K$ channel the S- and D-wave interactions and K^* resonances in the $K\pi$ channel are accounted for. So, one has:

$$a_{\bar{K}K;\pi^0} = A^{00}_{\bar{K}K;\pi^0} + F_2 A^{02}_{\bar{K}K;\pi^0},$$
$$a_{K\pi^0;\bar{K}} = A^{1/2,1}_{K\pi^0;\bar{K}}\,. \quad (5.90)$$

Here the isotopic spin $I = 1/2$. For these amplitudes one can get the following integral equations:

$$A^{1/2,1}_{K\pi^0;\bar{K}} = \lambda^{1/2,1}_{K\pi^0;\bar{K}}\left\{\hat{B}\frac{\hat{I}}{\hat{I}-\hat{b}\hat{B}}\right\}^{1/2,1}_{\pi^0K;\pi^0K}$$
$$+\int\limits_{(\mu_\pi+\mu_K)^2}^{\infty}\frac{ds'_{12}}{\pi}\frac{\rho_{\pi^0K}(s'_{12})N^{1/21}_{\pi^0K;\pi^0K}(s'_{12},s_{12})}{s'_{12}-s_{12}-i0}$$
$$\times\Bigg(\Big\langle a_{K\pi^0;\bar{K}}(s'_{13},z)\Big\rangle_J$$
$$+\Big\langle a_{\bar{K}K;\pi^0}(s'_{23},z)\Big\rangle_J\Bigg)\left\{\hat{B}\frac{\hat{I}}{\hat{I}-\hat{b}\hat{B}}\right\}^{1/2,1}_{\pi^0K;\pi^0K}. \quad (5.91)$$

The integral equation for the $A^{0J}_{\bar{K}K;\pi^0}$ amplitude has the following form:

$$A^{0J}_{\bar{K}K;\pi^0} = \lambda^{0J}_{\pi\pi;\pi^0}\left\{\hat{B}\frac{\hat{I}}{\hat{I}-\hat{b}\hat{B}}\right\}^{0J}_{\pi\pi;\bar{K}K}$$
$$+\int\limits_{4\mu_\pi^2}^{\infty}\frac{ds'_{12}}{\pi}\frac{\rho_{\pi\pi}(s'_{12})N^{0J}_{\pi\pi;\bar{K}K}(s'_{12},s_{12})}{s'_{12}-s_{12}}$$
$$\times\Bigg(\Big\langle\frac{2}{3}a^{(0)}_{\pi\pi;\pi^0}(s'_{13},z)\Big\rangle_J+\Big\langle\frac{20}{9}a^{(2)}_{\pi\pi;\pi^0}(s'_{23},z)\Big\rangle_J$$
$$+\Big\langle\frac{4}{3}a^{(1)}_{\pi\pi;\pi^0}(s'_{23},z)\Big\rangle_J\Bigg)\left\{\hat{B}\frac{\hat{I}}{\hat{I}-\hat{b}\hat{B}}\right\}^{0J}_{\pi\pi;\bar{K}K}$$

$$
\begin{aligned}
&+ \lambda^{0J}_{\eta\eta;\pi^0}\left\{\hat{B}\frac{\hat{I}}{\hat{I}-\hat{b}\hat{B}}\right\}^{0J}_{\eta\eta;\bar{K}K} \\
&+ \int\limits_{4\mu_\eta^2}^{\infty}\frac{ds'_{12}}{\pi}\frac{\rho_{\eta\eta}(s'_{12})N^{0J}_{\eta\eta;\eta\eta}(s'_{12},s_{12})}{s'_{12}-s_{12}-i0} \\
&\quad\times\Big\langle 2\, a_{\pi^0\eta;\eta}(s'_{13},z)\Big\rangle_J\left\{\hat{B}\frac{\hat{I}}{\hat{I}-\hat{b}\hat{B}}\right\}^{0J}_{\eta\eta;\bar{K}K} \\
&+ \lambda^{0J}_{\bar{K}K;\pi^0}\left\{\hat{B}\frac{\hat{I}}{\hat{I}-\hat{b}\hat{B}}\right\}^{0J}_{\bar{K}K;\bar{K}K} \\
&+ \int\limits_{4\mu_K^2}^{\infty}\frac{ds'_{12}}{\pi}\frac{\rho_{\bar{K}K}(s'_{12})N^{0J}_{\bar{K}K;\bar{K}K}(s'_{12},s_{12})}{s'_{12}-s_{12}-i0} \\
&\quad\times\bigg(\Big\langle a_{\pi^0K;\bar{K}}(s'_{13},z)\Big\rangle_J \\
&\qquad+\Big\langle a_{\pi^0\bar{K};K}(s'_{13},z)\Big\rangle_J\bigg)\left\{\hat{B}\frac{\hat{I}}{\hat{I}-\hat{b}\hat{B}}\right\}^{0J}_{\bar{K}K;\bar{K}K}. \qquad (5.92)
\end{aligned}
$$

Thereby, Eqs. (5.82)–(5.92) give us a set of equations for the coupled decay channels $\bar{p}p(IJ^{PC}=10^{-+})\to\pi^0\pi^0\pi^0$, $\eta\eta\pi^0$, $\bar{K}K\pi^0$. An analogous set of integral equations for the reactions $\bar{p}p$ $(IJ^{PC}=00^{-+})\to\eta\pi^0\pi^0$, $\bar{K}K\pi^0$ may be written in the same way.

5.4 Conclusion

To summarize, the propagators are given for mesons and baryons with arbitrary spins. That allows to write amplitudes for the production of excited resonances, high spin states included, and amplitudes with resonances in intermediate states.

As an example of application of this technique, reactions $\bar{p}p(IJ^{PC}=10^{-+},\text{at rest})\to\pi^0\pi^0\pi^0$, $\eta\eta\pi^0$, $\bar{K}K\pi^0$ are considered with accounting of all binary rescattering. In this way the spectral integral equations are obtained which take into account not only the S wave interactions but also interactions with higher angular momenta (P and D waves). These equations response to resonance production and can easely generalized for the non-resonance type of two-particle interactions.

Three-body spectral integral equations take into account not only pole singularities of amplitudes but the next-to-leading, the logarithmic ones, as well. This is an important point meaning the application of isobar models to fitting data. In multiparticle production reactions the logarithmic singularities can appear near the physical region and, since the amplitude in the singular point tends to infinity, it may imitate a resonance. This problem was discussed in [17, 18]. In the analysis of the $p\bar{p}$ annihilation data at rest for a more precise determination of the studied resonance parameters, one should include the logarithmic singularities into the fit, such an inclusion was carried out in [6]. The full set of singularities is given by Eqs. (5.82)–(5.92), the leading and next-to-leading singularities can be extracted from these equations by an iteration procedure.

References

[1] A.V. Anisovich, V.V. Anisovich, V.N. Markov, M.A. Matveev, and A.V. Sarantsev, J. Phys. G: Nucl. Part. Phys. **28**, 15 (2002).

[2] A.V. Anisovich, V.V. Anisovich, M.A. Matveev, V.A. Nikonov, J. Nyiri, A.V. Sarantsev, "Mesons and Baryons – Systematization and Methods of Analysis", World Scientific (2004).

[3] V.V. Anisovich, A.V. Sarantsev, D.V. Bugg, Nucl. Phys. A **537**, 501 (1992).

[4] D.V. Bugg, Phys. Rep. **397**, 257 (2004).

[5] E. Klempt and A. Zaitsev, Phys. Rep. **454**, 1 (2007).

[6] V.V. Anisovich, D.V. Bugg, A.V. Sarantsev, and B.S. Zou, Phys. Rev. D **50**, 4412 (1994).

[7] A.V. Anisovich, Yad. Fiz. **66**, 173 (2003) [Phys. Atom. Nucl. **66** , 172 (2003)].

[8] V.V. Anisovich et al. Phys. Lett. B **323**, 233 (1994).

[9] A.V. Anisovich, C.A. Baker, C.J. Batty *et al.*, Phys. Lett. B**449**, 114 (1999); B **472**, 168 (2000); B **476**, 15 (2000); B **477**, 19 (2000); B **491**, 40 (2000); B **491**, 47 (2000); B **496**, 145 (2000); B **507**, 23 (2001); B **508**, 6 (2001); B **513**, 281 (2001); B **517**, 273 (2001); B **542**, 8 (2002);
Nucl. Phys. A **651**, 253 (1999); A **662**, 319 (2000); A **662**, 344 (2000);
M.A. Matveev, AIP Conf. Proc. 717:72-76, 2004.

[10] A.V. Anisovich, C.A. Baker, C.J. Batty *et al.*, Phys. Lett. B **452**, 173 (1999); Phys. Lett. B **452**, 187 (1999); Phys. Lett. B **517**, 261 (2001).

[11] R.L. Jaffe and K. Johnson, Phys. Lett. B **60**, 201 (1976);
J.F. Donoghue, K. Johnson, and B.A. Li, Phys. Lett. B **99**, 416 (1981);
J. Paton and N. Isgur, Phys. Rev. D **31**, 2910 (1985);

[12] G. Parisi and R. Petronzio, Phys. Lett.B**94**, 51 (1980);
M. Consoli and J.H. Field, Phys. Rev. D**49**, 1293 (1994).

[13] G.S. Bali et al. Phys. Lett. B **309**, 378 (1993);
I. Chen et al. Nucl. Phys. **B34** (Proc. Suppl.) 357 (1994);
C.J. Morningstar, M.J. Peardon, Phys. Rev. D **60**, 034509 (2003).

[14] C. Amsler and F.E. Close, Phys. Lett. B **353**, 385 (1995);
V.V. Anisovich, Phys. Lett. B **364**, 195 (1995).

[15] V.V. Anisovich, Y.D. Prokoshkin, A.V. Sarantsev, Phys. Lett. **389**, 388 (1996); Z. Phys. **A357**, 123 (1997).

[16] V.V. Anisovich, JETP Lett. **80**, 845 (2004);
V.V. Anisovich and A.V. Sarantsev, JETP Lett. **81**, 417 (2005) ;
V.V. Anisovich, M.A. Matveev, J. Nyiri, A.V. Sarantsev, Int. J. Mod. Phys. **A 20**, 6327 (2005).

[17] V.V. Anisovich and L.G. Dakhno Phys. Lett. **10**, 221 (1964); V.V. Anisovich and L.G. Dakhno Nucl. Phys. **76**, 657 (1966).

[18] A.V. Anisovich and V.V. Anisovich Phys. Lett. **B345** , 321 (1995).

Chapter 6

Isobar model and partial wave analysis. D-matrix method and investigation of the meson-meson and meson-nucleon spectra

In this chapter we discuss hadron production reactions in the region of the isobar model (the region labeled as III in the Introduction). To be specific, we use as an example the partial wave analyses of the meson-meson scattering data and data from πN and γN collisions. The presented analyses are based on the K-matrix and dispersion relation D-matrix techniques.

The D-matrix method fully satisfies the unitarity and analyticity conditions and is suitable for a fast analysis of modern high statistic data.

The extraction of the physical information about hadrons from scattering reactions is related to a procedure which is called the partial wave analysis. The purpose of such an analysis is to extract partial wave amplitudes from the angular distributions of the collected data. Then every partial wave is a subject of further investigations with the aim of extracting amplitude poles and non-resonant contributions. Such an approach is usually called the energy independent partial wave analysis. However, in most cases such a two-step analysis is not possible: either there is no comprehensive information about the analyzed reaction, *e.g.* the absence of some polarization observables, or the angular distributions do not cover all angles and partial waves can not be unambiguously extracted, or partial waves should be extracted from complicated reactions with a multi-particle final state.

Recently an energy dependent partial wave analysis of a set of related reactions in terms of K-matrix technique became a very popular approach. In this chapter, we discuss the advantages of the D-matrix method which takes into account dispersion corrections neglected in the K-matrix approach.

In Section 6.1 we provide a comparative survey of formulae used in the K-matrix and D-matrix approaches. In Section 6.2 we compare the results of the partial wave analysis obtained in the framework of K-matrix and D-matrix techniques for isoscalar-scalar meson states. In Section 6.3 a similar comparision is performed for the analysis of the baryon spectrum. There we compare the results of the K- and D-matrix analyses of the pion-induced and photo production reactions.

6.1 The *K*-Matrix and *D*-Matrix Techniques

In the K-matrix approach real parts of loop diagrams are neglected, or better to say they are taken into account effectively as a renormalization of resonance masses. These real parts have left-hand side singularities and the amplitude can not be treated correctly in the low energy region where these singularities can play a notable role. In the D-matrix approach the real part is calculated directly and one can expect a proper treatment of the amplitude in the whole energy region.

Although the D-matrix method is theoretically better founded, the majority of the partial wave analyses had been made in the framework of the K-matrix approach and a good understanding of stability and limitations of the obtained results is an important task.

6.1.1 *K-matrix approach*

The K-matrix approach was introduced to satisfy directly the unitarity condition. The S-matrix for the transition between different final states can be expressed as

$$S = \left(I + i\hat{\rho}\hat{K}\right)\left(I - i\hat{\rho}\hat{K}\right)^{-1} = I + 2i\hat{\rho}\hat{K}\left(I - i\hat{\rho}\hat{K}\right)^{-1} \qquad (6.1)$$

where $\hat{\rho}$ is a diagonal matrix of phase volumes and $\hat{K}$ is the real matrix which describes resonant and non-resonant contributions.

For the partial wave amplitude $A(s)$ one obtains:

$$\hat{A} = \hat{K}\left(I - i\hat{\rho}\hat{K}\right)^{-1} = \hat{K} + \hat{K}i\hat{\rho}\hat{K} + \hat{K}i\hat{\rho}\hat{K}i\hat{\rho}\hat{K} + \ldots \qquad (6.2)$$

Or in the matrix form:

$$\hat{A} = \hat{A}\, i\,\hat{\rho}\hat{K} + \hat{K}. \qquad (6.3)$$

The factor $(I - i\hat{\rho}\hat{K})^{-1}$ describes the rescattering of particles in the channels directly taken into account. The elements of the K-matrix are parametrized as a sum of resonant terms and non-resonant contributions:

$$K_{ij} = \sum_{\alpha} \frac{g_i^{(\alpha)} g_j^{(\alpha)}}{M_\alpha^2 - s} + f_{ij} . \tag{6.4}$$

This form is defined by the symmetry condition and the condition that the scattering amplitude has only pole singularities of the first order.

As a rule the K-matrix amplitude includes explicitly only channels which contribute notably to the rescattering term. In cases when particles are produced from a weak channel, for example, the meson production in the $\gamma\gamma$ collision or proton-antiproton annihilation, the initial interaction can be taken into account only once. Then the transition amplitude is a vector with elements which describe the transition from the initial state to K-matrix channels:

$$\vec{A}(\bar{p}p) = \vec{P}\left(I - i\hat{\rho}\hat{K}\right)^{-1} = \vec{P} + \vec{P}i\hat{\rho}\hat{K} + \vec{P}i\hat{\rho}\hat{K}i\hat{\rho}\hat{K} + \dots . \tag{6.5}$$

Here the factor $(I - i\hat{\rho}\hat{K})^{-1}$ is the same as in the transition amplitude (6.2) and elements of the production vector P_j have a form similar to the K-matrix elements (6.4):

$$P_j = \sum_{\alpha} \frac{\Lambda_\alpha^P g_j^{(\alpha)}}{M_\alpha^2 - s} + G_j^P . \tag{6.6}$$

The first term in Eq.(6.6) refers to the production of resonances; the second one, G_j^P, to a non-resonant production.

An analog of the P-vector approach is the F-vector method which describes the transition amplitude from a K-matrix channel to a weak channel measured in the final state. Here the amplitude is also described by a vector with elements corresponding to the initial K-matrix channels:

$$\vec{A} = \vec{F} + K\left(I - i\hat{\rho}\hat{K}\right)^{-1} i\hat{\rho}\vec{F} \tag{6.7}$$

where

$$F_j = \sum_{\alpha} \frac{g_j^{(\alpha)} \Lambda_\alpha^F}{M_\alpha^2 - s} + G_j^F . \tag{6.8}$$

Again, the first term describes the decay of resonances; the second one, G_j^F, a non-resonant transition from the initial K-matrix state to the final state.

Using these equations, one can write an expression for the transition amplitude between two weak channels which are not included explicitly in the K-matrix. In this case the amplitude is a scalar function described as

$$A=\sum_{\alpha}\frac{\Lambda_\alpha^P\Lambda_\alpha^F}{M_\alpha^2-s}+R_{PF}+P\left(I-i\hat{\rho}\hat{K}\right)^{-1}i\hat{\rho}\vec{F}, \tag{6.9}$$

where R_{PF} describes a non-resonant transition between the initial and final states.

6.1.2 *Spectral integral equation for the K-matrix amplitude*

Let us introduce so called "bare" and "dressed" particles: the "bare" states correspond to the poles on the real-s axis which transformed due to rescattering of particles (*e.g.* mesons) into states dressed by "coats" of these states. In the K-matrix approach we deal with a "coat" formed by real particles – the contribution of virtual ones is included in the principal part of the loop diagram, $B(s)$, and is taken into account effectively by the renormalization of mass and couplings.

In terms of the dispersion relation technique the K-matrix amplitude can be considered as a solution of the spectral integral equation which is an analog of the Bethe-Salpeter equation [1] for the Feynman technique. This spectral integral equation is presented graphically in Fig. 6.1 and reads:

$$A_{ab}(s)=\int\frac{ds'}{\pi}\frac{A_{aj}(s,s')}{s'-s-i0}\rho_j(s')K_{jb}(s',s)+K_{ab}(s)\ . \tag{6.10}$$

Here $\rho_j(s')$ is the diagonal matrix of the phase volumes, $A_{aj}(s,s')$ is the off-shell transition amplitude from the channel a to channel j and $K_{jb}(s,s')$ is the off-shell elementary interaction. Let us remind that in the dispersion relation technique, just as in quantum mechanics, there is no energy conservation for the intermediate states.

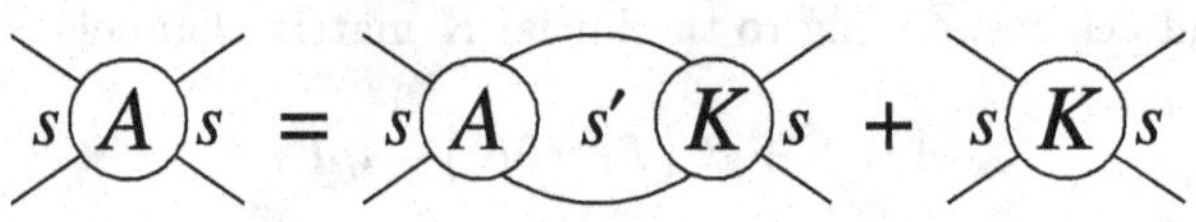

Fig. 6.1 Graphical representation of the spectral integral equation for the K-matrix amplitude.

The standard way of the transformation of Eq. (6.10) into the K-matrix form is the extraction of the imaginary and principal parts of the integral. The principal part has no singularities in the physical region and can be omitted (or taken into account by a renormalization of the K-matrix parameters):

$$\int \frac{ds'}{\pi} \frac{A_{aj}(s,s')}{s'-s-i0} \rho_j(s') K_{jb}(s',s) = P \int \frac{ds'}{\pi} \frac{A_{aj}(s,s')}{s'-s} \rho_j(s') K_{jb}(s',s) + iA_{aj}(s,s)\rho_j(s)K_{jb}(s) \to iA_{aj}(s,s)\rho_j(s)K_{jb}(s), \qquad (6.11)$$

and we obtain a standard K-matrix expression (6.3).

One of the easiest way to take into account the real part of the integral (6.11) (so called dispersion corrections) is to assume that the amplitude and K-matrix have a trivial dependence on s'. Such a case corresponds, for example, to a parametrization of the resonant couplings and non-resonant K-matrix terms as constants and regularization of the integral (6.11) which depends on the scattering channel only (*e.g.* a subtraction at a fixed point). In this case

$$\int \frac{ds'}{\pi} \frac{A_{aj}(s,s')}{s'-s-i0} \rho_j(s') K_{jb}(s',s) = A_{aj}(s,s) Re\, B_j(s) K_{jb}(s,s) + iA_{aj}(s,s)\rho_j(s)K_{jb}(s) \qquad (6.12)$$

where

$$Re\, B_j(s) = P \int \frac{ds'}{\pi} \frac{\rho_j(s')}{s'-s} \qquad (6.13)$$

And for the transition amplitude we obtain:

$$A = \hat{K}\left(I - \hat{Re}\, B\hat{K} - i\hat{\rho}\hat{K}\right)^{-1}$$

$$S = \left(I - \hat{Re}\, B\hat{K} + i\hat{\rho}\hat{K}\right)\left(I - \hat{Re}\, B\hat{K} - i\hat{\rho}\hat{K}\right)^{-1} \qquad (6.14)$$

Such an approach provides a correct continuation of the amplitude below the thresholds.

6.1.3 *D-matrix approach*

As we discussed above, the K-matrix approach can be considered as an effective way to calculate an infinite sum of the rescattering diagrams from the spectral integral equation. Here the rescattering diagrams can be divided into blocks which describe a transition from one channel into another. Thus the rank of the K-matrix is defined by the number of the channels taken explicitly into account. The key issue of the K-matrix approach is a factorization of such blocks which is automatically fulfilled for the imaginary part of the loop diagrams. The real part should be either omitted or introduced in the form to fulfil this factorization, like one discussed in the previous section. If the decay vertices and non-resonant terms have a non-trivial energy dependence, the real part can not be factorized into the K-matrix blocks and another approach should be used for the calculation of the amplitude. The most straightforward idea is to factorize the amplitude into blocks which describe a transition from one "bare" state to another. Such a factorization is automatically fulfilled for the pole terms. The treatment of non-resonant terms is the main problem of such approach and will be discussed in details below.

Let us introduce the block $D_{\alpha\beta}$ which describes a transition between the bare state α and the bare state β. For such a block one can write the following equation:

$$D_{\alpha\beta} = D_{\alpha\gamma} \sum_j B^j_{\gamma\delta} d_{\delta\beta} + d_{\alpha\beta}. \tag{6.15}$$

Or, in the matrix form:

$$\hat{D} = \hat{D}\hat{B}\hat{d} + \hat{d} \qquad \hat{D} = \hat{d}(I - \hat{B}\hat{d})^{-1}. \tag{6.16}$$

Here the $\hat{d}$ is a diagonal matrix of the propagators:

$$\hat{d} = diag\left(\frac{1}{M_1^2 - s}, \frac{1}{M_2^2 - s}, \ldots, \frac{1}{M_N^2 - s}, R_{N+1}, R_{N+2}\ldots\right), \tag{6.17}$$

where N is the number of resonant terms and R_α are propagators for effective bare states which describe non-resonant transitions. The elements of the $\hat{B}$-matrix are equal to:

$$\hat{B}_{\alpha\beta} = \sum_j B^j_{\alpha\beta} = \sum_j \int\limits_{(m_{1j}+m_{2j})^2}^{\infty} \frac{ds'}{\pi} \frac{g_j^{R(\alpha)} \rho_j(s', m_{1j}, m_{2j}) g_j^{L(\beta)}}{s' - s - i0}. \tag{6.18}$$

The $g_j^{R(\alpha)}$ and $g_j^{L(\alpha)}$ are right and left vertices for a transition from the bare state α to the channel j. The function B^j_{ab} depends on initial, intermediate and final states and allows us to introduce for every transition a specific energy dependence and regularization procedure.

For the resonance transition the right and left vertices are the same:

$$g_j^{R(\alpha)} = g_j^{L(\alpha)} = g_j^{(\alpha)}. \tag{6.19}$$

The easiest way to take into account the non-resonant transitions is to introduce for every non-resonant term a separate non-resonant bare state. A more compact way was suggested in [2]. There it was shown that two effective bare states can describe all non-resonant transitions from a fixed initial channel i to all final channels:

$$g_i^{L(N+1)} R_{N+1}\, g_j^{R(N+1)} + g_i^{L(N+2)} R_{N+2}\, g_j^{R(N+2)}, \qquad (6.20)$$

where:

$$\begin{aligned} &g_1^{L(N+1)} = 1 \qquad g_j^{R(N+1)} = f_{ij} \qquad R_{N+1} = 1, \\ &g_{j>1}^{L(N+2)} = f_{ij} \qquad g_1^{R(N+2)} = 1 \qquad R_{N+2} = 1 \\ &g_{j>1}^{L(N+1)} = 0 \qquad g_{j>1}^{R(N+2)} = 0 \qquad g_1^{L(N+2)} = 0\,. \end{aligned} \qquad (6.21)$$

Let us note that such a parametrization is not unique and assumes a specific dependence of non-resonant terms on s and s' for virtual energy violation transition. Another way is, for example, to introduce a separable interaction. However, we should stress that non-resonant contributions are mostly model dependent and as a rule have a large parametrization flexibility.

To take into account the non-resonant transitions from the second channel to other channels, two more effective bare states should be introduced. However, these states should not have vertices which correspond to the transition from/to the first channel.

The scattering amplitude between channels i and j which are taken into account in the rescattering has the form

$$A_{ij}(s) = g_i^{L(\alpha)}(s) D_{\alpha\gamma} g_j^{R(\gamma)}(s). \qquad (6.22)$$

The production amplitude from a weak channel not included in the rescattering (analog of the P-vector approach) can be written as:

$$A_j(s) = P_j(s,s) + \left(\sum_i \int\limits_{th_i}^{\infty} \frac{ds'}{\pi} \frac{P_i(s,s')\rho_i(s',m_{1i},m_{2i}) g_i^{L(\beta)}(s')}{s'-s-i0}\right) D_{\beta\gamma} g_j^{R(\gamma)}(s) \qquad (6.23)$$

where $th_i = (m_{1i}+m_{2i})^2$ and the vector P_j has the same parametrization as that in the K-matrix approach:

$$P_j(s,s') = \sum_{\alpha=1}^{N} \frac{\Lambda_\alpha^P(s) g_j^{R(\alpha)}(s')}{M_\alpha^2 - s} + G_j^P(s,s')\,. \qquad (6.24)$$

The decay into a weak channel is described as:

$$\begin{aligned} A_j(s) = {} & g_j^{L(\gamma)}(s) D_{\gamma\beta} \left(\sum_i \int\limits_{th_i}^{\infty} \frac{ds'}{\pi} \frac{g_i^{R(\beta)}(s')\rho_i(s',m_{1i},m_{2i}) F_i(s,s')}{s'-s-i0}\right) \\ & + F_j(s,s), \\ F_j(s',s) = {} & \sum_{\alpha=1}^{N} \frac{g_j^{L(\alpha)}(s')\Lambda_\alpha^F(s)}{M_\alpha^2 - s} + G_j^F(s',s)\,. \end{aligned} \qquad (6.25)$$

6.2 Meson-meson scattering

6.2.1 *K-matrix fit*

One of the most successful descriptions of the meson-meson S-wave interaction in the isoscalar sector ($IJ^{PC} = 00^{++}$) was obtained with the following K-matrix parametrization [3]:

$$K^{00}_{ab}(s) = \left(\sum_{\alpha} \frac{g_a^{(\alpha)} g_b^{(\alpha)}}{M_\alpha^2 - s} + f_{ab}\frac{1\ \text{GeV}^2 + s_0}{s + s_0}\right)\frac{s - s_A}{s + s_{A0}}\ . \tag{6.26}$$

Here the K-matrix describes the transition between 5 channels ($a, b =$ 1,2,3,4,5), where $1 = \pi\pi$, $2 = K\bar{K}$, $3 = \eta\eta$, $4 = \eta\eta'$ and $5 =$ multimeson states (four-pion state mainly at $\sqrt{s} < 1.6$ GeV). The $g_a^{(\alpha)}$ are coupling constants of the bare state α to meson channels; the parameters f_{ab} and s_0 describe the smooth part of the K-matrix elements ($s_0 > 1.5$ GeV^2). The factor $(s - s_A)/(s + s_{A0})$, where $s_A \sim (0.1 - 0.5)m_\pi^2$, describes Adler's zero in the two-pion channel. Although such a factor is demanded in the $\pi\pi$ channel only, it was useful to introduce it also in other channels to suppress the effect of the left-hand side false kinematic singularities in the K-matrix amplitude.

The standard form of the two-particle phase volume is

$$\rho_a(s, m_{1a}, m_{2a}) = \sqrt{\frac{(s - (m_{1a} + m_{2a})^2)(s - (m_{1a} - m_{2a})^2)}{s^2}}\,,$$
$$a = 1, 2, 3, 4 \tag{6.27}$$

where m_{1a} and m_{2a} are masses of the final particles. In the case of different masses this expression includes the term $\sqrt{s - (m_{1a} - m_{2a})^2}$ which in the K-matrix approach can be another source of false kinematic singularities on the first (physical) sheet: the loop diagram amplitude, $B(s)$, does not contain this type of singularities on the first sheet. Such a cancelation can be taken into account effectively by replacing the $\eta\eta'$ phase volume:

$$\sqrt{\frac{(s - (m_{1a} + m_{2a})^2)(s - (m_{1a} - m_{2a})^2)}{s^2}} \to \sqrt{\frac{s - (m_{1a} + m_{2a})^2}{s}}. \tag{6.28}$$

The 00^{++}-amplitude has also threshold singularities related to the four pion channel: the cut at the real s-axis started at $\sqrt{s} = 4m_\pi$ and cuts in the complex-s plane related to the production of vector and scalar particles: $\pi\pi\rho$ (at $\sqrt{s} = 2m_\pi + m_\rho$ with a complex mass m_ρ), $\rho\rho$ (at $\sqrt{s} = 2m_\rho$) and $f_0 f_0$. The phase space factor for the $\rho\rho$-state which contains 4π, $\pi\pi\rho$ and

$\rho\rho$ threshold singularities can be written as

$$\rho_{4\pi}(s) = \int\limits_{4m_\pi^2}^{(\sqrt{s}-2m_\pi)^2} \frac{ds_{12}}{\pi} \int\limits_{4m_\pi^2}^{(\sqrt{s}-\sqrt{s_{12}})^2} \frac{ds_{34}}{\pi} G_{in}^2(s, s_{12}, s_{34})\, \rho(s, \sqrt{s_{12}}, \sqrt{s_{34}})$$

$$\times \frac{G^2(s_{12})(s_{12}-4m_\pi^2)\rho(s_{12}, m_\pi, m_\pi)}{(s_{12}-M_\rho^2)^2+(M_\rho\Gamma_\rho)^2} \frac{G^2(s_{34})\,(s_{34}-4m_\pi^2)\rho(s_{34}, m_\pi, m_\pi)}{(s_{34}-M_\rho^2)^2+(M_\rho\Gamma_\rho)^2} \tag{6.29}$$

The form factors $G_{in}(s, s_{12}, s_{34})$, $G(s_{12})$, $G(s_{34})$ are introduced into (6.29) to provide the convergence of the integrals. This phase volume describes the production of $\rho\rho$ in the S-wave and P-wave production of pions in the ρ-meson decays. For example, in the analysis [4] the vertices $G(s_{ij})$ are parametrized as P-wave Blatt–Weisskopf form factors and $G_{in}(s, s_{12}, s_{34})=1$.

The proton-antiproton annihilation is described with the P-vector parametrized as:

$$P_a(s) = \sum_\alpha \frac{\Lambda_\alpha^P g_a^{(\alpha)}}{M_\alpha^2 - s} + G_a^P \frac{1\ \text{GeV}^2 + s_0}{s + s_0}\,. \tag{6.30}$$

Such a parametrization produced a very successful description of the meson-meson scattering and proton-antiproton annihilation data into three meson states.

Moreover, one can impose on the K-matrix couplings the relations which correspond to the decay of the $q\bar{q}$ states without notable changes in the data description.

These relations are based on the quark combinatoric and given in [5] or, in [6], in more detail. The couplings for channels $a = \pi\pi$, $K\bar{K}$, $\eta\eta$, $\eta\eta'$, calculated in leading terms of the $1/N_c$ expansion, are presented in Table 6.1. They depend only on the constant g which is universal for all nonet states, the mixing angle Φ which determines the proportion of the $n\bar{n} = (u\bar{u} + d\bar{d})/\sqrt{2}$ and $s\bar{s}$ components in the decaying $q\bar{q}$ state, and the $s\bar{s}$ production suppression parameter $\lambda \sim 0.5 - 0.7$. Two scalar-isoscalar states of the same nonet are orthogonal if:

$$\Phi^{(I)} - \Phi^{(II)} = \pm 90^o. \tag{6.31}$$

The equality of the coupling constants g and the fulfilment of the mixing angle relation (6.31) is a basis for the determination of mesons of a $q\bar{q}$-nonet.

For one of the K-matrix state located in the mass region 1300-1550 MeV it is necessary to introduce an admixture of the gluonic component decaying in the channels $a = \pi\pi$, $K\bar{K}$, $\eta\eta, \eta\eta'$ with the same couplings as the $q\bar{q}$-state

Table 6.1 Coupling constants given by quark combinatorics for $(q\bar{q})_{I=0}$ meson and glueball decays into two pseudoscalar mesons in the leading terms of the $1/N_c$ expansion. The Φ is the mixing angle for $n\bar{n} = (u\bar{u} + d\bar{d})/\sqrt{2}$ and $s\bar{s}$ states: $(q\bar{q})_{I=0} = n\bar{n}\cos\Phi + s\bar{s}\sin\Phi$. The Θ is the mixing angle for $\eta - \eta'$ mesons: $\eta = n\bar{n}\cos\Theta - s\bar{s}\sin\Theta$ and $\eta' = n\bar{n}\sin\Theta + s\bar{s}\cos\Theta$ with $\Theta \simeq 37^o$.

decay channel	$q\bar{q}$-meson decay coupling	gg state decay coupling	identity factor
$\pi^0\pi^0$	$g\ \cos\Phi/\sqrt{2}$	G	1/2
$\pi^+\pi^-$	$g\ \cos\Phi/\sqrt{2}$	G	1
K^+K^-	$g(\sqrt{2}\sin\Phi + \sqrt{\lambda}\cos\Phi)/\sqrt{8}$	$\sqrt{\lambda}G$	1
$K^0\bar{K}^0$	$g(\sqrt{2}\sin\Phi + \sqrt{\lambda}\cos\Phi)/\sqrt{8}$	$\sqrt{\lambda}G$	1
$\eta\eta$	$g(\cos^2\Theta\ \cos\Phi/\sqrt{2} + \sqrt{\lambda}\ \sin\Phi\ \sin^2\Theta)$	$G(\cos^2\Theta + \sqrt{\lambda}\sin^2\Theta)$	1/2
$\eta\eta'$	$g\sin\Theta\ \cos\Theta(\cos\Phi/\sqrt{2} - \sqrt{\lambda}\ \sin\Phi)$	$G(1-\lambda)\cos\Theta\sin\Theta$	1

but at a fixed mixing angle $\Phi \to \Phi_{glueball}$ which is determined by the value of λ, namely: $\Phi_{glueball} = \cos^{-1}\sqrt{2/(2+\lambda)}$. The corresponding couplings are given in Table 6.1 as well.

The quality of the K-matrix description for the set of meson-meson scattering and proton-antiproton annihilation reactions in terms of χ^2 is given in the first column of Table 6.2 and K-matrix parameters are listen in Table 6.3. The position of the poles for the 0^+0^{++} amplitude is provided in the first column of the Table 6.4.

Table 6.2 List of the reactions and χ^2 values for the K-matrix and D-matrix solutions: solutions 3,5 taking into account confinement interaction, solutions 1,2,4 without it.

Reaction	K-matrix	D-matrix	Reaction	K-matrix	D-matrix
The Crystal Barrel data (liquid H_2)			The Crystal Barrel data (gaseous H_2)		
$\bar{p}p \to \pi^0\pi^0\pi^0$	1.32	1.37	$\bar{p}p \to \pi^0\pi^0\pi^0$	1.39	1.44
$\bar{p}p \to \pi^0\eta\eta$	1.33	1.34	$\bar{p}p \to \pi^0\eta\eta$	1.31	1.34
$\bar{p}p \to \pi^0\pi^0\eta$	1.24	1.33	$\bar{p}p \to \pi^0\pi^0\eta$	1.20	1.22
The Crystal Barrel data (liquid H_2)			The Crystal Barrel data (liquid D_2)		
$\bar{p}p \to \pi^+\pi^0\pi^-$	1.54	1.46	$\bar{p}n \to \pi^0\pi^0\pi^-$	1.51	1.47
$\bar{p}p \to K_SK_S\pi^0$	1.09	1.10	$\bar{p}n \to \pi^-\pi^-\pi^+$	1.61	1.54
$\bar{p}n \to K^+K^-\pi^0$	0.98	1.00	$\bar{p}p \to K_SK_S\pi^0$	1.09	1.10
$\bar{p}n \to K_LK^\pm\pi^\mp$	0.78	0.79	$\bar{p}n \to K^+K^-\pi^0$	0.98	1.00
			$\bar{p}p \to K_SK_S\pi^-$	1.66	1.64
			$\bar{p}n \to K_SK^-\pi^0$	1.33	1.31
The GAMS and BNL data					
$\pi\pi \to (\pi^0\pi^0)_{S-wave}$	1.23	1.13	$\pi\pi \to (\eta\eta)_{S-wave}$	1.02	1.05
$\pi\pi \to (\eta\eta')_{S-wave}$	0.45	0.30	$\pi\pi \to (K\bar{K})_{S-wave}$	1.32	1.13
The CERN-Munich data: $Y_0^0 \ldots Y_6^1$			The K_{e4} decay data		
$\pi^-\pi^+ \to \pi^-\pi^+$	1.82	1.86	$\delta_0^0(\pi^-\pi^+ \to \pi^-\pi^+)$	1.51	1.02

Table 6.3 The $f_0^{\rm bare}$-resonances: masses M_n (in MeV units), decay coupling constants g_n (in GeV units, $g_{4\pi} \equiv g_5$), mixing angles (in degrees) defined as in Table 1, background terms f_n and confinement singularity term G/s^2 (factor G in GeV units). In all fits the position of the Adler zero was fixed at $s_A = 0.5\ m_\pi^2$.

	K-matrix	D-matrix		K-matrix	D-matrix
M_1	671	685	g_1	0.860	0.926
M_2	1205	1135	g_2	0.956	0.950
M_3	1560	1561	g_3	0.373	0.290
M_4	1210	1290	g_4	0.447	0.307
M_5	1816	1850	g_5	0.458	0.369
$g^{(1)}_{\eta\eta}$	-0.382	-0.213	$g^{(1)}_{\eta\eta'}$	-0.322	-0.500
$g^{(1)}_{4\pi}$	0	0	Φ_1	-74	-83
$g^{(2)}_{4\pi}$	0	0	Φ_2	6	-3
$g^{(3)}_{4\pi}$	0.638	0.534	Φ_3	9	5
$g^{(4)}_{4\pi}$	0.997	0.790	Φ_4	38	31
$g^{(5)}_{4\pi}$	-0.901	-0.862	Φ_5	-64	-71
$f_{\pi\pi\to\pi\pi}$	0.337	0.408	$f_{\pi\pi\to\eta\eta}$	0.389	0.438
$f_{\pi\pi\to K\bar K}$	0.212	0.036	$f_{\pi\pi\to\eta\eta'}$	0.394	0.518
$f_{\pi\pi\to 4\pi}$	-0.199	-0.101			

Table 6.4 The position of poles of the scalar-isoscalar amplitude.

Resonance	K-matrix	D-matrix	Resonance	K-matrix	D-matrix
σ-meson	420-i 395	407-i 281	$f_0(980)$	1014-i 31	1015-i 36
$f_0(1300)$	1302-i 180	1307-i 137	$f_0(1500)$	1487-i 58	1487-i 60
$f_0(1750)$	1738-i 152	1781-i 140			

6.2.2 *D-matrix fit*

The D-matrix parameters can be expressed in the same terms (bare masses and couplings) as parameters of a K-matrix fit. However, in this case the false K-matrix kinematic singularities are absent. Thus it is not necessary to introduce the regularization of the $\eta\eta'$ phase volume and we use the standard expression (6.27). It is also not necessary to introduce any regularization for the D-matrix elements at $s = 0$ and the term with the Adler zero is introduced in the $\pi\pi$ channel only. Technically, it can be done either by the modification of vertices or by the modification of the $\pi\pi$ phase volume:

$$\rho_1(s, m_\pi, m_\pi) = \frac{s - s_A}{s + s_{A0}}\sqrt{\frac{s - 4m_\pi}{s}} \tag{6.32}$$

In the K-matrix fit only non-resonant transitions between the $\pi\pi$ channel and other channels are needed for a good description of all fitted data.

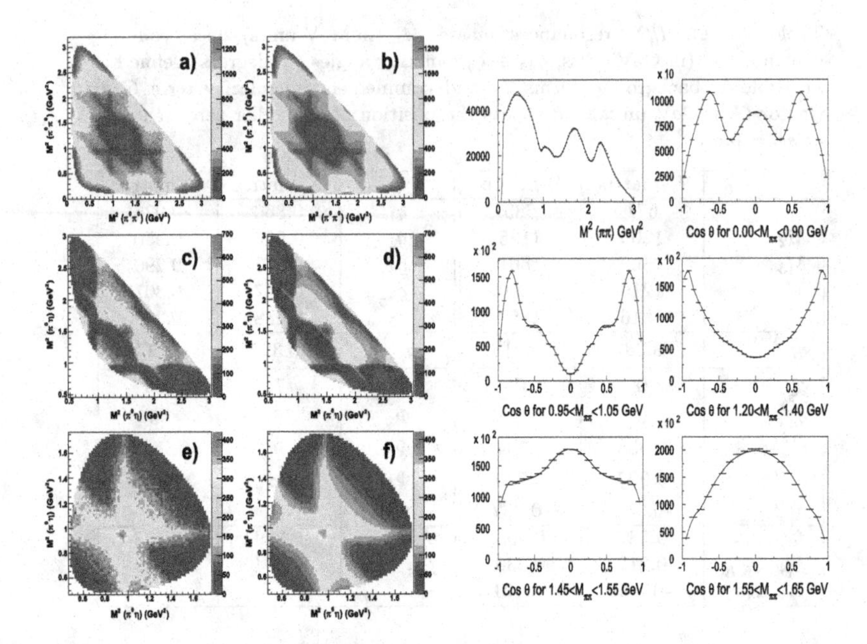

Fig. 6.2 Left panel:Description of the $\bar{p}p$ annihilation in liquid hydrogen into the $3\pi^0$ (a,b), $2\pi^0$ (c,d) and $\pi^0 2\eta$ (e,f) final states. The panels (a,c,e) show the experimental Dalitz plots and (b,d,f) the results of fit [2]. Right panel: Mass and angular projections for $\bar{p}p$ annihilation into $3\pi^0$ in liquid hydrogen.

It means that in the D-matrix fit two effective bare states should be introduced. The non-zero left and right vertices for these states can be chosen as

$$g_j^{L(N+1)} = f_{1j}\frac{1\ \text{GeV}^2 + s_0}{s+s_0} \qquad g_1^{R(N+1)} = 1 \qquad R_{N+1} = 1\,,$$
$$g_1^{L(N+2)} = 1 \qquad g_{j>1}^{R(N+2)} = f_{1j}\frac{1\ \text{GeV}^2 + s_0}{s+s_0} \qquad R_{N+2} = 1 \quad (6.33)$$

and

$$g_{j>1}^{R(N+1)} = g_{j>1}^{L(N+2)} = g_1^{R(N+2)} = 0 \quad (6.34)$$

Another alternative parametrization for the non-zero terms is [2]:

$$g_j^{L(N+1)} = f_{1j} \qquad g_1^{R(N+1)} = 1 \qquad R_{N+1} = \frac{1\ \text{GeV}^2 + s_0}{s+s_0}\,,$$
$$g_1^{L(N+2)} = 1 \qquad g_{j>1}^{R(N+2)} = f_{1j} \qquad R_{N+2} = \frac{1\ \text{GeV}^2 + s_0}{s+s_0}\,. \quad (6.35)$$

The elements of the $B^j_{\alpha\beta}$ can be calculated using one subtraction:

$$B^j_{\alpha\beta}(s) = B^j_{\alpha\beta}(M_j^2) + (s - M_j^2) \int\limits_{(m_{1j}+m_{2j})^2}^{\infty} \frac{ds'}{\pi} \frac{g_j^{R(\alpha)} \rho_j(s', m_{1j}, m_{2j}) g_j^{L(\beta)}}{(s' - s - i0)(s' - M_j^2)} . \tag{6.36}$$

The subtraction point can be taken at the corresponding two-particle threshold $M_j = (m_{1j} + m_{2j})$ and therefore the contribution of the loop diagram can be rewritten as:

$$B^j_{\alpha\beta}(s) = g_a^{R(\alpha)} \left(b^j + (s - M_j^2) \int\limits_{(m_{1j}+m_{2j})^2}^{\infty} \frac{ds'}{\pi} \frac{\rho_j(s', m_{1j}, m_{2j})}{(s' - s - i0)(s' - M_j^2)} \right) g_b^{L(\beta)} . \tag{6.37}$$

Here the parameters b^j depend on decay channels only and were optimized in the fit. Let us mention that in this case the D-matrix approach is identical to the K-matrix modified approach (6.12).

The D-matrix fit describes data base with a very similar quality as the K-matrix approach: the list of studied reactions as well as χ^2 for the K-matrix and D-matrix solutions is presented in Table 6.2. The pole positions of the scalar-isoscalar amplitude found in both solutions are given in Table 6.4 and bare masses and their couplings in Table 6.3.

Examples of the description of experimental Dalitz plots with D-matrix solution for proton-antiproton annihilation into three meson states are shown in Figs. 6.2,6.3 together with mass and angular projections for selected Dalitz plots. The description of the CERN-Munich data [7] is shown in Fig. 6.4 and the description of the S-wave intensities for $\pi\pi$ transition into different final states in Fig. 6.5.

It is seen from Table 6.3 that the masses of bare states are hardly changed from the K-matrix solution and most of the couplings are shifted by less than 20%. The positions of the amplitude poles above 900 MeV also changed very little: here both approaches can be successfully applied. The biggest change is observed for the lowest state: the pole in the mass region around 300-400 MeV. Indeed the influence of the dispersion corrections is expected to be largest here.

6.3 Partial wave analysis of baryon spectra in the frameworks of *K*-matrix and *D*-matrix methods

One of the main problem of the baryon spectrum is the problem of missing resonances. The classical quark model which is based on the interaction

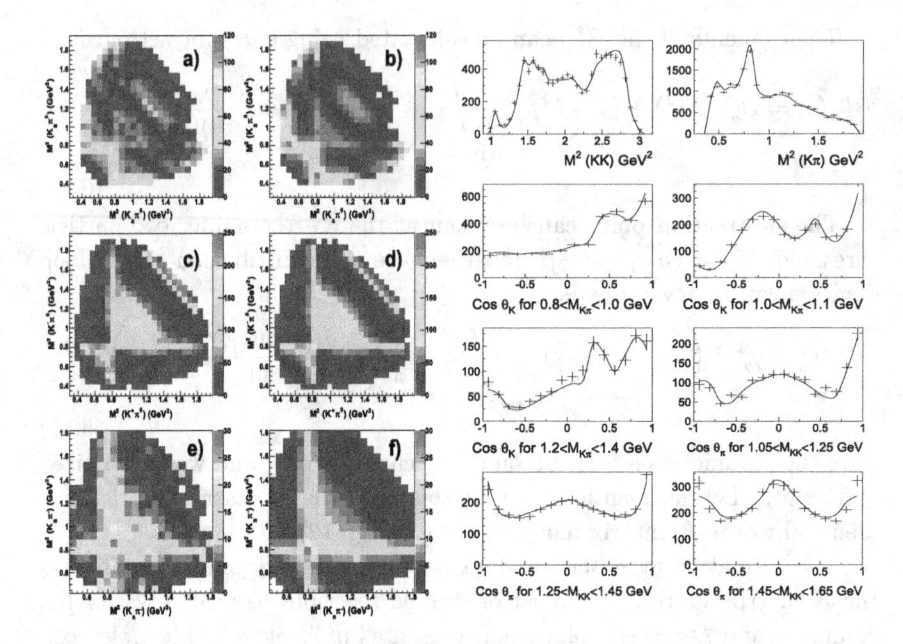

Fig. 6.3 Left panel: Description of the $\bar{p}p$ annihilation in liquid hydrogen into the $K_S K_S \pi^0$ (a,b), $K^+K^-\pi^0$ (c,d) and in liquid deuterium into $K_S K_S \pi^-$ (e,f) final states. The panels (a,d,e) show the experimental Dalitz plots and (b,d,f) the result of the fit [2]. Right panel: Mass and angular projections for $\bar{p}p$ annihilation into $K_S K_S \pi^0$ in liquid hydrogen.

of three quarks predicts a large density of the nucleon states with masses above 1900 MeV and Δ states with masses above 2000 MeV. Only a few of these states are observed experimentally. It could be that the predicted states escaped an identification due to a lack of the experimental data. Indeed, for a long time our knowledge of the baryon spectrum was based purely on the elastic πN scattering experiments performed more than 30 years ago. If a state has a small πN coupling it can not be solidly identified from these data. It should be mentioned that the classical quark model, indeed, predicts small couplings of radial excitations to the πN channel.

Nevertheless, the problem of "missing resonances" triggered a wide development of new theoretical approaches for the calculation of the baryon spectra. One of the simplest idea which reduces the number of the states is the so-called diquark model. In this model two quarks form a diquark with zero orbital momentum (S-wave) and then this diquark interacts with a spectator quark. An interesting result was obtained in the calculation of

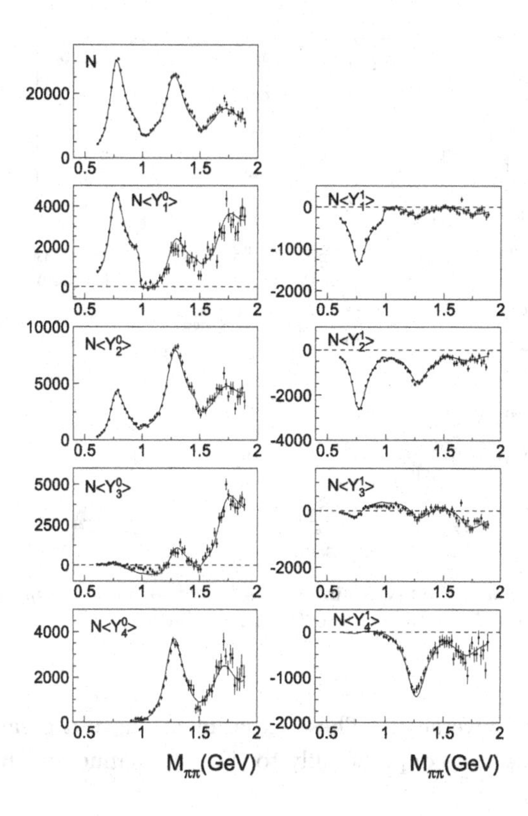

Fig. 6.4 Description of the CERN-Munich data with the solution 2.

the baryon spectrum in the AdS QCD approach. Although this approach is also based on the quar$K$-diquark interaction, it imposes different selection rules. In the case of the soft-wall approximation the AdS QCD predicts very well the baryon spectrum up to 2 GeV [8]. Another approach suggest a restoration of the chiral symmetry at large baryon masses. In this case the produced states should degenerate in parity and produced as doublets. Although this idea does not provide any numerical predictions for the spectrum, it is interesting to check it experimentally.

Though the number of states predicted by these models is less than that predicted by the classical quark model, it is still notably larger than the number of experimentally observed states. Recently a set of new experiments which study the baryon spectrum in the γN collision were launched in the USA (JLAB, CEBAF) and in Europe (CB-ELSA, Bonn University,

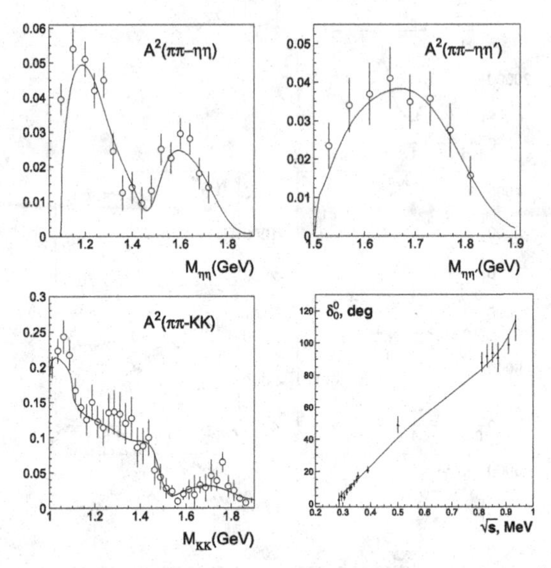

Fig. 6.5 Description of the S-wave intensities for $\pi\pi \to \eta\eta$, $\pi\pi \to \eta\eta'$ (GAMS), $\pi\pi \to KK$ (BNL) and the K_{e4} data with the D-matrix solution

MAMI, Mainz University). These experiments have a great potential to discover states which couple weakly to the πN channel but have a notable coupling to γN.

A combined analysis of pion-induced and photo-production reactions in the framework of the energy dependent approach can provide a stable result already from a restricted data base and has a large potential for the discovery of new states. Indeed, such characteristics of the resonances as mass, width, helicity coupling can be fixed from one reaction and automatically reduce the freedom in the fit of data sets with different final states. Here we demonstrate the difference between K-matrix and D-matrix approach based on the partial wave analysis performed by the Bonn-Gatchina group.

6.3.1 *Pion and photo induced reactions*

Baryon states can be classified by isospin, total spin and parity or by their decay into the πN system. In the last case the state is denoted as $L_{2I,2J}$ where L is the orbital momentum of the pion nucleon system, I is the isospin and J is the total spin of a baryon state. In turn, the decay properties of baryons into πN system can be divided into two groups: states with $J =$

L+1/2, where L is the orbital momentum of the meson-nucleon system, are called '+' states ($J^P = 1/2^-, 3/2^+, 5/2^-, \ldots$). The states with $J = L-1/2$ are called '-' states ($J^P = 1/2^+, 3/2^-, 5/2^+, \ldots$).

In the Bonn-Gatchina approach the K-matrix for the πN transition amplitude was parametrized as:

$$K_{ab}(s) = \sum_\alpha \frac{g_a^{(\alpha)} g_b^{(\alpha)}}{M_\alpha^2 - s} + f_{ab} \,. \tag{6.38}$$

The non-resonant contributions from f_{ab} were assumed to be constants except of the S_{11} channel where for elastic πN transition the following parametrization was used:

$$f_{ab} = \frac{a + b\sqrt{s}}{s + s_0} \,. \tag{6.39}$$

The pseudoscalar-meson nucleon phase volumes for decays of '+' and '-' states were calculated in [9]:

$$\rho^{(+)L}_{\mu N}(s) = \frac{\alpha_L}{2L+1}|\vec{k}|^{2L}\frac{m_N + k_{N0}}{2m_N}\rho(s, m_\nu, m_N)\,.$$
$$\rho^{(-)L}_{\mu N}(s) = \frac{\alpha_L}{L}|\vec{k}|^{2L}\frac{m_N + k_{N0}}{2m_N}\rho(s, m_\mu, m_N)\,. \tag{6.40}$$

where $\mu = \pi, \eta, K$ and $N = n, p, \Lambda, \Sigma$ and

$$\alpha_L = \prod_{l=1}^{L} \frac{2l-1}{l} = \frac{(2L-1)!!}{L!} \,. \tag{6.41}$$

The S_{11} and P_{11} partial waves were parametrized as a six channel K-matrix: πN, ηN, $K\Lambda$, $K\Sigma$, $\pi\Delta$ and $N(\pi\pi)_S$, where $(\pi\pi)_S$ is an effective description of the low energy $\pi\pi$ scalar isoscalar partial wave. The S_{31} and P_{31} partial waves were parametrized as a two-pole four channel (πN, $K\Sigma$, $\Delta\pi$, $N(1440)\pi$) K-matrix. In the P_{13} and P_{33} partial waves two channels with $\Delta\pi$ (orbital momentum $L = 1, 3$) and $D_{13}(1520)\pi$ with $L = 0$ were introduced in addition to two-body final states.

A successful description of the large set of pion induced and photo-production data was obtained with including non-resonant terms from πN channel only. The only exception is the S_{11} partial wave where the non-resonant terms from ηN channel to ηN and $\Delta\pi$ were included. It means that in the S_{11} channel 4 effective bare states should be included for the description of the non-resonant transitions in framework of the D-matrix approach:

$$\begin{array}{lll} g_j^{L(N+1)} = f_{1j} & g_1^{R(N+1)} = 1 & R_{N+1} = 1\,, \\ g_1^{L(N+2)} = 1 & g_{j>1}^{R(N+2)} = f_{1j} & R_{N+2} = 1 \\ g_{j>1}^{L(N+3)} = f_{2j} & g_2^{R(N+3)} = 1 & R_{N+3} = 1 \\ g_2^{L(N+4)} = 1 & g_{j>2}^{R(N+4)} = f_{1j} & R_{N+4} = 1 \end{array} \tag{6.42}$$

and

$$\begin{array}{l} g_{j>1}^{R(N+1)} = g_{j>1}^{L(N+2)} = g_1^{R(N+2)} = 0 \\ g_{j\neq 2}^{R(N+3)} = g_1^{L(N+3)} = g_{j\neq 2}^{L(N+4)} = g_{1,2}^{R(N+4)} = 0\,, \end{array} \tag{6.43}$$

while for other partial waves only two effective bare states were introduced.

The comparison in the description of the elastic data from the K-matrix and D-matrix fits of the GWU data [10] is shown in Fig. 6.6. The χ^2 was found to be practically the same and the difference is hardly seen on the plot.

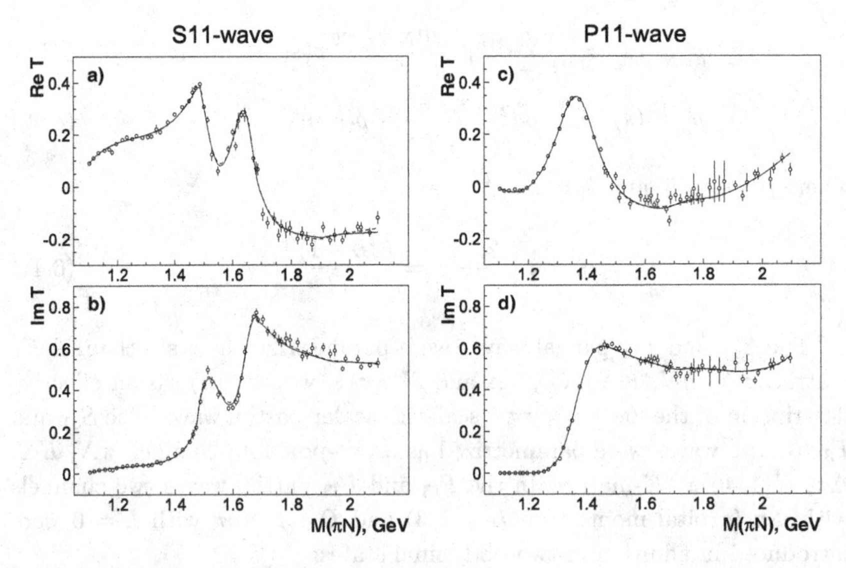

Fig. 6.6 Description of the πN elastic amplitudes with K-matrix solution BG2011-02 (solid lines) and D-matrix solution BG2011-02D (dashed lines). Left panel S_{11} and right-hand panel the P_{11} partial waves. The points with error bars are from George-Washington University energy independent solution [10].

The difference in intensity for the fit of the $\pi N \to K\Lambda$, $\gamma p \to \pi^0 p$ and $\gamma p \to \eta p$ reactions is demonstrated in Fig. 6.7. The deviations which are observed here are typical for the K-matrix solution with slightly different assumptions about behavior of the non-resonant contributions. The same effect was observed for all fitted data sets. One probably can say that the D-matrix solution provides a better description of the data near reaction production thresholds where we expect that the dispersion corrections play a notable role.

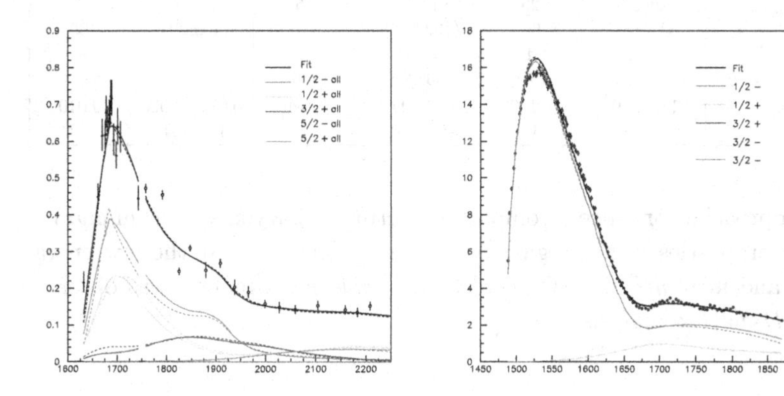

Fig. 6.7 Contribution of the leading partial waves to a) $\pi^- p \to K\Lambda$, b) $\gamma p \to \pi^0 p$ and c) $\gamma p \to \eta p$ total cross sections. The K-matrix solution BG2011-02 is shown with solid lines and D-matrix solution BG2011-02D with dashed lines.

The largest effect from the D-matrix fit was found for the S_{11} wave. The amplitude pole positions, elastic and inelastic residues in these poles are listed in Table 6.5. All these numbers are complex ones and we provide the absolute values and phases (in degrees). For the K-matrix solution the values are given with systematic errors calculated from class of solutions BG2011. These solutions differ by a contribution from weakly defined states and by the description of different (sometimes not fully compatible) data sets. For the lowest $N(1535)S_{11}$ state the values obtained in the framework of the D-matrix approach deviate at more than at one standard deviation. This is not a big surprise: this state is located very close to the ηN threshold and has a larger coupling to this channel. For the higher states the difference is much smaller and usually is less than one standard deviation.

We found even smaller differences in the description of the S_{31}, P_{11}, P_{31}, P_{33} and P_{13} partial waves in the framework of K-matrix and D-matrix approaches: all numbers obtained in the D-matrix fits were within a standard deviation from the K-matrix solutions. On this basis we conclude that

Table 6.5 Comparison of the properties of the S_{11} states from the K-matrix and D-matrix fits.

	$N(1535)S_{11}$		$N(1650)S_{11}$		$N(1890)S_{11}$	
	K-matrix	D-matrix	K-matrix	D-matrix	K-matrix	D-matrix
M_{pole}	1501±4	1494	1647±6	1651	1900±15	1905
Γ_{pole}	134±11	116	103±8	95	90^{+30}_{-15}	106
Elastic residue	31±4	25	24±3	23	1±1	1.5
Phase	-(29±5)°	-38°	-(75±12)°	-62°	–	–
$Res_{\pi N\to N\eta}$	28±3	25	15±3	15	4±2	5
Phase	-(76±8)°	-69°	(132±10)°	140	(40±20)°	42°
$Res_{\pi N\to\Delta\pi}$	7±4	4	11±3	12	–	–
Phase	(147±17)°	157°	-(30±20)°	-40	–	–
$A^{1/2}$ (GeV$^{-\frac{1}{2}}$)	0.116±0.010	0.107	0.033±0.007	0.029	0.012±0.006	0.010
Phase	(7±6)°	1°	-(9±15)°	0°	120±50°	150°

both approaches produce a compatible picture of baryon spectrum and resonance properties in the region of baryon resonances. At the low energy region and near ηN and $K\Lambda$ thresholds some improvement was observed in the D-matrix analysis.

References

[1] E. Salpeter and H.A. Bethe, Phys. Rev. **84**, 1232 (1951).

[2] A.V. Anisovich, V.A. Nikonov, A.V. Sarantsev, V.V. Anisovich, M.A. Matveev, T.O. Vulfs, K.V. Nikonov, J. Nyiri, Phys. Rev. **D 84**, 076001 (2011).

[3] V.V. Anisovich and A.V. Sarantsev, Int. J. Mod. Phys. A**24**, 2481, (2009); V.V. Anisovich and A.V. Sarantsev, Yad. Fiz. **72**, 1950 (2009) [Phys. Atom. Nucl. **72**, 1889 (2009)]; V.V. Anisovich and A.V. Sarantsev, Yad. Fiz. **72**, 1981 (2009) [Phys. Atom. Nucl. **72**, 1920 (2009)].

[4] D. V. Bugg, B. S. Zou and A. V. Sarantsev, Nucl. Phys. B **471** (1996) 59.

[5] V.V. Anisovich, Yu.D. Prokoshkin, and A.V. Sarantsev, Phys. Lett. B **389**, 388 (1996).

[6] A.V. Anisovich, V.V. Anisovich, M.A. Matveev, V.A. Nikonov, J. Nyiri and A.V. Sarantsev, *Mesons and Baryons*, World Scientific, Singapore (2008).

[7] B. Hyams et al., Nucl. Phys. **B64** (1973) 134.

[8] H. Forkel and E. Klempt, Phys. Lett. B **679**, 77 (2009).

[9] A. V. Anisovich and A. V. Sarantsev, Eur. Phys. J. A **30**, 427 (2006).

[10] http://gwdac.phys.gwu.edu/analysis/pin_analysis.html

Chapter 7

Reggeon-Exchange Technique for the Description of the Reactions of Two-Meson Diffractive Production

In this chapter we present the dispersion relation approach for a direct use of reggeon-exchange technique in the description of the reactions of two-meson diffractive production in three-meson processes at large energies. As an example we consider the reaction $\pi N \to two\,mesons + N$ at large energies of the initial pion and small momentum transfer to a nucleon. The reggeon exchange approach allows us to describe simultaneously distributions over M (invariant mass of two mesons) and t (momentum transfer squared to nucleons). Making use of this technique, the following resonances (as well as corresponding bare states), produced in the $\pi N \to \pi^0\pi^0 N$ reaction were studied [1]: $f_0(980)$, $f_0(1370)$, broad state $f_0(1200-1600)$, $f_0(1500)$, $f_0(1750)$, $f_2(1270)$, $f_2(1525)$, $f_2(1565)$, $f_2(2020)$, $f_4(2025)$. In the Appendices we present technical aspects of the fitting procedure.

7.1 Introduction

The study of the mass spectrum of hadrons and their properties is the key point for the understanding of colour particle interactions at large distances. But even the meson sector, though less complicated than the baryon one, is far from being completely understood. And the situation is only partly connected with the lack of data. In the lower mass region there is a lot of data taken from the proton–antiproton annihilation at rest (Crystal Barrel, Obelix), from the $\gamma\gamma$ interaction (L3), from the proton–proton central collisions (WA102), from J/ψ decay (Mark III, BES), from D- and B-meson decays (Focus, D0, BaBar, Belle, Cleo C) and from $\pi N \to two\,mesons + N$ reactions with high-energy pion beams (GAMS, VES, E852). Most of these data are of high statistics, thus allowing us to determine resonance properties with a high accuracy (though, let us emphasize, in the reactions $\pi N \to two\,mesons + N$ polarized-target data are lacking).

Nevertheless, in many cases there are significant contradictions between analyses performed by different groups. The ambiguities originate from two circumstances.

First, in the discussed sectors the analyses of data taken from a single experiment cannot provide us with a unique solution. A unique solution can be obtained only from the combined analysis of a large set of data taken in different experiments.

Second, there are some simplifications inherent in many analyses. The unitarity was neglected frequently even when the amplitudes were close to the unitarity limit. A striking example is that up to now there is no proper K-matrix parametrization of the 1^{--} and 2^{++} waves which are considered by many physicists as mostly understood ones. As to multiparticle final states, only a few analyses have ever considered the contributions of triangle or box singularities to the measured cross sections. However, these contributions can simulate the resonant behavior of the studied distributions, especially in the threshold region (for more detail, see [2] and references therein).

In the analysis of meson spectra in high-energy reactions $\pi N \to$ *two mesons* $+ N$, many results are related to the decomposition of the cross sections into natural and unnatural amplitudes based on certain models developed for the two-pion production at small momenta transferred, (*e.g.*, see [3, 4, 5]). However, as was discussed by the cited authors, a direct application of these methods at large momenta transferred to the analysis of data may lead to a wrong result. In addition, the $\pi N \to$ *two mesons* $+ N$ data were discussed mostly in terms of t-channel particle exchange, though without proper analysis of the t-channel exchange amplitudes.

A decade ago our group performed a combined analysis of data on proton–antiproton annihilation at rest into three pseudoscalar mesons, together with the data on two-meson S waves extracted form the $\pi N \to \pi\pi N$, $\eta\eta N$, $K\bar{K}$ and $\eta\eta' N$ reactions [6, 7, 8]. The analysis has been carried out in the framework of the K-matrix approach which preserves unitarity and analyticity of the amplitude in the two-meson physical region. Although the two-meson data extracted from the reaction $\pi N \to$ *two mesons* $+ N$ at small momentum transfer appeared to be highly compatible with those found in proton–antiproton annihilation, we have faced a set of problems, describing the $\pi N \to$ *two mesons* $+ N$ data at large momentum transfer. As we have seen now, the problems were owing to the use of partial-wave decomposition which was performed by the E852 Collaboration and showed a huge signal at 1300 MeV in the S wave.

The plan of our presentation is as follows. The analysis of the $\pi N \to$ *two mesons*$+N$ data is based on the t-channel reggeized exchanges. For the $\pi N \to$ *two mesons* $+ N$ reactions, the data at small and large momentum transfers are included. Here, as the first step, we perform the analysis in the framework of the K-matrix parametrization for all fitting channels (K-matrix approach insures the unitarity and analyticity in the physical region). At the next stage, we use the dispersion relation method (D-matrix approach) for two-meson amplitudes satisfying these requirements in the whole complex plane.

We present the method for the analysis of the πN interactions based on the t-channel reggeized exchanges supplemented by a study of the proton–antiproton annihilation at rest. The method is applied to a combined analysis of the $\pi N \to \pi^0\pi^0 N$ data taken by E852 at small and large momentum transfers and Crystal Barrel data on the proton–antiproton annihilation at rest into three neutral pseudoscalar mesons. The even waves, which contributed to this set of data, are parametrized within the K-matrix approach. To check a strong S-wave signal around 1300 MeV, which has been reported by the E852 Collaboration from the analysis of data at large momentum transfers, is a subject of a particular interest in the present analysis.

We present the results of the new K-matrix analysis of two-meson spectra in the scalar, $J^P = 0^+$, and tensor, $J^P = 2^+$, sectors: these sectors need a particular attention because just here we meet with the low-lying glueballs, $f_0(1200-1600)$ and $f_2(2000)$. The situation with the tensor glueball is rather transparent allowing us to make a definite conclusion about the gluonium structure of $f_2(2000)$, while the status of the broad state $f_0(1200 - 1600)$ requires a special discussion: this state is nearly flavor-blind but the corresponding pole of the amplitude dives deeply into the complex M plane. It is definitely seen only in the analysis of a large number of different reactions in broad intervals of mass spectra (for example, see [2] and references therein).

7.2 Meson–nucleon collisions at high energies: peripheral two-meson production in terms of reggeon exchanges

The two-meson production reactions $\pi p \to \pi\pi n$, $K\bar{K}n$, $\eta\eta n$, $\eta\eta' n$ at high energies and small momentum transfers to the nucleon are used for obtaining the S-wave amplitudes $\pi\pi \to \pi\pi$, $K\bar{K}$, $\eta\eta$, $\eta\eta'$ at $|t| < 0.2\,(\text{GeV/c})^2$ because, as commonly believed, the π exchange dominates this wave at such momentum transferred. At larger momentum transfers, $|t| \gtrsim 0.2\,(\mathrm{GeV}/c)^2$, we observe definitely a change of the regime in the S-wave production — a significant contribution of other reggeons is possible (a_1 exchange, daughter-π and daughter-a_1 exchanges). Nevertheless, the study of the two-meson production processes at $|t| \sim 0.5 - 1.5\,(\mathrm{GeV}/c)^2$ looks promising, for at such momentum transfers the contribution of the broad resonance (the scalar glueball $f_0(1200-1600)$) vanishes. Therefore, the production of other resonances (such as the $f_0(980)$ and $f_0(1300)$) appears practically without background – this is important for finding out their characteristics as well as a mechanism of their production.

7.2.1 *K-matrix and D-matrix approaches*

What we know about the reactions $\pi p \to \pi\pi n$, $K\bar{K}n$, $\eta\eta n$, $\eta\eta' n$ allows us to suggest that a consistent analysis of the peripheral two-meson production in terms of reggeon exchanges may be a good tool for studying meson resonances. Note that the investigation of two-meson scattering amplitudes by means of the reggeon-exchange expansion of the peripheral two-meson production amplitudes was proposed long ago [9] but was not used because of the lack of data until now.

7.2.1.1 *K-matrix approach*

The K-matrix amplitude of the peripheral production of two mesons with total angular momentum J reads

$$\left(\bar{\psi}_N(k_3)\hat{G}_R\psi_N(p_2)\right)R(s_{\pi N},t)\sum_J \widehat{K}^{(J)}_{\pi R(t)}(s)\frac{1}{1-i\hat{\rho}_J(s)\widehat{K}_J(s)}Q^{(J)}(k_1,k_2). \tag{7.1}$$

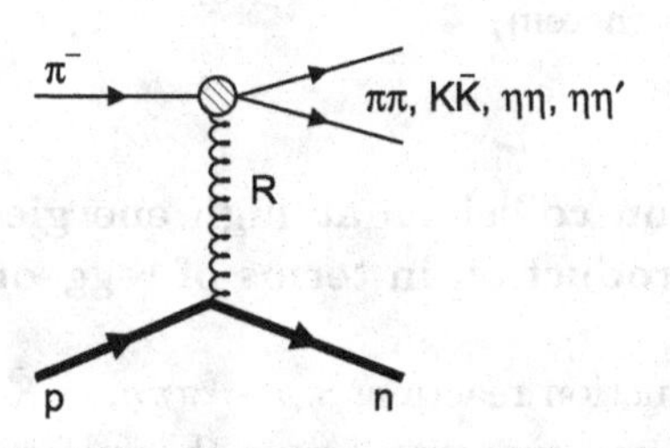

Fig. 7.1 Diagram for the production of two mesons ($\pi\pi$, $K\bar{K}$, $\eta\eta$, $\eta\eta'$) in $\pi^- p$ collisions due to reggeon (R) exchange.

This formula is illustrated by Fig. 7.1 for the production of $\pi\pi$, $K\bar{K}$, $\eta\eta$, $\eta\eta'$ systems. Here the factor $(\bar{\psi}_N(k_3)\hat{G}_R\psi_N(p_2))$ stands for the reggeon–nucleon vertex, and $\hat{G}_R$ is the spin operator; $R(s_{\pi N},t)$ is the reggeon propagator depending on the total energy squared of colliding particles, $s_{\pi N} = (p_1+p_2)^2$, and the momentum transfer squared $t = (p_2-k_3)^2$, while the factor $\widehat{K}_{\pi R(t)}[1-i\hat{\rho}_J(s)\widehat{K}_J(s)]^{-1}$ is related to the block of two-meson production; $s \equiv M^2 = (k_1+k_2)^2$, and $\hat{\rho}(s)$ is the phase space matrix. In the reactions $\pi p \to \pi\pi n$, $K\bar{K}n$, $\eta\eta n$, $\eta\eta' n$, 4π (one can approximate $4\pi \to \sigma\sigma$), the factor $\widehat{K}_{\pi R(t)}(s)[1-i\hat{\rho}_J(s)\widehat{K}_J(s)]^{-1}$ describes transitions $\pi R(t) \to \pi\pi$, $K\bar{K}$, $\eta\eta$, $\eta\eta'$: in this way the block $\widehat{K}_{\pi R(t)}$ is associated with the prompt meson production, and $[1-i\hat{\rho}_J(s)\widehat{K}_J(s)]^{-1}$ is the K-matrix factor for meson rescattering (of the type of $\pi\pi \to \pi\pi$, $\pi\pi \to K\bar{K}$, $K\bar{K} \to \eta\eta$, and so on). The prompt-production block for transition $\pi R \to b$ (where a and b run over $a,b = \pi\pi$, $K\bar{K}$, $\eta\eta$, $\eta\eta'$, $\sigma\sigma$, ...) is parametrized with singular (pole) and smooth terms [6, 8, 10]:

$$\left(\widehat{K}^{(J)}_{\pi R(t)}(s)\right)_{\pi R,b} = \sum_n \frac{G^{(n)}_{\pi R}(t)g^{(n)}_b}{\mu_n^2 - s} + f^{(J)}_{\pi R,b}(t,s),$$

$$\left(\widehat{K}_J(s)\right)_{a,b} = \sum_n \frac{g^{(n)}_a g^{(n)}_b}{\mu_n^2 - s} + f^{(J)}_{a,b}(s). \tag{7.2}$$

The pole singular term, $G^{(n)}_{\pi R}(t)g^{(n)}_b/(\mu_n^2 - s)$, determines the bare state: here $G^{(n)}_{\pi R}(t)$ is the bare-state production vertex while the parameters $g^{(n)}_b$ and μ_n are the coupling and the mass of the bare state – they are the same as in the partial-wave transition amplitudes $\pi\pi \to \pi\pi$, $K\bar{K}$, $\eta\eta$, $\eta\eta'$, $\sigma\sigma$, ... The smooth term $f_{\pi R,b}$ stands for the background production of mesons. The $G^{(n)}_{\pi R}(t)$, $f_{\pi R,b}$, $g^{(n)}_b$, μ_n are free parameters of the fitting procedure, while the characteristics of resonances are determined by poles of the K-matrix amplitude (remind that the position of poles is given by zeros of the amplitude denominator, $det|1-i\hat{\rho}_J(s)\widehat{K}_J(s)| = 0$).

The phase space matrix, $\hat{\rho}_J(s)$ is diagonal:

$$\hat{\rho}_J(s) = \mathrm{diag}\left(\rho^{(J)}_{\pi\pi}(s),\ \rho^{(J)}_{K\bar{K}}(s),\ \rho^{(J)}_{\eta\eta}(s),\ \rho^{(J)}_{\eta\eta'}(s),\ \rho^{(J)}_{\sigma\sigma}(s)\right). \tag{7.3}$$

The factor $Q^{(J)}(k_1,k_2)$ in (7.1) is related to momenta of particles in the meson block. It depends on J and the type of the reggeon.

Below we explain in detail the method of analysis of meson spectra using as an example the reactions $\pi N \to \pi\pi N$, $K\bar{K}N$, $\eta\eta N$, $\eta\eta' N$, $\pi\pi\pi\pi N$.

7.2.1.2 *D-matrix approach.*

A simple way to turn to D-matrix representation, starting from the K-matrix, is to replace in Eq. (7.1) (i) the phase space factor $\hat{\rho}_J(s)$ to the universal loop diagram $\hat{B}_J(s)$ and K-matrix elements by D-matrix ones:

$$i\,\hat{\rho}_J(s) \to \hat{B}_J(s),$$
$$\hat{B}_J(s) = \text{diag}\Big(B^{(J)}_{\pi\pi}(s),\, B^{(J)}_{K\bar{K}}(s),\, B^{(J)}_{\eta\eta}(s),\, B^{(J)}_{\eta\eta'}(s),\, B^{(J)}_{\sigma\sigma}(s)\Big) \quad (7.4)$$

and

$$\Big(\widehat{K}^{(J)}_{\pi R(t)}(s)\Big)_{\pi R,b} \to \Big(\widehat{d}^{(J)}_{\pi R(t)}(s)\Big)_{\pi R,b} = \sum_n \frac{g^{(n)}_{\pi R}(t)g^{(n)}_b}{\mu_n^2 - s} + f^{(J)}_{\pi R,b}(t,s),$$
$$\Big(\widehat{K}_J(s)\Big)_{a,b} \to \Big(\widehat{d}_J(s)\Big)_{a,b} = \sum_n \frac{g^{(n)}_a g^{(n)}_b}{\mu_n^2 - s} + f^{(J)}_{a,b}(s) \quad (7.5)$$

The non-pole terms can be considered as a contribution of distant poles.

So, the D-matrix amplitude of the peripheral production of two mesons with total angular momentum J reads:

$$\Big(\bar{\psi}_N(k_3)\hat{G}_R\psi_N(p_2)\Big)R(s_{\pi N},t)\sum_J \widehat{d}^{(J)}_{\pi R(t)}(s)\frac{1}{1-\widehat{B}_J(s)\widehat{d}_J(s)}Q^{(J)}(k_1,k_2)\,. \quad (7.6)$$

Let us present an example of the universal loop diagram for $\pi\pi$ state, in this example we make an subtraction in $s = 4m^2$:

$$B^{(J=0)}_{\pi\pi}(s) = b_{\pi\pi} + (s-4m_\pi^2)\int\limits_{4m_\pi^2}^{\infty}\frac{ds'}{\pi}\frac{1}{16\pi}\sqrt{\frac{s'-4m_\pi^2}{s'}}\frac{1}{(s'-4m_\pi^2)(s'-s-i0)}$$
$$= b_{\pi\pi} + \frac{1}{16\pi}\sqrt{\frac{s-4m_\pi^2}{s}}\left[\frac{1}{\pi}\ln\frac{\sqrt{s}-\sqrt{s-4m_\pi^2}}{\sqrt{s}+\sqrt{s_\pi-4m^2}} + i\right],$$
$$s > 4m_\pi^2. \quad (7.7)$$

Eq. (7.7) presents the loop diagram on the first sheet of the complex s-plane. Below the $\pi\pi$ threshold, $B_{\pi\pi}(s)$ is real, $\text{Im}B^{(J=0)}_{\pi\pi}(s) = 0$, and analytical at $s \to 0$:

$$B^{(J=0)}_{\pi\pi}(s\to 0) \simeq b_{\pi\pi} + \frac{i}{16\pi}\sqrt{\frac{4m_\pi^2}{s}}\left[\frac{1}{\pi}\left(-i\pi + 2i\sqrt{\frac{s}{4m_\pi^2}}\right) + i\right], \quad (7.8)$$

On the first sheet the only singularity of $B(s)$ is the threshold one, at $s = 4m^2$. The singularity $s = 0$, being absent on the first sheet, is located on the second sheet which can be reached from (7.7) at $\sqrt{s-4m^2} \to -i\sqrt{4m^2-s}$ with decreasing s.

Certainly, one can use other presentations of the loop diagram, for example, without subtraction but with a universal cutting factor in the integrand.

A more detailed discussion of the connection between the D-matrix and K-matrix representations can be found in Appendix A.

7.2.2 *Reggeized pion-exchange trajectories for the waves* $J^{PC} = 0^{++}, 1^{--}, 2^{++}, 3^{--}, 4^{++}$

Here we present the technique of the analysis of high-energy reaction $\pi^- p \to$ *mesons* $+ n$, with the production of mesons in the $J^{PC} = 0^{++}, 1^{--}, 2^{++}, 3^{--}, 4^{++}$ states at small and moderate momenta transferred to the nucleon. The partial-wave decomposition of the produced meson states describes directly the measured cross sections without using the published moment expansions (which usually were done under some simplifying assumptions – it is discussed below in more detail).

7.2.2.1 *Kinematics for reggeon exchange amplitudes*

First, we consider the process $\pi^- p \to \pi\pi + n$ in the c.m. system of the reaction and present the momenta of the incoming and outgoing particles (below we use the notation for four-vectors: $x = (x_0, \mathbf{x}_\perp, x_z)$.

We have for the incoming particles:

$$\begin{aligned}
\text{pion momentum}: &\quad p_1 = (p_z + \frac{m_\pi^2}{2p_z}, 0, p_z)\,, \\
\text{proton momentum}: &\quad p_2 = (p_z + \frac{m_N^2}{2p_z}, 0, -p_z)\,, \\
\text{total energy squared}: &\quad s_{\pi N} = (p_1 + p_2)^2\,.
\end{aligned} \tag{7.9}$$

Here we have performed an expansion over the large momentum p_z. Likewise, we write for the outgoing particles:

$$\begin{aligned}
&\text{meson momenta}\,(i = 1, 2): \quad k_i = (k_{iz} - \frac{m_i^2 + k_{i\perp}^2}{2k_{iz}}, \mathbf{k}_{i\perp}, k_{iz})\,, \\
&\text{total momentum of mesons}: \\
&P = k_1 + k_2 = p_1 - q = (p_z + \frac{s + m_\pi^2 + 2q_\perp^2}{4p_z}, \mathbf{q}_\perp, p_z - \frac{s - m_\pi^2}{4p_z})\,, \\
&\text{proton momentum}: \\
&k_3 = p_2 + q = (p_z - \frac{s - m_\pi^2 + 2q_\perp^2}{4p_z}, -\mathbf{q}_\perp, -p_z + \frac{s - m_\pi^2}{4p_z})\,, \\
&\text{energy squared of mesons}: \quad s = P^2 = (k_1 + k_2)^2\,.
\end{aligned} \tag{7.10}$$

The relative momenta of mesons in the initial and final states read:

$$p = \frac{1}{2}(p_1 + q), \quad k = \frac{1}{2}(k_1 - k_2). \tag{7.11}$$

The momentum squared transferred to the nucleon is comparatively small: $t \equiv q^2 \sim m_N^2 << s_{\pi N}$ where

$$q = (-\frac{s + m_\pi^2 + 2q_\perp^2}{4p_z}, -\mathbf{q}_\perp, \frac{s - m_\pi^2}{4p_z}). \tag{7.12}$$

Neglecting $O(1/p_z^2)$-terms, one has $q \simeq (0, -\mathbf{q}_\perp, 0)$ and $q^2 \simeq -q_\perp^2$.

7.2.2.2 *Amplitude with leading and daughter pion trajectory exchanges*

The amplitude with t-channel pion trajectory exchanges can be written as follows:

$$A^{(\pi-\text{traj.})}_{\pi p \to \pi\pi n} = \sum_{R(\pi_j)} A\Big(\pi R(\pi_j) \to \pi\pi\Big) R_{\pi_j}(s_{\pi N}, q^2)\left(\varphi_n^+(\boldsymbol{\sigma}\mathbf{q}_\perp)\varphi_p\right) g_{pn}^{(\pi_j)}(t). \tag{7.13}$$

The summation is carried out over the leading and daughter trajectories. Here, $A(\pi R(\pi_j) \to \pi\pi)$ is the transition amplitude for the meson block in Fig. 7.1, $g_{pn}^{(\pi_j)}$ is the reggeon–NN coupling and $R_{\pi_j}(s_{\pi N}, q^2)$ is the reggeon propagator:

$$R_{\pi_j}(s_{\pi N}, q^2) = \exp\left(-i\frac{\pi}{2}\alpha_\pi^{(j)}(q^2)\right) \frac{(s_{\pi N}/s_{\pi N0})^{\alpha_\pi^{(j)}(q^2)}}{\sin\left(\frac{\pi}{2}\alpha_\pi^{(j)}(q^2)\right)\Gamma\left(\frac{1}{2}\alpha_\pi^{(j)}(q^2)+1\right)}. \tag{7.14}$$

The π–reggeon has a positive signature, $\xi_\pi = +1$. Following [2, 11, 12, 13], we use for pion trajectories:

$$\alpha_\pi^{(\text{leading})}(q^2) \simeq -0.015 + 0.72q^2, \quad \alpha_\pi^{(\text{daughter}-1)}(q^2) \simeq -1.10 + 0.72q^2, \tag{7.15}$$

where the slope parameters are given in $(\text{GeV}/c)^{-2}$ units. The normalization parameter $s_{\pi N0}$ is of the order of 2–20 GeV^2. To eliminate the poles at $q^2 < 0$ we introduce Gamma-functions in the reggeon propagators (recall that $1/\Gamma(x) = 0$ at $x = 0, -1, -2, \ldots$).

For the nucleon–reggeon vertex $\hat{G}_{pn}^{(\pi)}$ we use in the infinite momentum frame the two-component spinors φ_p and φ_n (see, for example, [2, 11, 14]):

$$g_\pi(\bar{\psi}(k_3)\gamma_5\psi(p_2)) \longrightarrow \left(\varphi_n^+(\boldsymbol{\sigma}\mathbf{q}_\perp)\varphi_p\right) g_{pn}^{(\pi)}(t). \tag{7.16}$$

As to the meson–reggeon vertex, we use the covariant representation [2, 11, 15]. For the production of two pseudoscalar particles, it reads:

$$A\Big(\pi R(\pi_j)\to\pi\pi\Big)=\sum_J A^{(J)}_{\pi R(\pi_j)\to\pi\pi}(s)X^{(J)}_{\mu_1\dots\mu_J}(p^\perp)$$
$$\times(-1)^J O^{\mu_1\dots\mu_J}_{\nu_1\dots\nu_J}(\perp P)X^{(J)}_{\nu_1\dots\nu_J}(k^\perp)\,\xi_J\,,$$
$$\xi_J=\frac{16\pi(2J+1)}{\alpha_J},\qquad \alpha_J=\prod_{n=1}^{J}\frac{2n-1}{n}\,. \qquad (7.17)$$

Let us remind that the notation $O^{\mu_1\dots\mu_J}_{\nu_1\dots\nu_J}(\perp P)$ means that the operator $O^{\mu_1\dots\mu_J}_{\nu_1\dots\nu_J}$ acts in the space orthogonal to P.

The angular-momentum operators are constructed of momenta $p^\perp$ and $k^\perp$ which are orthogonal to the momentum of the two-pion system $P=k_1+k_2$. Recall: $g^\perp_{\mu\nu}=g_{\mu\nu}-P_\mu P_\nu/P^2$, $k^\perp_\mu=(k_1-k_2)_\nu g^\perp_{\mu\nu}/2$ and $p^\perp_\mu=(p_1+q)_\nu g^\perp_{\mu\nu}/2$. The coefficient ξ_J normalizes the angular-momentum operators, so that the unitarity condition appears in a simple form (for technical details in calculation meson spectra see Appendices B and C).

In the K-matrix representation, the amplitude $A^{(J)}_{\pi R(\pi_j)\to\pi\pi}(s)$, being determined by (7.1), reads:

$$A^{(J)}_{\pi R(\pi_j)\to\pi\pi}(s)=\sum_{a=\pi\pi,K\bar K,\dots}\widehat{K}^{(J)}_{\pi R(t),a}(s)\left(\frac{1}{1-i\hat\rho_J(s)\widehat{K}_J(s)}\right)_{a,\pi\pi}. \qquad (7.18)$$

For the D-matrix representation, following (7.6), we write:

$$A^{(J)}_{\pi R(\pi_j)\to\pi\pi}(s)=\sum_{a=\pi\pi,K\bar K,\dots}\widehat{d}^{(J)}_{\pi R(t),a}(s)\left(\frac{1}{1-\widehat{B}_J(s)\widehat{d}_J(s)}\right)_{a,\pi\pi}. \qquad (7.19)$$

The above consideration demonstrates that the procedure of reforming the K-matrix formulae into the D-matrix ones is a standard one, so below we present the K-matrix formulae only.

7.2.2.3 *The t-channel π_2 exchange*

The π_2-exchange is needed for the description of the differential cross sections (see Appendix C). The effects appear owing to the interference in the two-meson production amplitude.

The π_2-exchange amplitude is written as:

$$\sum_a A_{\alpha\beta}\Big(\pi R(\pi_2)\to\pi\pi\Big)\varepsilon^{(a)}_{\alpha\beta}R_{\pi_2}(s_{\pi N},q^2)\frac{\varepsilon^{(a)+}_{\alpha'\beta'}}{s^2_{\pi N}}$$
$$\times X^{(2)}_{\alpha'\beta'}(k_3^{\perp q})\left(\varphi^+_n(\boldsymbol{\sigma}\mathbf{q}_\perp)\varphi_p\right)g^{(\pi_2)}_{pn}(t), \qquad (7.20)$$

where $A_{\alpha\beta}\Big(\pi R(\pi_2) \to \pi\pi\Big)$ is the meson block of the amplitude related to the π_2-reggeized t-channel transition, $g_{pn}^{(\pi_2)}$ is the reggeon–pn vertex, $R_{\pi_2}(s_{\pi N}, q^2)$ is the reggeon propagator, and $\varepsilon_{\alpha\beta}^{(a)}$ is the polarization tensor for the 2^{-+} state. Let us remind that k_3 is the momentum of the outgoing nucleon and $k_{3\mu}^{\perp q} = g_{\mu\nu}^{\perp q} k_{3\nu}$ with $g_{\mu\nu}^{\perp q} = g_{\mu\nu} - q_\mu q_\nu / q^2$.

The π_2 particles are located on the pion trajectories and are described by a similar reggeized propagator. But in the meson block, the 2^{-+} state exchange leads to vertices different from those in the 0^{-+} exchange, so it is convenient to single out these contributions. Therefore, we use for $R_{\pi_2}(s_{\pi N}, q^2)$ the propagator given by (7.14) but with an eliminated $\pi(0^{-+})$ contribution:

$$R_{\pi_2}(s_{\pi N}, q^2) = \exp\left(-i\frac{\pi}{2}\alpha_\pi^{(\text{leading})}(q^2)\right) \times \frac{(s_{\pi N}/s_{\pi N0})^{\alpha_\pi^{(\text{leading})}(q^2)}}{\sin\left(\frac{\pi}{2}\alpha_\pi^{(\text{leading})}(q^2)\right)\Gamma\left(\frac{1}{2}\alpha_\pi^{(\text{leading})}(q^2)\right)}. \tag{7.21}$$

Taking into account that

$$\sum_{a=1}^{5} \varepsilon_{\alpha\beta}^{(a)} \varepsilon_{\alpha'\beta'}^{(a)+} = \frac{1}{2}\left(g_{\alpha\alpha'}^{\perp q} g_{\beta\beta'}^{\perp q} + g_{\beta\alpha'}^{\perp q} g_{\alpha\beta'}^{\perp q} - \frac{2}{3} g_{\alpha\beta}^{\perp q} g_{\alpha'\beta'}^{\perp q}\right), \tag{7.22}$$

one obtains:

$$\frac{X_{\alpha'\beta'}^{(2)}(k_3^{\perp q})}{2s_{\pi N}^2}\left(g_{\alpha\alpha'}^{\perp q} g_{\beta\beta'}^{\perp q} + g_{\beta\alpha'}^{\perp q} g_{\alpha\beta'}^{\perp q} - \frac{2}{3} g_{\alpha\beta}^{\perp q} g_{\alpha'\beta'}^{\perp q}\right) = \frac{3}{2}\frac{k_{3\alpha}^{\perp q} k_{3\beta}^{\perp q}}{s_{\pi N}^2} - \frac{4m_N^2 - q^2}{8s_{\pi N}^2}\left(g_{\alpha\beta} - \frac{q_\alpha q_\beta}{q^2}\right). \tag{7.23}$$

In the large-momentum limit of the initial pion, the second term in (7.24) is always small and can be neglected, while the convolution of $k_{3\alpha}^{\perp q} k_{3\beta}^{\perp q}$ with the momenta of the meson block results in the term $\sim s_{\pi N}^2$. Hence, the amplitude for π_2 exchange can be rewritten as follows:

$$A_{\pi p \to \pi\pi n}^{(\pi_2-\text{exchange})} = \frac{3}{2} A_{\alpha\beta}(\pi R(\pi_2) \to \pi\pi) \frac{k_{3\alpha}^{\perp q} k_{3\beta}^{\perp q}}{s_{\pi N}^2} \times R_{\pi_2}(s_{\pi N}, q^2)\left(\varphi_n^+(\boldsymbol{\sigma}\mathbf{q}_\perp)\varphi_p\right) g_{pn}^{(\pi_2)}. \tag{7.24}$$

A resonance with spin J and fixed parity can be produced owing to the π_2 exchange with three angular momenta $L = J - 2$, $L = J$ and $L = J + 2$, so we have

$$A_{\alpha\beta}(\pi R(\pi_2) \to \pi\pi) = \sum_J A^{(J)}_{+2}(s) X^{(J+2)}_{\alpha\beta\mu_1\ldots\mu_J}(p^\perp)$$
$$\times(-1)^J O^{\mu_1\ldots\mu_J}_{\nu_1\ldots\nu_J}(\perp P) X^{(J)}_{\nu_1\ldots\nu_J}(k^\perp)\xi_J$$
$$+\sum_J A^{(J)}_0(s) O^{\alpha\beta}_{\chi\tau}(\perp q) X^{(J)}_{\chi\mu_2\ldots\mu_J}(p^\perp)(-1)^J O^{\tau\mu_2\ldots\mu_J}_{\nu_1\nu_2\ldots\nu_J}(\perp P) X^{(J)}_{\nu_1\ldots\nu_J}(k^\perp)\xi_J$$
$$+\sum_J A^{(J)}_{-2}(s) X^{(J-2)}_{\mu_3\ldots\mu_J}(p^\perp)(-1)^J O^{\alpha\beta\mu_3\ldots\mu_J}_{\nu_1\nu_2\nu_3\ldots\nu_J}(\perp P) X^{(J)}_{\nu_1\ldots\nu_J}(k^\perp)\xi_J\,. \quad (7.25)$$

The sum of the two terms presented in (7.13) and (7.24) gives us an amplitude with a full set of the π_j-meson exchanges.

Let us emphasize an important point: in the K-matrix representation the amplitudes $A^{(J)}_{\pi R(\pi_j)\to\pi\pi}(s)$ (Eq. (7.17), $j = leading, daughter\text{-}1$) and $A^{(J)}_{+2}(s)$, $A^{(J)}_0(s)$, $A^{(J)}_{-2}(s)$ (Eq. (7.25)) differ only due to the prompt-production K-matrix block (the term $\widehat{K}_{\pi R(t)}(s)$ in (7.1)), while the final-state-interaction factor ($[1 - i\hat{\rho}(s)\widehat{K}(s)]^{-1}$ in (7.1)) is the same for each J.

7.2.3 *Amplitudes with a_J-trajectory exchanges*

Here we present formulae for leading and daughter a_1 trajectories and the leading a_2 trajectory.

7.2.3.1 *Amplitudes with leading and daughter a_1-trajectory exchanges*

The amplitude with t-channel a_1 exchanges is a sum of leading and daughter trajectories:

$$A^{(a_1-\text{trajectories})}_{\pi p\to\pi\pi n} = \sum_{a_1^{(j)}} A\left(\pi R(a_1^{(j)}) \to \pi\pi\right)$$
$$\times\, R_{a_1^{(j)}}(s_{\pi N}, q^2) i \left(\varphi^+_n (\boldsymbol{\sigma}\mathbf{n}_z)\varphi_p\right) g^{(a_{1j})}_{pn}(t)\,, \quad (7.26)$$

where $g^{(a_{1j})}_{pn}$ is the reggeon–NN coupling and the reggeon propagator $R_{a_1^{(j)}}(s_{\pi N}, q^2)$ has the form:

$$R_{a_1^{(j)}}(s_{\pi N}, q^2) = i\exp\left(-i\frac{\pi}{2}\alpha^{(j)}_{a_1}(q^2)\right)\frac{(s_{\pi N}/s_{\pi N0})^{\alpha^{(j)}_{a_1}(q^2)}}{\cos\left(\frac{\pi}{2}\alpha^{(j)}_{a_1}(q^2)\right)\Gamma\left(\frac{1}{2}\alpha^{(j)}_{a_1}(q^2)+\frac{1}{2}\right)}\,. \quad (7.27)$$

Recall that the a_1 trajectories have a negative signature, $\xi_\pi = -1$. Here we take into account the leading and first daughter trajectories which are linear and have a universal slope parameter [11, 12, 13] (see also Appendix D):

$$\alpha_{a_1}^{(\text{leading})}(q^2) \simeq -0.10 + 0.72q^2, \quad \alpha_{a_1}^{(\text{daughter}-1)}(q^2) \simeq -1.10 + 0.72q^2. \tag{7.28}$$

As previously, the normalization parameter $s_{\pi N0}$ is of the order of 2–20 GeV^2, and the Gamma-functions in the reggeon propagators are introduced in order to eliminate the poles at $q^2 < 0$.

For the nucleon–reggeon vertex we use two-component spinors in the infinite momentum frame, φ_p and φ_n, so the vertex reads $(\varphi_n^+ i(\boldsymbol{\sigma}\mathbf{n}_z)\varphi_p)\, g_{pn}^{(a_1)}$, where $\mathbf{n}_z$ is the unit vector directed along the nucleon momentum in the c.m. frame of colliding particles.

At fixed partial wave $J^{PC} = J^{++}$, the $\pi R(a_1^j)$ channel (j = *leading*, *daughter*-1) is characterized by two angular momenta $L = J + 1$, $L = J - 1$, therefore we have two amplitudes for each J:

$$\begin{aligned} A\Big(\pi R(a_1^{(j)}) \to \pi\pi\Big) = \sum_J \epsilon_\beta^{(-)} \Big[& A^{(J+)}_{\pi a_1^{(j)} \to \pi\pi}(s) X^{(J+1)}_{\beta\mu_1\ldots\mu_J}(p^\perp) \\ & + A^{(J-)}_{\pi a_1^{(j)} \to \pi\pi}(s) Z_{\mu_1\ldots\mu_J,\beta}(p^\perp)\Big] \\ & \times (-1)^J O^{\mu_1\ldots\mu_J}_{\nu_1\ldots\nu_J}(\perp P) X^{(J)}_{\nu_1\ldots\nu_J}(k^\perp), \end{aligned} \tag{7.29}$$

where the polarization vector $\epsilon_\beta^{(-)} \sim n_\beta^{(-)}$; the GLF-vectors [16] defined in the c.m. system of the colliding particles are

$$n_\beta^{(-)} = (1,0,0,-1)/2p_z, \quad n_\beta^{(+)} = (1,0,0,1)/2p_z \tag{7.30}$$

with $p_z \to \infty$.

The products of Z and X operators can be expressed through vectors $V_\beta^{(J+)}$ and $V_\beta^{(J-)}$:

$$\begin{aligned} & X^{(J+1)}_{\beta\mu_1\ldots\mu_J}(p^\perp)(-1)^J X_{\mu_1\ldots\mu_J}(k^\perp) = \alpha_J(\sqrt{-p_\perp^2})^{J+1}(\sqrt{-k_\perp^2})^J V_\beta^{(J+)}, \\ & V_\beta^{(J+)} = \frac{1}{J+1}\left[P'_{J+1}(z)\frac{p_\beta^\perp}{\sqrt{-p_\perp^2}} - P'_J(z)\frac{k_\beta^\perp}{\sqrt{-k_\perp^2}}\right], \\ & Z_{\mu_1\ldots\mu_J,\beta}(p^\perp)(-1)^J X^{(J)}_{\mu_1\ldots\mu_J}(k^\perp) = \alpha_J(\sqrt{-p_\perp^2})^{J-1}(\sqrt{-k_\perp^2})^J V_\beta^{(J-)}, \\ & V_\beta^{(J-)} = \frac{1}{J}\left[P'_{J-1}(z)\frac{p_\beta^\perp}{\sqrt{-p_\perp^2}} - P'_J(z)\frac{k_\beta^\perp}{\sqrt{-k_\perp^2}}\right]. \end{aligned} \tag{7.31}$$

Here $k_\perp^2$, $p_\perp^2$ and z are defined as: $k_\perp^2 = (k^\perp k^\perp)$, $p_\perp^2 = (p^\perp p^\perp)$, $z = \Big(-(k^\perp p^\perp)\Big)/\Big(\sqrt{-k_\perp^2}\sqrt{-p_\perp^2}\Big)$.

7.2.3.2 *The amplitude with a_2-trajectory exchange*

The amplitude with t-channel a_2-trajectory exchange reads:

$$A^{(a_2)}_{\pi p\to\pi\pi n} = \sum_a A_{\alpha\beta}\left(\pi R(a_2)\to\pi\pi\right)\varepsilon^{(a)}_{\alpha\beta} R_{a_2}(s_{\pi N},q^2)\frac{\varepsilon^{(a)+}_{\alpha'\beta'}}{s^2_{\pi N}}$$
$$\times X^{(2)}_{\alpha'\beta'}(k_3^{\perp q})\Big(\bar\psi(k_3)\psi(p_2)\Big) g^{(a_2)}_{pn}(q^2)\,, \qquad (7.32)$$

where $g^{(a_2)}_{pn}$ is the reggeon–NN coupling and the reggeon propagator $R_{a_2}(s_{\pi N},q^2)$ has the form:

$$R_{a_2}(s_{\pi N},q^2) = \exp\left(-i\frac{\pi}{2}\alpha_{a_2}(q^2)\right)\frac{\left(s_{\pi N}/s_{\pi N0}\right)^{\alpha_{a_2}(q^2)}}{\sin\left(\frac{\pi}{2}\alpha_{a_2}(q^2)\right)\Gamma\left(\frac{1}{2}\alpha_{a_2}(q^2)\right)}. \qquad (7.33)$$

Recall that the leading a_2 trajectory has a positive signature, $\xi_\pi = +1$, it is linear with the following slope parameter [11, 12, 13]:

$$\alpha_{a_2}(q^2) = 0.45\pm0.05 + (0.72\pm0.05)q^2\,. \qquad (7.34)$$

As previously, the normalization parameter $s_{\pi N0}$ is of the order of 2–20 GeV^2, and the Gamma-function in the reggeon propagator is introduced in order to eliminate the poles at $q^2 < 0$.

Using Eqs. (7.22), (7.24), we obtain:

$$A^{(a_2)}_{\pi p\to\pi\pi n} = \frac{3}{2}A_{\alpha\beta}\left(\pi R(a_2)\to\pi\pi\right)\frac{k^{\perp q}_{3\alpha}k^{\perp q}_{3\beta}}{s^2_{\pi N}}$$
$$\times\, R_{a_2}(s_{\pi N},q^2)\Big(\bar\psi(k_3)\psi(p_2)\Big) g^{(a_2)}_{pn}(q^2)\,. \qquad (7.35)$$

Due to the a_2 exchange, the resonance with spin J can be produced from orbital momentum either $J-1$ or $J+1$. Thus,

$$A_{\alpha\beta}(\pi R(a_2)\to\pi\pi) = \sum_J\left(A^{(J)}_{-1}(s)T^{(J-1)}_{\alpha\beta} + A^{(J)}_{+1}(s)T^{(J+1)}_{\alpha\beta}\right), \qquad (7.36)$$

where

$$T^{(J-1)}_{\alpha\beta} = \varepsilon_{\xi\alpha\tau\eta}\frac{P_\eta}{\sqrt{s}}X^{(J-1)}_{\xi\mu_3\ldots\mu_J}(p^\perp)O^{\tau\beta\mu_3\ldots\mu_J}_{\nu_1\ldots\nu_J}(\perp P)(-1)^J X^{(J)}_{\nu_1\ldots\nu_J}(k^\perp),$$
$$T^{(J+1)}_{\alpha\beta} = \varepsilon_{\xi\alpha\tau\eta}\frac{P_\eta}{\sqrt{s}}X^{(J+1)}_{\xi\beta\mu_2\ldots\mu_J}(p^\perp)O^{\tau\mu_2\ldots\mu_J}_{\nu_1\ldots\nu_J}(\perp P)(-1)^J X^{(J)}_{\nu_1\ldots\nu_J}(k^\perp). \qquad (7.37)$$

Taking into account that the tensors $T^{(J\pm1)}_{\alpha\beta}$ convolute with the symmetrical tensor $k^{\perp q}_{3\alpha}k^{\perp q}_{3\beta}$, we obtain:

$$
\begin{aligned}
T^{(J-1)}_{\alpha\beta}k^{\perp q}_{3\alpha}k^{\perp q}_{3\beta} &= \frac{\varepsilon_{p\alpha kP}}{\sqrt{s}}\frac{\alpha_{J-1}}{J(J-1)}\frac{\left(\sqrt{p^2_\perp k^2_\perp}\right)^{J-1}}{\sqrt{-p^2_\perp}} \\
&\times\left(P''_J(z)\frac{k^\perp_\beta}{\sqrt{-k^2_\perp}}-P''_{J-1}\frac{p^\perp_\beta}{\sqrt{-p^2_\perp}}\right)k^{\perp q}_{3\alpha}k^{\perp q}_{3\beta}, \\
T^{(J+1)}_{\alpha\beta}k^{\perp q}_{3\alpha}k^{\perp q}_{3\beta} &= -\frac{\alpha_{J+1}}{J}\varepsilon_{p\alpha kP}\frac{\left(\sqrt{p^2_\perp k^2_\perp}\right)^{J-1}}{\sqrt{s}}p^\perp_\beta P'_J(z)k^{\perp q}_{3\alpha}k^{\perp q}_{3\beta} \\
&-\frac{p^2_\perp(J-1)\alpha_J}{(J+1)\alpha_{J-1}}T^{(J-1)}_{\alpha\beta}k^{\perp q}_{3\alpha}k^{\perp q}_{3\beta}\,.
\end{aligned}
\qquad (7.38)
$$

7.2.3.3 *Calculations in the Godfrey–Jackson system*

In the c.m. system of the produced mesons, which is used for the calculation of the meson block (the GJ system), we write:

$$
\epsilon^{(-)}_\beta=\frac{1}{s_{\pi N}}\left(k_{3\mu}-\frac{q_\mu}{2}\right). \qquad (7.39)
$$

In this system the momenta are as follows:

$$
\begin{aligned}
&p^{\perp P}_1\equiv p_\perp=(0,0,0,p),\quad p^2=\frac{(s+m^2_\pi-t)^2}{4s}-m^2_\pi,\quad k^2=\frac{s}{4}-m^2_\pi, \\
&k^{\perp P}_1\equiv k_\perp=(0,k\sin\Theta\cos\varphi,k\sin\Theta\sin\varphi,k\cos\Theta), \\
&q=(q_0,0,0,p),\qquad q_0=(s-m^2_\pi+t)/(2\sqrt{s})\,, \\
&k_3=(k_{30},k_{3x},0,k_{3z}),\quad k_{30}=(s_{\pi N}-s-m^2_n)/(2\sqrt{s}), \\
&k_{3z}=(2k_{30}q_0-t)/(2p)\,.
\end{aligned}
\qquad (7.40)
$$

Recall that we use the notation $A=(A_0,A_x,A_y,A_z)$ and $\cos\Theta\equiv z=-(k^\perp p^\perp)/(\sqrt{-k^2_\perp}\sqrt{-p^2_\perp})$.

For the a_1 exchange the convolutions $V^{(J+)}_\beta(k_{3\beta}-\frac{q_\beta}{2})$, $V^{(J-)}_\beta(k_{3\beta}-\frac{q_\beta}{2})$ give us the amplitude for the transition $\pi R(a^{(j)}_1)$ into two pions (in a GJ-system the momentum $\mathbf{k}_3$ is usually situated in the (xz)-plane). We write the amplitude in the form

$$
A(\pi R(a^{(j)}_1)\to\pi\pi)=\sum_J\alpha_J p^{J-1}k^J\left(W^{(J)}_0(s)Y^0_J(\Theta,\varphi)+W^{(J)}_1(s)\mathrm{Re}Y^1_J(\Theta,\varphi)\right) \qquad (7.41)
$$

where the coefficients $W_0^{(J)}(s)$, $W_1^{(J)}(s)$ are easily calculated:

$$W_0^{(J)} = \sum_i -N_{J0}\left(k_{3z} - \frac{|\mathbf{p}|}{2}\right)\left(|\mathbf{p}|^2 A^{(J+)}_{\pi a_1^{(i)}\to\pi\pi} - A^{(J-)}_{\pi a_1^{(i)}\to\pi\pi}\right),$$

$$W_1^{(J)} = \sum_i -\frac{N_{J1}}{J(J+1)}k_{3x}\left(|\mathbf{p}|^2 J\, A^{(J+)}_{\pi a_1^{(i)}\to\pi\pi} + (J+1)A^{(J-)}_{\pi a_1^{(i)}\to\pi\pi}\right). \quad (7.42)$$

For a_2 exchange, one has:

$$T^{(J-1)}_{\alpha\beta}k^{\perp q}_{3\alpha}k^{\perp q}_{3\beta} = \frac{\alpha_{J-1}}{J}p^{J-1}k^J k_{3x}\Big[(k_{3z}-\frac{p}{2})N_{1J}\,\mathrm{Im}\,Y^1_J(\Theta,\varphi) - \frac{k_{3x}}{2}\frac{N_{2J}}{J-1}\mathrm{Im}\,Y^2_J(\Theta,\varphi)\Big] \quad (7.43)$$

For the amplitude with orbital momentum $J+1$, we write:

$$T^{(J+1)}_{\alpha\beta}k^{\perp q}_{3\alpha}k^{\perp q}_{3\beta} = -\alpha_{J+1}p^{J+1}k^J\left(k_{3z}-\frac{p}{2}\right)\frac{N_{1J}}{J}\mathrm{Im}\,Y^1_J(\Theta,\varphi) - \frac{p_\perp^2(J-1)\alpha_J}{(J+1)\alpha_{J-1}}T^{(J-1)}_{\alpha\beta}k^{\perp q}_{3\alpha}k^{\perp q}_{3\beta}\,. \quad (7.44)$$

The final expression for the a_2-exchange amplitude can be written as follows:

$$A^{(a_2)}_{\pi p\to\pi\pi n} = \frac{3k_{3x}}{2s^2_{\pi N}}\sum_J p^{J-1}k^J\Big[W^{1J}_{a_2}\mathrm{Im}\,Y^1_J(\Theta,\varphi) + W^{2J}_{a_2}\mathrm{Im}\,Y^2_J(\Theta,\varphi)\Big] \times R_{a_2}(s_{\pi N},q^2)\Big(\bar\psi(k_3)\psi(p_2)\Big)g^{(a_2)}_{pn}(q^2)\,, \quad (7.45)$$

where

$$W^{1J}_{a_2} = \frac{N_{1J}}{J}\left(k_{3z}-\frac{p}{2}\right)\Big[-p^2\alpha_{J+1}A^{(J)}_{+1} + \left(\frac{\alpha_{J-1}}{J-1}A^{(J)}_{-1} + \frac{p^2\alpha_J}{J+1}A^{(J)}_{+1}\right)(J-1)\Big]$$

$$W^{2J}_{a_2} = -\frac{N_{2J}}{J}\frac{k_{3x}}{2}\Big[\frac{\alpha_{J-1}}{J-1}A^{(J)}_{-1} + \frac{p^2\alpha_J}{J+1}A^{(J)}_{+1}\Big]. \quad (7.46)$$

For the unpolarized cross section, the amplitude related to a_2 exchange does not interfere with either π, π_2 or a_1-exchange amplitudes. If the highest moments are small in the cross section, one can assume that the combination in front of Y^2_n is close to 0. Then

$$W^{1J}_{a_2} = -N_{1J}\left(k_{3z}-\frac{p}{2}\right)p^2\alpha_{J+1}A^{(J)}_{+1},$$

$$W^{2J}_{a_2} = 0, \quad (7.47)$$

and, as a result, we have:

$$A^{(a_2)}_{\pi p\to\pi\pi n} = -\frac{3k_{3x}}{2s^2_{\pi N}}\sum_J\frac{\xi_J}{J}p^{J+1}k^J\left[N_{1J}\left(k_{3z}-\frac{p}{2}\right)\alpha_{J+1}A^{(J)}_{+1}\mathrm{Im}\,Y^1_J(\Theta,\varphi)\right] \times R_{a_2}(s_{\pi N},q^2)\Big(\bar\psi(k_3)\psi(p_2)\Big)g^{(a_2)}_{pn}(q^2)\,. \quad (7.48)$$

7.2.3.4 *Partial-wave decomposition*

The partial-wave amplitude $\pi R(a_1^{(j)}) \to \pi\pi$ with fixed J^{++} is presented in the K-matrix form:

$$A^{(L=J\pm1,J^{++})}_{\pi R(a_1^{(j)}),\pi\pi}(s) = \sum_b K^{(L=J\pm1,J^{++})}_{\pi R(a_1^{(j)}),\, b}(s,q^2)\left[\frac{\hat{I}}{\hat{I} - i\hat{\rho}(s)\hat{K}^{(J^{++})}(s)}\right]_{b,\pi\pi}, \tag{7.49}$$

where $K^{(L=J\pm1,J^{++})}_{\pi R(a_1^{(j)}),b}(s,q^2)$ is the following vector ($b = \pi\pi$, $K\bar{K}$, $\eta\eta$, $\eta\eta'$, $\pi\pi\pi\pi$):

$$K^{(L=J\pm1,J^{++})}_{\pi R(a_1^{(j)}),\, b}(s,q^2) = \left(\sum_\alpha \frac{G^{(L=J\pm1,J^{++},\,\alpha)}_{\pi R(a_1^{(j)})}(q^2)\, g_b^{(J^{++},\,\alpha)}}{M_\alpha^2 - s} + F^{(J^{L=J\pm1,J^{++}})}_{\pi R(a_1^{(j)}),\, b}(q^2)\frac{1\ \mathrm{GeV}^2 + s_{R0}}{s + s_{R0}}\right)\frac{s - s_A}{s + s_{A0}}\,. \tag{7.50}$$

Here $G^{(L=J\pm1,J^{++},\,\alpha)}_{\pi R(a_1^{(j)})}(q^2)$ and $F^{(J^{L=J\pm1,++})}_{\pi R(a_1^{(j)}),\, b}(q^2)$ are the q^2-dependent reggeon form factors.

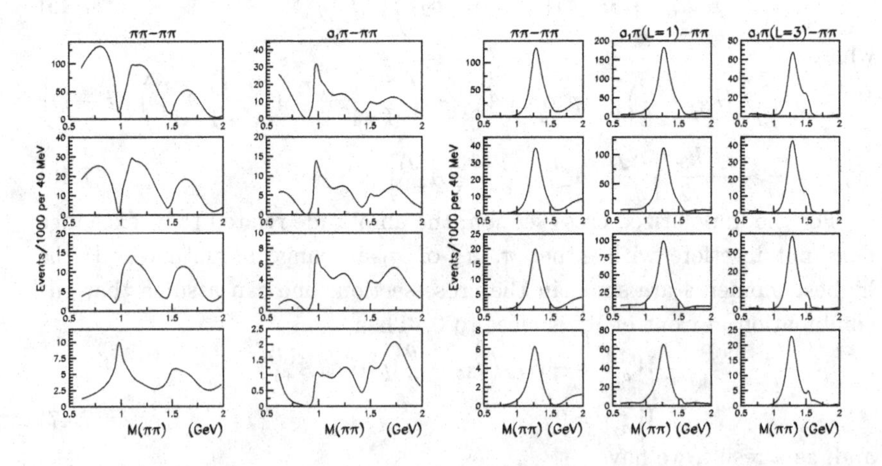

Fig. 7.2 Solution I. The contributions of S-wave (two left columns) and D-wave (three right columns) to Y_{00} moment integrated over t intervals. First line: $-0.1 < t < -0.01$ GeV^2, second line: $-0.2 < t < -0.1$ GeV^2, third line: $-0.4 < t < -0.2$ GeV^2 and the bottom line: $-0.4 < t < -1.5$ GeV^2.

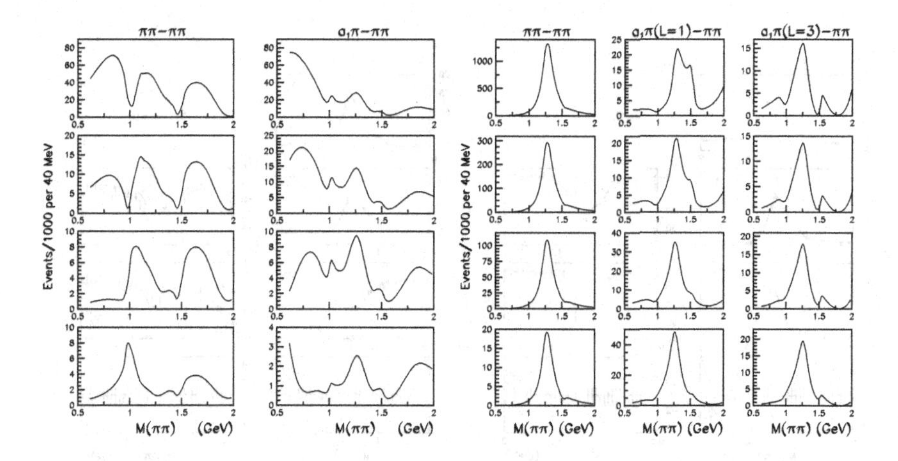

Fig. 7.3 Solution II. The contributions of S-wave (two left columns) and D-wave (three right columns) to Y_{00} moment integrated over t intervals. First line: $-0.1 < t < -0.01$ GeV^2, second line: $-0.2 < t < -0.1$ GeV^2, third line: $-0.4 < t < -0.2$ GeV^2 and the bottom line: $-0.4 < t < -1.5$ GeV^2.

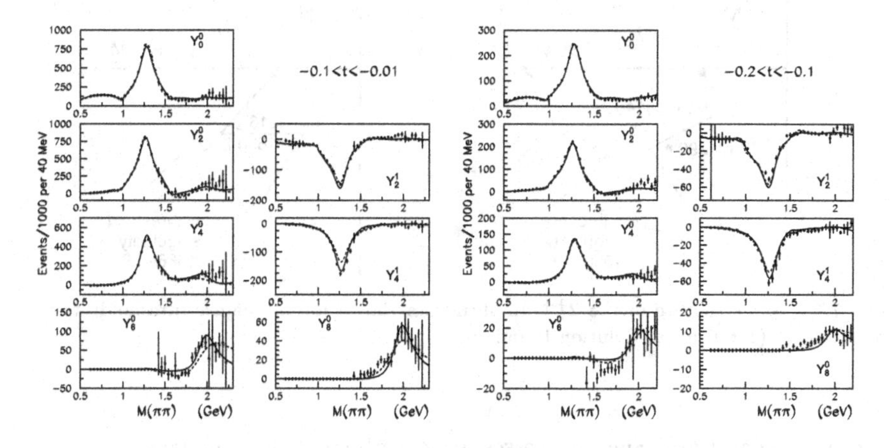

Fig. 7.4 The description of the moments extracted at $-0.1 < t < -0.01$ GeV^2 (the left two columns) and $-0.2 < t < -0.1$ GeV^2 (the right two columns). Dashed curves correspond to the Solution I and full curves to the Solution II.

7.2.4 $\pi^- p \to K\bar{K} n$ reaction with exchange by ρ-meson trajectories

In the case of the production of a $K\bar{K}$ system the resonance in this channel can have isospins $I = 0$ and $I = 1$, with even spin (production of states of

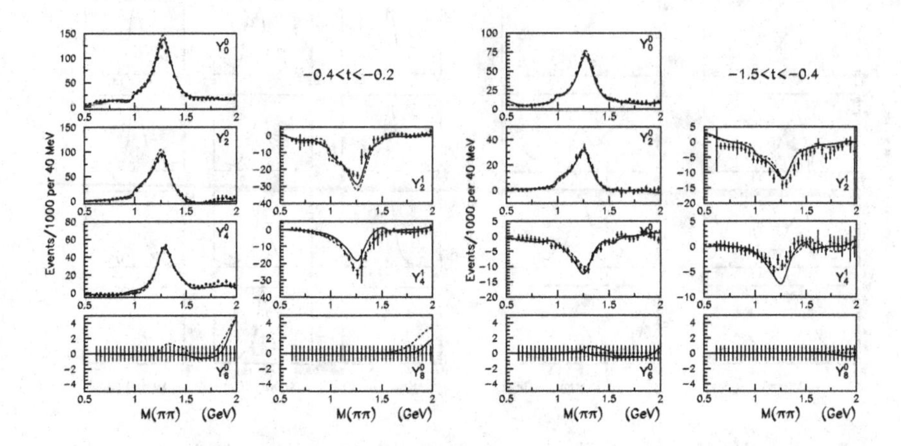

Fig. 7.5 The description of the moments extracted at $-0.4 < t < -0.2$ GeV^2 (two left columns) and $-1.5 < t < -0.4$ GeV^2 (two right columns). Dashed curves correspond to the Solution I and full lines to the Solution II.

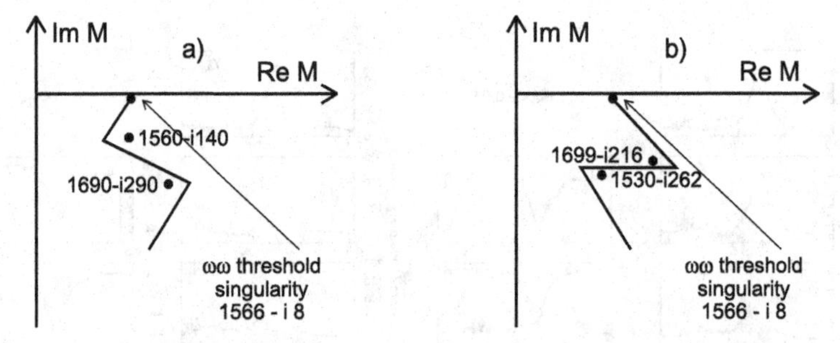

Fig. 7.6 Pole structure of the 2^{++}-amplitude in the region of the $\omega\omega$-threshold: the resonance $f_2(1560)$ in (a) Solution I, and (b) Solution II

the types ϕ and a_0). Such processes are described by ρ exchanges.

7.2.4.1 *Amplitude with exchanges of ρ-meson trajectories*

The amplitude with t-channel ρ-meson exchanges is written as follows:

$$A^{(\rho-\text{trajectories})}_{\pi p \to K\bar{K}n} = \sum_{\rho_j} A\Big(\pi R(\rho_j) \to K\bar{K}\Big) R_{\rho_j}(s_{\pi N}, q^2)\hat{g}^{(\rho_j)}_{pn}, \quad (7.51)$$

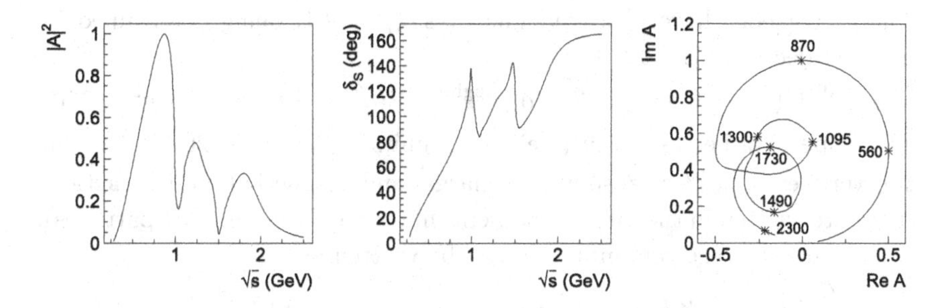

Fig. 7.7 From left to right: (a) The $\pi\pi \to \pi\pi$ S-wave amplitude squared, (b) the amplitude phase and (c) the Argand diagram for the S-wave amplitude $\pi\pi \to \pi\pi$.

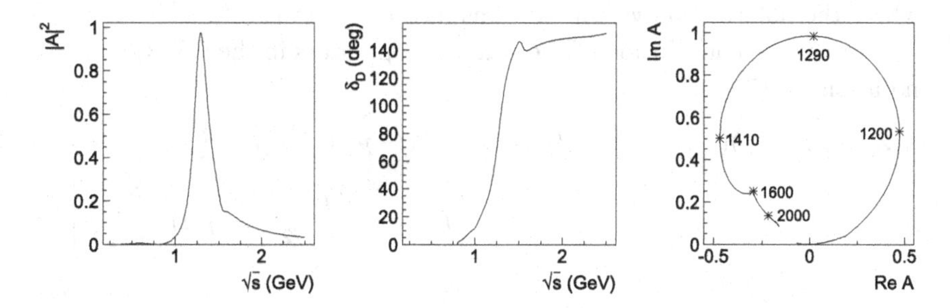

Fig. 7.8 From left to right: The $\pi\pi \to \pi\pi$ D-wave amplitude squared, the amplitude phase and the Argand diagram for the amplitude.

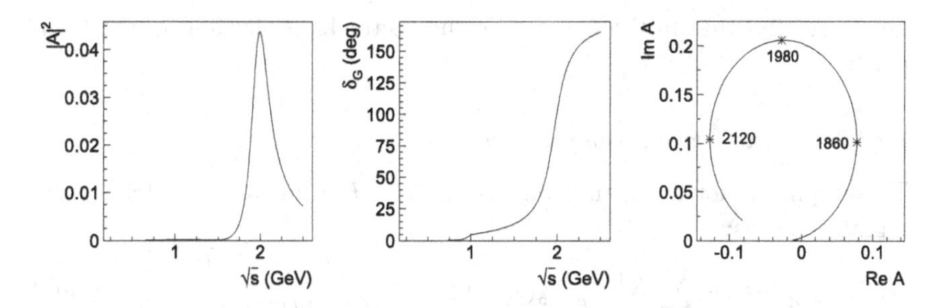

Fig. 7.9 From left to right: The $\pi\pi \to \pi\pi$ G-wave amplitude squared, the amplitude phase and the Argand diagram for the amplitude.

where the reggeon propagator $R_{\rho_j}(s_{\pi N}, q^2)$ and the reggeon–nucleon vertex $\hat{g}_{pn}^{(\rho_j)}$ read, respectively:

$$R_{\rho_j}(s_{\pi N}, q^2) = \exp\left(-i\frac{\pi}{2}\alpha_\rho^{(j)}(q^2)\right)\frac{(s_{\pi N}/s_{\pi N0})^{\alpha_\rho^{(j)}(q^2)}}{\sin\left(\frac{\pi}{2}\alpha_\rho^{(j)}(q^2)\right)\Gamma\left(\frac{1}{2}\alpha_\rho^{(j)}(q^2)+1\right)},$$

$$\hat{g}_{pn}^{(\rho_j)} = g_{pn}^{(\rho_j)}(1)(\varphi_n^+\varphi_p) + g_{pn}^{(\rho_j)}(2)\left(\varphi_n^+\frac{i}{2m_N}(\mathbf{q}_\perp[\mathbf{n}_z, \boldsymbol{\sigma}])\varphi_p\right). \quad (7.52)$$

The ρ_j reggeons have positive signatures, $\xi_\rho = +1$, being determined by linear trajectories [11, 12, 13]:

$$\alpha_\rho^{(\mathrm{leading})}(q^2) \simeq 0.50 + 0.83q^2, \alpha_\rho^{(\mathrm{daughter}-1)}(q^2) \simeq -0.75 + 0.83q^2. \quad (7.53)$$

The slope parameters are in $(\mathrm{GeV}/c)^{-2}$ units, $s_{\pi N0} \sim 2 - 20\ \mathrm{GeV}^2$. The two vertices in $\hat{g}_{pn}^{(\rho_j)}$ correspond to charge- and magnetic-type interactions (they are written in the infinite momentum frame of the colliding particles).

The meson–reggeon amplitude can be written as

$$A\Big(\pi R(\rho_j) \to K\bar{K}\Big) = \sum_J \varepsilon_{\beta\epsilon^{(-)}pP} Z_{\mu_1\mu_2\ldots\mu_J,\beta}(p^\perp) A^{(J^{++})}_{\pi R_\rho(q^2),K\bar{K}}(s) \times X^{(J)}_{\mu_1\mu_2\ldots\mu_J}(k^\perp)(-1)^J\,, \quad (7.54)$$

where the polarization vector $\epsilon_\beta^{(-)}$ was introduced in (7.39).

We use the convolution of the Z and X operators in the GJ-system (see notations in (7.40):

$$Z_{\mu_1\ldots\mu_J,\beta}(p^\perp)(-1)^J X^{(J)}_{\mu_1\ldots\mu_J}(k^\perp) = \frac{\alpha_J}{J}(\sqrt{-p_\perp^2})^{J-1}(\sqrt{-k_\perp^2})^J \times \left[P'_{J-1}(z)\frac{p_\beta^\perp}{\sqrt{-p_\perp^2}} - P'_J(z)\frac{k_\beta^\perp}{\sqrt{-k_\perp^2}}\right]. \quad (7.55)$$

The convolution of the spin–momentum operators in (7.54) gives:

$$A(\pi\rho_j \to \pi\pi) = \sum_J \frac{\alpha_J}{J} p^J k^J k_{3x}\sqrt{s} N_{j1} \mathrm{Im}\, Y_J^1(\Theta,\varphi) A^{(J^{++})}_{\pi R_\rho(q^2),K\bar{K}}(s). \quad (7.56)$$

Let us remind that in the GJ-system the vector $\mathbf{k}_3$ is situated in the (xz)-plane.

7.2.4.2 *Partial-wave decomposition*

The amplitude for the transition $\pi R_{\rho_j}(q^2) \to K\bar{K}$ in the K-matrix representation reads:

$$A^{(J^{++})}_{\pi R(\rho_j),K\bar{K}}(s) = \sum_b K^{(J^{++})}_{\pi R(\rho_j),\,b}(s,q^2)\left[\frac{\hat{I}}{\hat{I} - i\hat{\rho}(s)\hat{K}^{(J^{++})}(s)}\right]_{b,K\bar{K}}, \quad (7.57)$$

where $K^{(J^{++})}_{\pi R(\rho_j),b}(s,q^2)$ is the following vector ($b = \pi\pi$, $K\bar{K}$, $\eta\eta$, $\eta\eta'$, $\pi\pi\pi\pi$):

$$K^{(J^{++})}_{\pi R(\rho_j),\,b}(s,q^2) = \left(\sum_\alpha \frac{G^{(J^{++},\,\alpha)}_{\pi R(\rho_j)}(q^2) g_b^{(J^{++},\,\alpha)}}{M_\alpha^2 - s} + F^{(J^{++})}_{\pi R(\rho_j),\,b}(q^2)\frac{1\ \mathrm{GeV}^2 + s_{R0}}{s + s_{R0}}\right)\frac{s - s_A}{s + s_{A0}}. \quad (7.58)$$

Here $G^{(J^{++},\alpha)}_{\pi R(\rho_j)}(q^2)$ and $F^{(J^{++})}_{\pi R(\rho_j),b}(q^2)$ are the reggeon q^2-dependent form factors.

7.3 Results of the fit

The leading terms from the π-exchange trajectory can contribute only to the moments with $m = 0$, while the a_1 exchange can contribute to the moments up to $m = 2$. The characteristic feature of the a_1 exchange is that moments with $m = 2$ are suppressed compared to moments with $m = 1$ by the ratio k_{3x}/k_{3z} which is small for the system of two final mesons propagating with a large momentum in the beam direction.

The amplitudes defined by the π and a_1 exchanges are orthogonal if the nucleon polarization is not measured. This is due to the fact that the pion-trajectory states are defined by the singlet combination of the nucleon spins while the a_1-trajectory states are defined by the triplet combination. This effect is not taken into account for the S-wave contribution in (7.118) which can lead to a misidentification of this wave at large momenta transferred.

The π_2 particle is situated on the pion trajectory and therefore should be described by the reggeized pion exchange. However, the π_2 exchange has next-to-leading-order contributions with spherical functions at $m \geq 1$. The interference of such amplitudes with the pion exchange can be important (especially at small t) and is taken into account in the present analysis.

To reconstruct the total cross section of the reaction $\pi^- p \to \pi^0\pi^0 n$ which is not available to us we have used two partial-wave decompositions provided by the E852 collaboration [17]. The cross section was reconstructed by Eq. (7.118) and decomposed over moments. The two partial-wave decompositions produced very close results for the moments and we included the small differences between them as systematical errors.

The $\pi^- p \to \pi^0\pi^0 n$ moments can be described successfully with only π, a_1 and π_2 leading trajectories taken into account and a simple assumption about the t dependence of form factor for all partial waves. Moreover, we have found two solutions which differ by their contributions from these exchanges. Such an ambiguity is likely to be connected with the lack of polarization data and can be resolved by data from future experiments.

Let us present here more details.

These two solutions differ by the fraction of the π, a_1 and π_2 exchanges already in the region of small energy transferred. The first solution has a very large, practically dominant contribution from the a_1 exchange to the

D wave (see Fig. 7.2). The contribution from the a_1 exchange to the S wave is small. In this solution there is no notable signal from the $f_0(1300)$ state either at small or at large energy transferred. If $f_0(1300)$ is excluded from this solution, only the description of the Crystal Barrel and GAMS data is deteriorated while the description of the E852 data has the same quality.

In the second solution, the D wave at small energies transferred is dominantly produced from the π exchange. The fraction of a_1 exchange at $|t| < 0.1$ is about $2.5 - 3\%$. At large energy transferred, like in Solution I, the contribution from a_1 exchange becomes comparable and even dominant. The S wave has a well known structure at small $|t|$. At intermediate energies the contribution from the a_1 exchange becomes dominant and a signal from the $f_0(1300)$ state is well seen in this wave. At very large $|t|$ ($-1.5 < t < -0.4$ GeV^2) the contribution from a_1 exchange is rather small. The dominant contribution comes from the $f_0(980)$ state produced from π exchange. Here, our analysis is in contradiction with the result reported by the E852 collaboration which observed a strong S-wave signal around 1300 MeV in this t interval. However, the contribution from $f_0(1300)$ at intermediate energies transferred is important for the description of data with this solution. If this state is excluded from the fit, the description is notably deteriorated. This subject is considered in the following section in detail.

The Krakow group reported from the analysis of the polarized data that at small t the dominant contribution comes from the π exchange [18]. They point out that the second solution is possibly a physical one. However, the final conclusion can be made only after including these (yet unavailable to us) data in the present combined analysis which uses reggeon exchanges.

The description of the moments at small and large $|t|$ for the two solutions is shown in Figs. 7.4 and 7.5, correspondingly. The second solution produces a systematically better overall description except for the Y_4^1 moments at large energies transferred.

The S wave was fitted to 5 poles in the 5-channel K-matrix, described in detail in the previous sections. The parameters for the first solution are very close to those for the second one, *e.g.* the parametrization given in Table 7.1 describes both solutions, and the given errors cover a marginal change in both descriptions.

The D wave was fitted to 4 poles in the 5-channel ($\pi\pi$, $K\bar{K}$, $\eta\eta$, $\omega\omega$ and 4π) K matrix. The position of the first two D-wave poles was found to be $1270 - i97$ MeV and $1530 - i72$ MeV which corresponds to the well-known resonances $f_2(1270)$ and $f_2(1525)$. The third state has a Flatté-structure near the $\omega\omega$ threshold and is defined by two poles on the sheets defined by the $\omega\omega$ cut. Due to the fact that we do not fit directly the $\omega\omega$ production data these positions cannot be defined unambiguously. For example, in the framework of the Solution I (dominant a_1 exchange in the D wave) we found at least two solutions for the pole structure in the region of 1560 MeV. In the first the pole is situated at $1565 - i140$ MeV on the sheet above the $\omega\omega$ threshold and $1690 - i290$ MeV on the sheet below the $\omega\omega$ threshold. In the other solution the position of the pole is $1530 - i262$ and $1699 - i216$, correspondingly. The closest physical region is for both poles at the beginning of the $\omega\omega$ threshold $M \sim$1570 MeV, where they form a relatively narrow (220–250 MeV) structure which is called the $f_2(1560)$ state, see Fig. 7.6. A similar situation was observed in the Solution II. The K matrix D-wave parameters for the Solution II are given in Table 7.2.

The fourth D-wave K-matrix pole, $f_2^{\rm bare}(1980)$ cannot be rigidly fixed by the present data. The position of the corresponding pole is also not stable: one can easily increase the mass of the pole with the simultaneous increase of the width, spoiling only slightly the description of data. Because of that we consider this pole as some effective contribution of resonances located above 1900 MeV.

The $\pi\pi \to \pi\pi$ S-wave elastic amplitude for the second solution is shown in Fig. 7.7. The structure of the amplitude is well known, it is defined by the destructive interference of the broad component with $f_0(980)$ and $f_0(1500)$. Neither $f_0(1300)$ nor $f_0(1750)$ provide a strong change of the amplitudes. However, this is hardly a surprise: both these states are relatively broad and dominantly inelastic.

The $\pi\pi \to \pi\pi$ D-wave elastic amplitude is shown in Fig. 20. The amplitude squared is dominated by the $f_2(1270)$ state. Neither of $f_2(1560)$ and $f_2(1510)$ (which are included into the $K\bar{K}$ channel of the K matrix) show a meaningful structure in the amplitude squared. The K-matrix parameters found in the solution are given in Table 7.2.

7.3.1 *The $f_0(1300)$ state*

In the Solution II the fit of the E852 data shows a large contribution from the $f_0(1300)$ state to Y_0^0 moment due to a_1 exchange at $-0.2<t<-0.1$ and $-0.4<t<-0.2$ GeV^2. At very small ($-0.1<t<-0.01$ GeV^2) and large ($-1.5<t<-0.4$ GeV^2) energy transferred the contribution of this state to the Y_0^0 moment is less pronounced. If the K-matrix pole which corresponds

Table 7.1 Masses and couplings (in GeV units) for the S-wave K-matrix poles (f_0^{bare} states) as well as the amplitude pole positions (given in MeV). The II sheet is defined under the $\pi\pi$ and 4π cuts, the IV sheet is under $\pi\pi$, 4π, $K\bar{K}$ and $\eta\eta$ cuts, and the V sheet is determined by $\pi\pi$, 4π, $K\bar{K}$, $\eta\eta$ and $\eta\eta'$ cuts.

	$\alpha=1$	$\alpha=2$	$\alpha=3$	$\alpha=4$	$\alpha=5$
M	$0.720^{+0.50}_{-0.080}$	$1.220^{+0.040}_{-0.030}$	1.210 ± 0.030	$1.550^{+0.030}_{-0.020}$	1.850 ± 0.040
$g_0^{(\alpha)}$	$0.760^{+0.080}_{-0.060}$	0.820 ± 0.090	0.470 ± 0.050	0.360 ± 0.050	0.440 ± 0.050
$g_5^{(\alpha)}$	0	0	0.850 ± 0.100	0.570 ± 0.070	-0.900 ± 0.070
φ_α	-(60 ± 12)	28 ± 12	30 ± 14	8 ± 15	-(52 ± 14)
	$a=\pi\pi$	$a=K\bar{K}$	$a=\eta\eta$	$a=\eta\eta'$	$a=4\pi$
f_{1a}	0.180 ± 0.120	0.150 ± 0.100 $f_{ba}=0$	0.240 ± 0.100 $b=2,3,4,5$	0.300 ± 0.100	0.000 ± 0.060
			Pole position		
II sheet	1030^{+30}_{-10} $-i(35^{+10}_{-16})$				
III sheet	850^{+80}_{-50} $-i(100\pm25)$				
IV sheet		1290 ± 50 $-i(170^{+20}_{-40})$	1486 ± 10 $-i(57\pm5)$	1510 ± 130 $-i(800^{+100}_{-150})$	
V sheet					1800 ± 60 $-i(200\pm30)$

to the $f_0(1300)$ state is excluded from the fit (all couplings are put to zero) the total χ^2 changes rather appreciably.

For the description of the E852 data the main effect is seen, as expected, for the second and third t intervals. The comparison of the solutions with and without $f_0(1300)$ for these t regions is shown in Fig. 7.10. Here the description of the Y_4^1 moment is systematically worse for the fit where $f_0(1300)$ is excluded. The χ^2 per data points change for this moment from 1.84 to 3.63 for the $-0.2<t<-0.1$ GeV^2 interval and from 2.07 to 4.90 for the $-0.4<t<-0.2$ GeV^2 interval. The fit without $f_0(1300)$ produces a worse description also for Y_2^0 and Y_4^0. At intervals of small and large t the description has the same quality and can hardly be distinguished on the pictures. The contribution of the S wave to the moment Y_0^0 from this solution is shown in Fig. 7.11. It is seen that an appreciable contribution from a_1 exchange at the mass region 1300–1500 MeV is needed and the

Table 7.2 Masses and couplings (in GeV units) for D-wave K-matrix poles ($f_2^{\rm bare}$ states) for the Solution II. The III sheet is defined by $\pi\pi$ and 4π and $K\bar{K}$ cuts, the IV sheet by $\pi\pi$, 4π, $K\bar{K}$ and $\omega\omega$ cuts. The values marked by $*$ were fixed in the fit.

	$\alpha=1$	$\alpha=2$	$\alpha=3$	$\alpha=4$
M	1.286 ± 0.025	1.540 ± 0.015	1.560 ± 0.020	$2.200^{+0.300}_{-0.200}$
$g^{(\alpha)}_{\pi\pi}$	0.920 ± 0.020	-0.05 ± 0.080	0.280 ± 0.100	-0.30 ± 0.15
$g^{(\alpha)}_{\eta\eta}$	0.420 ± 0.060	0.27 ± 0.15	0.400 ± 0.200	$1.2\pm0.6^*$
$g^{(\alpha)}_{4\pi}$	-0.150 ± 0.200	0.370 ± 0.150	1.170 ± 0.450	1.0 ± 0.4
$g^{(\alpha)}_{\omega\omega}$	0^*	0^*	0.540 ± 0.150	-0.05 ± 0.2
	$a=\pi\pi$	$a=\eta\eta$	$a=\omega\omega$	$a=4\pi$
f_{1a}	0.03 ± 0.15	-0.11 ± 0.10	0^*	0^*
f_{2a}	-0.11 ± 0.10	-1.8 ± 0.60	0^*	0^*
		$f_{ba}=0$	$b=3,4,5$	
		Pole position		
III sheet	1.270 ± 0.008 $-i\,0.097\pm0.008$	1.530 ± 0.012 -i 0.064 ± 0.010		
III sheet			1.690 ± 0.015 $-i\,0.290\pm0.020$	
IV sheet			1.560 ± 0.015 $-i\,0.140\pm0.020$	

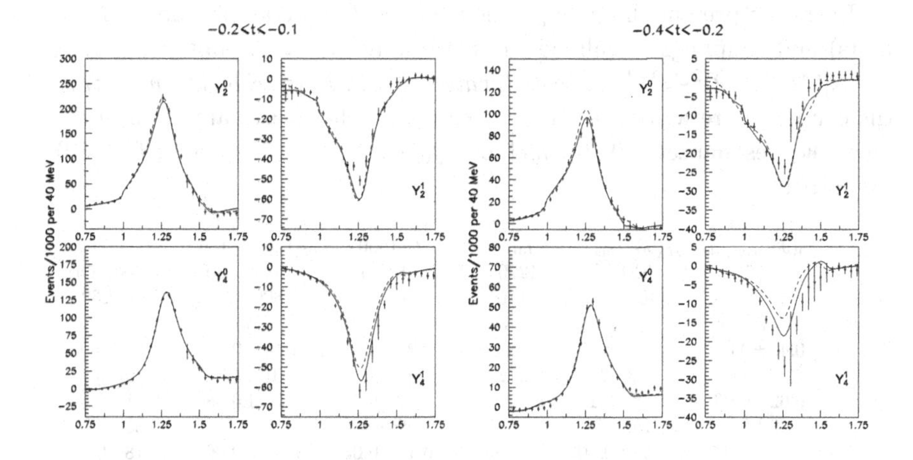

Fig. 7.10 The description of the moments extracted at $-0.1<t<-0.2$ GeV^2 (two left columns) and $-0.4<t<-0.2$ GeV^2 (two right columns). Solid curves correspond to the Solution II and dashed line to the Solution II(-) with excluded $f_0(1300)$.

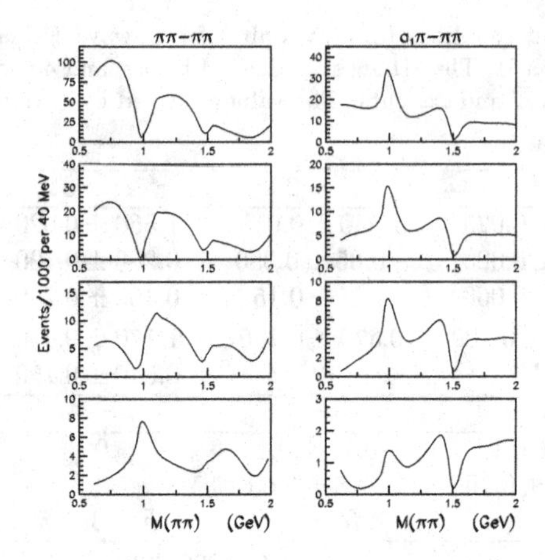

Fig. 7.11 Solution II(-) with $f_0(1300)$ excluded from the fit. The contributions of S-wave to Y_{00} moment integrated over different t intervals. First line: $t < -0.1$ GeV^2, second line: $-0.1 < t < -0.2$ GeV^2, third line: $-0.2 < t < -0.4$ GeV^2 and the bottom line: $-1.5 < t < -0.4$ GeV^2.

fit tries to simulate it (although not very successfully) by an interference between the broad component and the $f_0(1500)$ state.

Below we present the pole positions of the S-wave amplitude (in MeV units) and couplings calculated as pole residues (in GeV units): $A_{a\to b} \simeq G_a G_b [(M - i\Gamma/2)^2 - s]^{-1} + smooth\, terms$ with $a, b = \pi\pi, K\bar{K}, \eta\eta, \eta\eta', \pi\pi\pi\pi$; the couplings are written as $G_a = g_a \exp(i\varphi_a)$, the phases are given in degrees. For resonances $f_0(980)$, $f_0(1300)$, $f_0(1500)$, $f_0(1200-1600)$, $f_0(1750)$ we obtain:

	$f_0(980)_{\text{1st pole}}$	$f_0(980)_{\text{2nd pole}}$	$f_0(1300)$	$f_0(1500)$	$f_0(1200-1600)$	$f_0(1750)$
M	1030^{+30}_{-10}	850^{+80}_{-50}	1290 ± 50	1486 ± 10	1510 ± 130	1800 ± 60
$\Gamma/2$	35^{+10}_{-16}	100 ± 25	170^{+20}_{-40}	57 ± 5	800^{+100}_{-150}	200 ± 30
Sheet	II	III	IV	IV	IV	V
$g_{\pi\pi}$	0.42 ± 0.07	0.39 ± 0.05	0.28 ± 0.08	0.24 ± 0.05	0.82 ± 0.06	0.55 ± 0.05
$\varphi_{\pi\pi}$	-71 ± 8	45 ± 7	27 ± 10	65 ± 8	10 ± 12	15^{+6}_{-15}
$g_{K\bar{K}}$	0.62 ± 0.06	0.68 ± 0.12	0.15 ± 0.05	0.17 ± 0.04	0.84 ± 0.08	0.11 ± 0.04
$\varphi_{K\bar{K}}$	3 ± 8	155 ± 6	35 ± 15	48 ± 8	2 ± 10	55 ± 20
$g_{\eta\eta}$	0.51 ± 0.07	0.58 ± 0.10	0.14 ± 0.06	0.10 ± 0.03	0.40 ± 0.06	0.18 ± 0.05
$\varphi_{\eta\eta}$	10 ± 8	157 ± 10	57 ± 8	96 ± 6	16 ± 7	40 ± 12
$g_{\eta\eta'}$	0.42 ± 0.08	0.46 ± 0.12	0.17 ± 0.07	0.18 ± 0.06	0.14 ± 0.05	0.35 ± 0.07
$\varphi_{\eta\eta'}$	18 ± 8	160 ± 10	75 ± 15	143 ± 15	80 ± 17	18 ± 6
$g_{4\pi}$	0.16 ± 0.05	0.29 ± 0.10	0.80 ± 0.15	0.47 ± 0.08	1.30 ± 0.20	0.85 ± 0.20
$\varphi_{4\pi}$	25^{+8}_{-15}	155 ± 12	205 ± 12	156 ± 10	5 ± 12	150 ± 14

(7.59)

7.4 Summary for isoscalar resonances

We developed a method for the analysis of the reactions $\pi N \to$ *two mesons* $+ N$ at large energies of the initial pion. The approach is based on the use of the reggeized exchanges that allow us to analyze simultaneously the data obtained at small and large momentum transfers. Here the method is applied to the analysis of the $\pi^- N \to \pi^0\pi^0 N$ data measured by the E852 experiment. The inclusion of the Crystal Barrel data on the proton-antiproton annihilation at rest into the $3\pi^0$, $\pi^0\eta\eta$ and $\pi^0\pi^0\eta$ channels helps to reduce ambiguities in the isoscalar sector and investigate the properties of the isovector scalar and tensor states.

As a result of the analysis the K-matrix parameters of the isoscalar–scalar and isoscalar–tensor states was obtained up to the invariant mass 2 GeV and pole positions of corresponding amplitudes are defined.

7.4.1 *Isoscalar–scalar sector*

In the scalar sector the contribution of the $f_0(1300)$ is necessary to get a consistent description for the data set analyzed:

$$\text{Pole position of } f_0(1300): \qquad M = 1290 \pm 50 - i(170^{+20}_{-40}) \quad \text{MeV}. \tag{7.60}$$

According to our fit, the strong signal in the $\pi\pi$ spectrum in the region 1300 MeV is formed by two contributions, by $f_0(1300)$ (dominantly the a_1 reggeized exchange) and $f_2(1275)$ (the π and a_1 reggeized exchanges).

The position of the $f_0(980)$ is defined very well. The resonance reveals a double pole structure around the $K\bar{K}$ threshold.

$$\begin{aligned}
&\text{Pole positions of } f_0(980): \\
&\text{II sheet (under } \pi\pi \text{ and } \pi\pi\pi\pi \text{ cuts)}: \quad M = 1030^{+30}_{-10} - i(35^{+10}_{-16})\text{MeV}, \\
&\text{III sheet(under } \pi\pi,\ \pi\pi\pi\pi \text{ and K}\bar{\text{K}} \text{ cuts)}: \quad M = 850^{+80}_{-50} - i(200 \pm 50)\text{MeV}.
\end{aligned} \tag{7.61}$$

The $f_0(1500)$ is defined from the combined fit with a good accuracy:

$$\text{Pole position of } f_0(1500): \qquad M = 1486 \pm 10 - i(57 \pm 5) \quad \text{MeV}. \tag{7.62}$$

The broad state $f_0(1200-1600)$ (the scalar glueball descendant) gives a contribution in $\pi\pi$ scattering amplitudes in the region up to 2 GeV; the following pole position is found

$$\text{Pole position of } f_0(1200\!-\!1600): \qquad M = (1510 \pm 130) - i(800^{+100}_{-150}) \quad \text{MeV}. \tag{7.63}$$

The $f_0(1750)$ is a dominantly $s\bar{s}$ state [2] and is needed to describe $\pi\pi \to \pi\pi$ and $\pi\pi \to \eta\eta$ amplitudes above 1750 MeV.

$$\text{Pole position of } f_0(1750): \qquad M = 1800 \pm 60 - i(200 \pm 30) \quad \text{MeV.} \tag{7.64}$$

Parameters of this state differ from that observed by the BES [19] and WA102 [20] collaborations (denoted as $f_0(1710)$); one should, however, have in mind that in the case of strong interferences characteristics of a peak in the data does not correspond to the resonance position.

7.4.2 *Isoscalar–tensor sector*

The D wave reveals the resonances $f_2(1275)$, $f_2(1525)$, $f_2(1565)$, and $f_2(1980)$ with the following pole positions:

$$\begin{aligned} f_2(1275): &\qquad M = 1270 \pm 8 - i(97 \pm 8)\,\text{MeV}, \\ f_2(1525): &\qquad M = 1530 \pm 12 - i(64 \pm 10)\,\text{MeV}, \\ f_2(1565)\ (\text{2nd Solution}): &\qquad M_I = 1690 \pm 15 - i(290 \pm 20)\,\text{MeV}, \\ &\qquad M_{II} = 1560 \pm 15 - i(140 \pm 20)\,\text{MeV}, \end{aligned} \tag{7.65}$$

In the case of $f_2(1565)$ the K-matrix fit can be obtained only with the large coupling of this state to $\omega\omega$ (and, possibly, to $\rho\rho$) channel (note that this result is in a very good agreement with the analysis of the proton-antiproton annihilation into $\omega\omega\pi$ [21]). The large coupling to $\omega\omega$ leads to the double pole structure of $f_2(1565)$, see Fig. 7.6.

The state $f_2(1980)$ can not be identified unambiguously from the present data due to its large inelasticity. It plays the role of some broad contribution needed for the description of the πN data

7.4.3 *Isoscalar sector $J^{PC} = 4^{++}$*

For the description of high moments in the $\pi N \to \pi^0\pi^0 N$ data a contribution from a 4^{++} state is needed. This state is identified as $f_4(2025)$. Due to the lack of data at high masses this state was fitted as a two-channel ($\pi\pi$ and 4π) one-pole K matrix.

$$\begin{array}{cccc} M\ (\text{GeV}) & g_{\pi\pi} & g_{4\pi} & f_{\pi\pi\to\pi\pi} \\ 1.970 \pm 30 & 0.550 \pm 0.050 & 0.490 \pm 0.080 & -0.025 \pm 0.050 \end{array} \tag{7.66}$$

Here, as previously, masses and couplings are in GeV units. The position of the pole is equal to $(1966 \pm 25) - i\,(130 \pm 20)$. The amplitude phase and the Argand diagram for the isoscalar 4^{++} state is shown in Fig. 7.9. The $\pi\pi \to \pi\pi$ 4^{++} amplitude has a peak at 1995 MeV and is slightly asymmetrical: the half height is reached at the mass 1880 and 2165 MeV. The branching ratio of the $\pi\pi$ channel at the pole position is $20 \pm 3\%$ which is in agreement with the PDG value within the error.

***

Assignment of mesons to $q\bar{q}$-nonets is presented in Appendix E.

7.5 Appendix A. D-matrix technique in the two-meson production reactions

We consider here specific points of the D-matrix technique applied to two-meson states: $\pi\pi$, $K\bar{K}$, $\eta\eta$ and so on. Connection of the D-matrix technique with that in the K-matrix approach is emphasized.

7.5.1 *D-matrix in one-channel and two-channel cases*

We present below the one-channel amplitude in the D-matrix form and then introduce the second channel thus passing to the two-channel case.

7.5.1.1 *One-channel amplitude*

For the one-channel case, let it be $1 = \pi\pi$, the amplitude is:

$$A(s) = \frac{d(s)}{1 - B(s)d(s)}, \tag{7.67}$$

where, recall, $d(s)$ is the D-matrix term and $B(s)$ is the $\pi\pi$-loop diagram:

$$d(s) \to d_{11}(s) = \sum_{\alpha} \frac{g_1^{(\alpha)} g_1^{(\alpha)}}{M_\alpha^2 - s}, \tag{7.68}$$

$$B(s) \to B_1(s) = b_1 + \int\limits_{4m_\pi^2}^{\infty} \frac{ds'}{\pi} \frac{\rho_1(s')}{\beta_1(s')} \frac{s - s_1}{(s'-s_1)(s'-s-i0)} \equiv R_1(s) + i\frac{\rho_1(s)}{\beta_1(s)}.$$

At $s > 4m_\pi^2$ the function $R_1(s)$ is real, the imaginary part is equal $\rho_1(s)/\beta_1(s)$ where $\rho_1(s)$ is the $\pi\pi$ the phase space factor and $\beta_1(s)$ is the universal cutting factor for the $\pi\pi$ loop. For easy calculations one can use for $\beta_1(s)$ a polynomial form: an example with $\beta_1(s) = 1$ is given in Eq. (7.7), and we have a bit more cumbersome expression for $\beta_1(s) = s + s_\beta$ with $s_\beta > 0$.

Equation (7.67) can be re-written in the K-matrix form:

$$A(s)=\frac{K(s)}{1-i\rho_1(s)K(s)},\quad K(s)=\frac{d(s)}{\beta_1(s)}\frac{1}{1-R_1(s)}. \tag{7.69}$$

The K-matrix form underlines the fulfillment of unitarity constraint for scattering amplitude.

7.5.1.2 *Two-channel amplitude*

Let us include into the amplitude the next channel, $2=K\bar{K}$. We do that for the amplitude $\pi\pi\to\pi\pi$ by replacing in the above written formulae:

$$d_{11}(s)\to d_{11}(s)+d_{12}(s)\frac{B_2(s)}{1-B_2(s)d_{22}(s)}d_{21}(s), \tag{7.70}$$

where

$$d_{ij}(s)=\sum_\alpha\frac{g_i^{(\alpha)}g_j^{(\alpha)}}{M_\alpha^2-s},$$

$$B_2(s)=b_2+\int\limits_{4m_K^2}^{\infty}\frac{ds'}{\pi}\frac{\rho_2(s')}{\beta_2(s')}\frac{s-s_2}{(s'-s_2)(s'-s-i0)}\equiv R_2(s)+i\frac{\rho_2(s)}{\beta_2(s)}. \tag{7.71}$$

In the $(2=K\bar{K})$-channel we have the same poles as in the $(1=\pi\pi)$-channel but with different couplings, the $K\bar{K}$ loop diagram may have its own subtraction point, s_2, and the cutting factor, $\beta_2(s)$.

The amplitudes $A_{11}(s)$ and $A_{12}(s)$ read:

$$\begin{aligned}A_{11}(s)&=\left[d_{11}(s)+d_{12}(s)\frac{B_2(s)}{1-B_2(s)d_{22}(s)}d_{21}(s)\right]\\&\times\left[1-B_1(s)\Big(d_{11}(s)+d_{12}(s)\frac{B_2(s)}{1-B_2(s)d_{22}(s)}d_{21}(s)\Big)\right]^{-1}\\&=\frac{d_{11}(s)+B_2(s)\left[d_{12}(s)d_{21}(s)-d_{11}(s)d_{22}(s)\right]}{1-B_1(s)d_{11}(s)-B_2(s)d_{22}(s)-B_1(s)B_2(s)\left[d_{12}(s)d_{21}(s)-d_{11}(s)d_{22}(s)\right]}\end{aligned} \tag{7.72}$$

and

$$\begin{aligned}A_{12}(s)&=d_{12}(s)\frac{B_2(s)}{1-B_2(s)d_{22}(s)}\\&\times\left[1-B_1(s)\Big(d_{11}(s)+d_{12}(s)\frac{B_2(s)}{1-B_2(s)d_{22}(s)}d_{21}(s)\Big)\right]^{-1}\\&=\frac{d_{12}(s)}{1-B_1(s)d_{11}(s)-B_2(s)d_{22}(s)-B_1(s)B_2(s)\left[d_{12}(s)d_{21}(s)-d_{11}(s)d_{22}(s)\right]}\end{aligned} \tag{7.73}$$

Two other amplitudes, $A_{21}(s)$ and $A_{22}(s)$, are written by changing in (7.72) and (7.73) the indices $1 \rightleftharpoons 2$:

$$A_{21}(s) = A_{12}(s)|_{1\rightleftharpoons 2} = A_{12}(s), \quad A_{22}(s) = A_{11}(s)|_{1\rightleftharpoons 2}. \tag{7.74}$$

The amplitudes $A_{ij}(s)$ $(ij = 1,2)$ can be rewritten in the K-matrix form:

$$\begin{aligned}
A_{11}(s) &= \frac{K_{11}(s) + i\rho_2(s)a(s)}{1 - i\rho_1(s)K_{11}(s) - i\rho_2(s)K_{22}(s) - i\rho_1(s)i\rho_2(s)a(s)} \\
A_{12}(s) &= \frac{K_{12}(s)}{1 - i\rho_1(s)K_{11}(s) - i\rho_2(s)K_{22}(s) - i\rho_1(s)i\rho_2(s)a(s)} \\
a(s) &= K_{12}(s)K_{21}(s) - K_{11}(s)K_{22}(s)
\end{aligned} \tag{7.75}$$

with

$$\begin{aligned}
K_{12}(s) &= \frac{1}{\Delta(s)}d_{12}(s), \\
K_{11}(s) &= \frac{1}{\Delta(s)}\left[d_{11}(s) + R_1(s)\Big(d_{12}(s)d_{21}(s) - d_{11}(s)d_{22}(s)\Big)\right], \\
\Delta(s) &= 1 - R_1(s)d_{11}(s) - R_2(s)d_{22}(s) - R_1(s)R_2(s) \\
&\qquad \times\left[d_{12}(s)d_{21}(s) - d_{11}(s)d_{22}(s)\right]
\end{aligned}$$

The K-matrix form shows us that the D-matrix amplitudes obey the unitarity condition.

The analogous operation can be performed for three-channel amplitudes and, in a similar way, for amplitudes with an arbitrary number of channels.

7.6 Appendix B. Elements of the reggeon exchange technique in the two-meson production reactions

Here we present the details for the partial wave analysis of a two-meson system produced in the high-energy πN interaction when the two-meson production occurs due to reggeon exchanges. The reggeon exchange approach is a good tool for studying hadron binary reactions and processes with diffractive production of hadrons at high energies (see [2], Chapters 2 and 6). Interference effects in the amplitudes of the type $\pi N \to two\,mesons + N$ provide valuable information on the contributions of resonances with different quantum numbers.

7.6.1 *Angular momentum operators for two-meson systems.*

As in [22], we use angular momentum operators $X^{(L)}_{\mu_1\ldots\mu_L}(k^\perp)$, $Z^{\alpha}_{\mu_1\ldots\mu_L}(k^\perp)$ and the projection operator $O^{\mu_1\ldots\mu_L}_{\nu_1\ldots\nu_L}(\perp P)$. Let us recall their definition.

The operators are constructed from the relative momenta $k^\perp_\mu$ and tensor $g^\perp_{\mu\nu}$. Both of them are orthogonal to the total momentum of the system:

$$k^\perp_\mu = \frac{1}{2}g^\perp_{\mu\nu}(k_1-k_2)_\nu = k_{1\nu}g^{\perp P}_{\nu\mu} = -k_{2\nu}g^{\perp P}_{\nu\mu}, \quad g^\perp_{\mu\nu} = g_{\mu\nu} - \frac{P_\mu P_\nu}{s}. \tag{7.76}$$

The operator for $L=0$ is a scalar (we write $X^{(0)}(k^\perp)=1$), and the operator for $L=1$ is a vector, $X^{(1)}_\mu = k^\perp_\mu$. The operators $X^{(L)}_{\mu_1\ldots\mu_L}$ for $L\geq 1$ can be written in the form of a recurrency relation:

$$\begin{aligned}
X^{(L)}_{\mu_1\ldots\mu_L}(k^\perp) &= k^\perp_\alpha Z^{\alpha}_{\mu_1\ldots\mu_L}(k^\perp) \equiv k^\perp_\alpha Z_{\mu_1\ldots\mu_L,\alpha}(k^\perp), \\
Z^{\alpha}_{\mu_1\ldots\mu_L}(k^\perp) &\equiv Z_{\mu_1\ldots\mu_L,\alpha}(k^\perp) = \frac{2L-1}{L^2}\Big(\sum_{i=1}^{L} X^{(L-1)}_{\mu_1\ldots\mu_{i-1}\mu_{i+1}\ldots\mu_L}(k^\perp)g^\perp_{\mu_i\alpha} \\
&\quad -\frac{2}{2L-1}\sum_{\substack{i,j=1\\ i<j}}^{L} g^\perp_{\mu_i\mu_j}X^{(L-1)}_{\mu_1\ldots\mu_{i-1}\mu_{i+1}\ldots\mu_{j-1}\mu_{j+1}\ldots\mu_L\alpha}(k^\perp)\Big).
\end{aligned} \tag{7.77}$$

We have a convolution equality $X^{(L)}_{\mu_1\ldots\mu_L}(k^\perp)k^\perp_{\mu_L} = k^2_\perp X^{(L-1)}_{\mu_1\ldots\mu_{L-1}}(k^\perp)$, with $k^2_\perp \equiv k^\perp_\mu k^\perp_\mu$, and the tracelessness property of $X^{(L)}_{\mu\mu\mu_3\ldots\mu_L}=0$. On this basis, one can write down the normalization condition for orbital angular operators:

$$\int\frac{d\Omega}{4\pi}X^{(L)}_{\mu_1\ldots\mu_L}(k^\perp)X^{(L')}_{\mu_1\ldots\mu_L}(k^\perp) = \alpha_L k^{2L}_\perp, \quad \alpha_L = \prod_{l=1}^{L}\frac{2l-1}{l}, \tag{7.78}$$

where the integration is performed over spherical variables $\int d\Omega/(4\pi)=1$.

Iterating Eq. (7.77), one obtains the following expression for the operator $X^{(L)}_{\mu_1\ldots\mu_L}$ at $L\geq 1$:

$$\begin{aligned}
X^{(L)}_{\mu_1\ldots\mu_L}(k^\perp) = \alpha_L\Big[&k^\perp_{\mu_1}k^\perp_{\mu_2}k^\perp_{\mu_3}k^\perp_{\mu_4}\ldots k^\perp_{\mu_L} \\
&-\frac{k^2_\perp}{2L-1}\Big(g^\perp_{\mu_1\mu_2}k^\perp_{\mu_3}k^\perp_{\mu_4}\ldots k^\perp_{\mu_L} + g^\perp_{\mu_1\mu_3}k^\perp_{\mu_2}k^\perp_{\mu_4}\ldots k^\perp_{\mu_L}+\ldots\Big) \\
&+\frac{k^4_\perp}{(2L-1)(2L-3)}\Big(g^\perp_{\mu_1\mu_2}g^\perp_{\mu_3\mu_4}k^\perp_{\mu_5}k^\perp_{\mu_6}\ldots k^\perp_{\mu_L} \\
&+g^\perp_{\mu_1\mu_2}g^\perp_{\mu_3\mu_5}k^\perp_{\mu_4}k^\perp_{\mu_6}\ldots k^\perp_{\mu_L}+\ldots\Big)+\ldots\Big].
\end{aligned} \tag{7.79}$$

For the projection operators, one has:

$$O = 1, \qquad O^{\mu}_{\nu}(\perp P) = g^{\perp}_{\mu\nu},$$
$$O^{\mu_1\mu_2}_{\nu_1\nu_2}(\perp P) = \frac{1}{2}\left(g^{\perp}_{\mu_1\nu_1}g^{\perp}_{\mu_2\nu_2} + g^{\perp}_{\mu_1\nu_2}g^{\perp}_{\mu_2\nu_1} - \frac{2}{3}g^{\perp}_{\mu_1\mu_2}g^{\perp}_{\nu_1\nu_2}\right). \tag{7.80}$$

For higher states, the operator can be calculated using the recurrent expression:

$$O^{\mu_1\ldots\mu_L}_{\nu_1\ldots\nu_L} = \frac{1}{L^2}\Bigg(\sum_{i,j=1}^{L} g^{\perp}_{\mu_i\nu_j} O^{\mu_1\ldots\mu_{i-1}\mu_{i+1}\ldots\mu_L}_{\nu_1\ldots\nu_{j-1}\nu_{j+1}\ldots\nu_L} \tag{7.81}$$
$$- \frac{4}{(2L-1)(2L-3)}\sum_{\substack{i<j\\k<m}}^{L} g^{\perp}_{\mu_i\mu_j} g^{\perp}_{\nu_k\nu_m} O^{\mu_1\ldots\mu_{i-1}\mu_{i+1}\ldots\mu_{j-1}\mu_{j+1}\ldots\mu_L}_{\nu_1\ldots\nu_{k-1}\nu_{k+1}\ldots\nu_{m-1}\nu_{m+1}\ldots\nu_L}\Bigg).$$

The projection operators obey the relations:

$$O^{\mu_1\ldots\mu_J}_{\nu_1\ldots\nu_J}(\perp P)X^{(J)}_{\nu_1\ldots\nu_J}(k^{\perp}) = X^{(J)}_{\mu_1\ldots\mu_J}(k^{\perp}),$$
$$O^{\mu_1\ldots\mu_J}_{\nu_1\ldots\nu_J}(\perp P)k_{\nu_1}k_{\nu_2}\ldots k_{\nu_J} = \frac{1}{\alpha_J}X^{(J)}_{\mu_1\ldots\mu_J}(k^{\perp}). \tag{7.82}$$

Hence, the product of the two $X^J(k_\perp)$ operators results in the Legendre polynomials as follows:

$$X^{(J)(k_\perp)}_{\mu_1\ldots\mu_J}(p^{\perp})(-1)^J O^{\mu_1\ldots\mu_J}_{\nu_1\ldots\nu_J}(\perp P)X^{(J)}_{\nu_1\ldots\nu_J}(k^{\perp}) = \alpha_J\left(\sqrt{-p_\perp^2}\sqrt{-k_\perp^2}\right)^J P_J(z), \tag{7.83}$$

where $z \equiv (-p^{\perp}k^{\perp})/(\sqrt{-p_\perp^2}\sqrt{-k_\perp^2})$.

7.6.2 *Reggeized pion exchanges*

To be definite, we present here formulae which lead to differential cross-section moment expansion in processes initiated by reggeized pions situated on the trajectories $R(\pi_j)$. Recall that the index j labels different trajectories, which are leading and daughter ones, as well as trajectories generated by a reggeization of the t-channel states with $J > 0$ (in multiparticle processes, like $\pi N \to two\,mesons + N$, the vertices $\pi R(\pi_j)$ are different for states with $J = 0$ and $J > 0$, so it is convenient to separate their contributions).

7.6.2.1 *Production of two mesons at small $|t|$ – the hypothesis of dominant pion exchange*

Under this hypothesis, the amplitude for the $\pi\pi$ production block is written as follows:

$$A(\pi R(\pi_j)\to\pi\pi)\to A(\pi\pi\to\pi\pi)=16\pi\sum_J A^J_{\pi\pi\to\pi\pi}(s)(2J+1)N^0_J Y^0_J(z,\varphi),$$

$$Y^m_J(z,\varphi)=\frac{1}{N^m_J}P^m_J(z)e^{im\varphi},\quad N^m_J=\sqrt{\frac{4\pi}{2J+1}\frac{(J+m)!}{(J-m)!}}. \tag{7.84}$$

Below we consider the resonance decay in the set of channels with two pseudoscalar mesons in the final states, $\pi\pi\to c=\pi\pi, K\bar{K}, \eta\eta,\ldots$ Generalizing this consideration, we give the same freedom for the initial state. Therefore, we denote the initial and final ($I=0$)-channel states as a,c with $a,c=\pi\pi, K\bar{K}, \eta\eta$, and so on. Then the transition amplitudes are denoted as

$$X^{(J)}_{\mu_1\ldots\mu_J}(p^\perp)\,A^J_{a\to c}(s)(-1)^J O^{\mu_1\ldots\mu_J}_{\nu_1\ldots\nu_J}(\perp P)X^{(J)}_{\nu_1\ldots\nu_J}(k^\perp_c)\xi_J,$$

$$\xi_J=\frac{16\pi(2J+1)}{\alpha_J}, \tag{7.85}$$

where $k^\perp_{c\mu}=\frac{1}{2}(k^\perp_{c1\nu}-k^\perp_{c2\nu})g^\perp_{\nu\mu}$.

The unitarity condition for the transition amplitudes reads:

$$\mathrm{Im}A^J_{a\to c}(s)=\sum_b\frac{2\sqrt{-k^2_{b\perp}}}{\sqrt{s}}A^J_{a\to b}(s)A^{J*}_{b\to c}(s)(-k^2_{c\perp})^J\,, \tag{7.86}$$

where index b refers to the intermediate states ($b=\pi\pi$, $K\bar{K}$, $\eta\eta,\ldots$). The unitarity condition is fulfilled by using for $A^J_{a\to c}(s)$ the K-matrix form:

$$A^J_{a\to c}(s)=\sum_b\hat{K}^J_{ab}\left(\frac{I}{I-i\hat{\rho}^J(s)\hat{K}^J}\right)_{bc}, \tag{7.87}$$

where $\hat{\rho}$ is a diagonal matrix with elements $\rho^J_{bb}(s)=2\sqrt{-k^2_{b\perp}}(-k^2_{b\perp})^J/\sqrt{s}$.

We parametrize the elements of the K-matrix in the following form:

$$K^J_{ab}=\sum_n\frac{1}{B_J(-k^2_{a\perp},r_n)}\left(\frac{g^{n(J)}_a g^{n(J)}_b}{M^2_n-s}\right)\frac{1}{B_J(-k^2_{b\perp},r_n)}$$
$$+\frac{f^{(J)}_{ab}}{B_J(-k^2_{a\perp},r_0)B_J(-k^2_{b\perp},r_0)}. \tag{7.88}$$

The index n refers to a set of resonances, the resonance couplings g^n_c are constants, and f_{ac} is a non-resonance term. The form factors $B_J(-k^2_\perp,r)$

are introduced to compensate the divergence of the relative momentum factor at large energies. Such form factors are known as the Blatt–Weisskopf factors depending on the radius of the state r_n. For non-resonance transition, the radius r_0 is taken to be much larger than that for resonance contributions.

7.6.2.2 *Calculation routine for the reggeized pion*

In case of the two-meson production, the initial-state K-matrix element is called the P-vector: $K^J_{\pi R(\pi_j),b} \equiv P^J_{\pi R(\pi_j),b}$ (recall that index j labels the leading and daughter trajectories). We write for the two-meson production amplitude initiated by the pion–reggeon exchange the following representation:

$$A^J_{\pi R(\pi_j),c}(s) = \sum_b P^J_{\pi R(\pi_j),b}\left(\frac{I}{I - i\hat{\rho}^J(s)\hat{K}^J}\right)_{bc}. \tag{7.89}$$

The P-vector is parametrized in the form similar to Eq. (7.88):

$$P^J_{\pi R(\pi_j),c} = \sum_n \frac{1}{B_J(-p^2_\perp, r_n)}\left(\frac{G^{n(J)}_{\pi R(\pi_j)}(q^2)g^{n(J)}_c}{M^2_n - s}\right)\frac{1}{B_J(-k^2_{c\perp}, r_n)}$$
$$+ \frac{F^{(J)}_{\pi R(\pi_j),c}(q^2)}{B_J(-p^2_\perp, r_0)B_J(-k^2_{c\perp}, r_0)}. \tag{7.90}$$

The product of the two amplitudes is equal to:

$$A(\pi R(\pi_j)\to\pi\pi)A^*(\pi R(\pi_k)\to\pi\pi) = (16\pi)^2\sum_J Y^0_J(z,\varphi) \tag{7.91}$$
$$\times\sum_{J_1J_2} d^{\,000}_{J_1J_2J}A^{J_1}_{\pi R(\pi_j)\to\pi\pi}(s)A^{J_2}_{\pi R(\pi_k)\to\pi\pi}(s)(2J_1+1)(2J_2+1)N^0_{J_1}N^0_{J_2},$$

where the coefficients d^{iji+j}_{nmk} are given below. Averaging over the polarizations of the initial nucleons and summing over polarization of the final ones, we get $\mathrm{Sp}[(\boldsymbol{\sigma}\mathbf{q}_\perp)(\boldsymbol{\sigma}\mathbf{q}_\perp)] \simeq -q^2 = -t$. So, we obtain for the total amplitude squared that is a sum over reggeon propagators $R_{\pi_j}(s_{\pi N}, q^2)$:

$$|A^{(\text{pion trajectories})}_{\pi p\to\pi\pi n}|^2 = \sum_{R(\pi_j)R(\pi_k)} A(\pi R(\pi_j)\to\pi\pi)A^*(\pi R(\pi_k)\to\pi\pi)$$
$$\times R_{\pi_j}(s_{\pi N}, q^2)R^*_{\pi_k}(s_{\pi N}, q^2)(-t)(g^{(\pi)}_{pn})^2. \tag{7.92}$$

The final expression reads:

$$N(M,t)\langle Y^0_J\rangle = \frac{\rho(s)\sqrt{s}}{\pi|\mathbf{p}_2|^2 s_{\pi N}}\sum_{R(\pi_j)R(\pi_k)} R_{\pi_j}(s_{\pi N}, q^2)R^*_{\pi_k}(s_{\pi N}, q^2)(-t)(g^{(\pi)}_{pn})^2$$
$$\times\sum_{J_1J_2} d^{\,000}_{J_1J_2J}A^{J_1}_{\pi\pi_j\to\pi\pi}(s)A^{J_2}_{\pi R(\pi_k)\to\pi\pi}(s)(2J_1+1)(2J_2+1)N^0_{J_1}N^0_{J_2}. \tag{7.93}$$

Here $N(M,t)$ is a normalizing coefficient which is to be determined from experimental data.

(i) Spherical functions

Let us present here some relations for the spherical functions used in the calculations:

$$Y_l^m(\Theta,\varphi) = \frac{1}{N_l^m} P_l^m(z) e^{im\varphi}, \quad N_l^m = \sqrt{\frac{4\pi}{2l+1} \frac{(n+m)!}{(l-m)!}},$$
$$P_l^m(z) = (-1)^m (1-z^2)^{\frac{m}{2}} \frac{d^m}{dz^m} P_l(z), \tag{7.94}$$

where $z = \cos\Theta$. We have the following convolution rule for two spherical functions:

$$Y_n^i(\Theta,\varphi) Y_m^j(\Theta,\varphi) = \sum_{k=0}^{n+m} d_{n,m,k}^{i,j,i+j} Y_k^{i+j}(\Theta,\varphi), \tag{7.95}$$

$d_{n,m,k}^{i,j,i+j}$ is defined also in [22], Eq. (8).

7.6.2.3 *Calculations related to the expansion of the differential cross section $\pi p \to \pi\pi + N$ over spherical functions for the reggeized π_2-exchange*

Here we present formulas which refer to the calculation routine related to the reggeized π_2-exchange.

The convolution of angular momentum operators can be expressed through Legendre polynomials and their derivatives:

$$X^{(J+2)}_{\alpha\beta\mu_1\ldots\mu_J}(p^\perp)(-1)^J O^{\mu_1\ldots\mu_J}_{\nu_1\ldots\nu_J}(\perp P) X^{(J)}_{\nu_1\ldots\nu_J}(k^\perp) =$$
$$= \frac{2\alpha_J\left(\sqrt{-k_\perp^2}\right)^J\left(\sqrt{-p_\perp^2}\right)^{J+2}}{3(J+1)(J+2)}$$
$$\times\left(X^{(2)}_{\mu\nu}(p^\perp)\frac{P''_{J+2}}{-p_\perp^2} + X^{(2)}_{\mu\nu}(k^\perp)\frac{P''_J}{-k_\perp^2} - \frac{3\,P''_{J+1}}{\sqrt{-k_\perp^2}\sqrt{-p_\perp^2}} k^\perp_\mu p^\perp_\nu\right) O^{\alpha\beta}_{\mu\nu}(\perp q),$$
$$O^{\alpha\beta}_{\chi\tau}(\perp q) X^{(J)}_{\chi\mu_2\ldots\mu_J}(p^\perp)(-1)^J O^{\tau\mu_2\ldots\mu_J}_{\nu_1\nu_2\ldots\nu_J}(\perp P) X^{(J)}_{\nu_1\ldots\nu_J}(k^\perp) =$$
$$= \frac{2\alpha_{J-1}}{3J^2}\left(\sqrt{-k_\perp^2}\right)^J\left(\sqrt{-p_\perp^2}\right)^J$$
$$\times\left(X^{(2)}_{\mu\nu}(p^\perp)\frac{P''_J}{-p_\perp^2} + X^{(2)}_{\mu\nu}(k^\perp)\frac{P''_J}{-k_\perp^2} - \frac{P'_J + 2zP''_J}{\sqrt{-k_\perp^2}\sqrt{-p_\perp^2}} k^\perp_\mu p^\perp_\nu\right) O^{\alpha\beta}_{\mu\nu}(\perp q),$$

$$X^{(J-2)}_{\mu_3\ldots\mu_J}(p^\perp)(-1)^J O^{\alpha\beta\mu_3\ldots\mu_J}_{\nu_1\nu_2\nu_3\ldots\nu_J}(\perp P)X^{(J)}_{\nu_1\ldots\nu_J}(k^\perp) =$$
$$= \frac{2\alpha_{J-2}\left(\sqrt{-k_\perp^2}\right)^J\left(\sqrt{-p_\perp^2}\right)^{J-2}}{3(n-1)n}$$
$$\times\left(X^{(2)}_{\mu\nu}(p^\perp)\frac{P''_{J-2}}{-p_\perp^2} + X^{(2)}_{\mu\nu}(k^\perp)\frac{P''_J}{-k_\perp^2} - \frac{3\,P''_{J-1}}{\sqrt{-k_\perp^2}\sqrt{-p_\perp^2}}k^\perp_\mu p^\perp_\nu\right)O^{\alpha\beta}_{\mu\nu}(\perp q). \tag{7.96}$$

Let us remind that $p^\perp_\mu = p_{1\nu}g^{\perp P}_{\nu\mu}$ and $k^\perp_\mu = k_{1\nu}g^{\perp P}_{\nu\mu}$. Therefore, the amplitude $A_{\alpha\beta}(\pi R(\pi_2)\to\pi\pi)$ can be rewritten as:

$$A_{\alpha\beta}(\pi R(\pi_2)\to\pi\pi) = \frac{2}{3}\sum_J\left[\frac{X^{(2)}_{\alpha\beta}(p^\perp)}{-p_\perp^2}\left(C^{(J)}_1 P''_{J+2}A^{(J)}_{+2}(s)\right.\right.$$
$$\left.+C^{(J)}_2 P''_J A^{(J)}_0(s) + C^{(J)}_3 P''_{J-2}A^{(J)}_{-2}(s)\right)$$
$$+\frac{X^{(2)}_{\alpha\beta}(k^\perp)}{-k_\perp^2}P''_J\left(C^{(J)}_1 A^{(J)}_{+2}(s) + C^{(J)}_2 A^{(J)}_0(s) + C^{(J)}_3 A^{(J)}_{-2}(s)\right)$$
$$-\frac{O^{\alpha\beta}_{\mu\nu}k^\perp_\mu p^\perp_\nu}{\sqrt{-k_\perp^2}\sqrt{-p_\perp^2}}\left(3C^{(J)}_1 P''_{J+1}A^{(J)}_{+2}(s) + C^{(J)}_2(P'_J + 2zP''_J)A^{(J)}_0(s)\right.$$
$$\left.\left.+3C^{(J)}_3 P''_{j-1}A^{(J)}_{-2}(s)\right)\right], \tag{7.97}$$

where

$$C^{(J)}_1 = \frac{16\pi(2J+1)}{(J+1)(J+2)}\left(\sqrt{-k_\perp^2}\right)^J\left(\sqrt{-p_\perp^2}\right)^{J+2},$$
$$C^{(J)}_2 = \frac{16\pi(2J+1)}{J(2J-1)}\left(\sqrt{-k_\perp^2}\right)^J\left(\sqrt{-p_\perp^2}\right)^J,$$
$$C^{(J)}_3 = \frac{16\pi(2J+1)}{(2J-1)(2J-3)}\left(\sqrt{-k_\perp^2}\right)^J\left(\sqrt{-p_\perp^2}\right)^{J-2}. \tag{7.98}$$

In the amplitude with the $X^{(2)}_{\alpha\beta}(p^\perp)$ structure there is no $m=1$ component. This amplitude should be taken effectively into account by the π trajectory. The second amplitude has the same angular dependence $P''_J(z)$ and works for resonances with $J\geq 2$. In the first approximation it is reasonable to use the third term only, which has the smallest power of $p_\perp^2$.

The third amplitude has the following angular dependence:

$$P''_{J+1}(z),\qquad P'_J + 2zP''_J,\qquad P''_{J-1}. \tag{7.99}$$

The first and second angular dependences are the same for $J = 1, 2$ and differ only at $n \geq 3$, when the third term appears. Therefore, in the first approximation one can use only the second term which has a lower order of $p_\perp^2$ to fit the data. Thus, the $R(\pi_2)$-exchange amplitude can be approximated as:

$$A_{\alpha\beta}(\pi R(\pi_2) \to \pi\pi) \simeq \frac{2}{3}\sum_J \left[\frac{X^{(2)}_{\alpha\beta}(k^\perp)}{-k_\perp^2} P''_J C_3^{(J)} A_{-2}^{(J)}(s) - \right.$$
$$\left. - \frac{O^{\alpha\beta}_{\mu\nu} k^\perp_\mu p^\perp_\nu}{\sqrt{-k_\perp^2}\sqrt{-p_\perp^2}} C_2^{(J)} (P'_J + 2zP''_J) A_0^{(J)}(s) \right] . \quad (7.100)$$

The convolution of operators in (7.100) with $k^{\perp q}_{3\alpha} k^{\perp q}_{3\beta}$ in the GJ system gives:

$$k^{\perp q}_{3\alpha} k^{\perp q}_{3\beta} X^{(2)}_{\alpha\beta}(k_\perp^2) = |\mathbf{k}|^2 k^{\perp q}_{3z} \left(k^{\perp q}_{3z} P_2(z) + 3k_{3x} z \cos\varphi \sin\Theta \right),$$
$$k^{\perp q}_{3\alpha} k^{\perp q}_{3\beta} O^{\alpha\beta}_{\mu\nu} k^\perp_\mu p^\perp_\nu = \frac{1}{3} |\mathbf{k}||\mathbf{p}| k^{\perp q}_{3z} \left(2k^{\perp q}_{3z} z + 3k^{\perp q}_{3x} \cos\varphi \sin\Theta \right), \quad (7.101)$$

and the total amplitude is equal to:

$$A^{R(\pi_2)}_{\pi p \to \pi\pi n} = \frac{1}{s_{\pi N}^2} \sum_J \left(V_1^J A_{-2}^{(J)}(s) - V_2^J A_0^{(J)} \right) R_{\pi_2}(s_{\pi N}, q^2) \varphi_n^+ (\boldsymbol{\sigma} \mathbf{p}_\perp) \varphi_p g^{(\pi_2)}_{pn}, \quad (7.102)$$

where

$$V_1^{(J)} = C_3^{(J)} k^{\perp q}_{3z} \left(k^{\perp q}_{3z} P_2(z) + 3k_{3x} z \cos\varphi \sin\Theta \right) P''_J,$$
$$V_2^{(J)} = \frac{1}{3} C_2^{(J)} k^{\perp q}_{3z} (P'_J + 2zP''_J) \left(2k^{\perp q}_{3z} z + 3k^{\perp q}_{3x} \cos\varphi \sin\Theta \right). \quad (7.103)$$

For $J = 1$ the first vertex is equal to 0; for the next ones we have:

$$V_2^{(1)} = \frac{1}{3} C_2^{(1)} k^{\perp q}_{3z} \left(2k^{\perp q}_{3z} Y_1^0 N_1^0 - 3k^{\perp q}_{3x} \mathrm{Re}\, Y_1^1 N_1^1 \right),$$
$$V_1^{(2)} = C_3^{(2)} k^{\perp q}_{3z} \left(k^{\perp q}_{3z} Y_2^0 N_2^0 - k_{3x} \mathrm{Re}\, Y_2^1 N_2^1 \right),$$
$$V_2^{(2)} = \frac{1}{3} C_2^{(2)} k^{\perp q}_{3z} \left(12k^{\perp q}_{3z} Y_2^0 N_2^0 + 6k^{\perp q}_{3z} Y_0^0 N_0^0 - 9k^{\perp q}_{3x} \mathrm{Re}\, Y_2^1 N_2^1 \right),$$
$$V_1^{(3)} = C_3^{(3)} k^{\perp q}_{3z} \left(9k^{\perp q}_{3z} Y_3^0 N_3^0 + 18k^{\perp q}_{3z} Y_1^0 N_1^0 - 6k_{3x} \mathrm{Re} Y_3^1 N_3^1 \right.$$
$$\left. - 9k_{3x} \mathrm{Re}\, Y_1^1 N_1^1 \right),$$
$$V_2^{(3)} = \frac{1}{3} C_2^{(3)} k^{\perp q}_{3z} \left(30k^{\perp q}_{3z} Y_3^0 N_3^0 + 42k^{\perp q}_{3z} Y_1^0 N_1^0 - 15k^{\perp q}_{3x} \mathrm{Re}\, Y_3^1 N_3^1 \right.$$
$$\left. - 18k^{\perp q}_{3x} \mathrm{Re} Y_1^1 N_1^1 \right),$$

$$V_1^{(4)} = C_3^{(4)} k_{3z}^{\perp q} \Big(18 k_{3z}^{\perp q} Y_4^0 N_4^0 + 20 k_{3z}^{\perp q} Y_2^0 N_2^0 + 7 k_{3z}^{\perp q} Y_0^0 N_0^0$$
$$- 9 k_{3x} \mathrm{Re}\, Y_4^1 N_4^1 - 15 k_{3x} \mathrm{Re}\, Y_2^1 N_2^1 \Big),$$
$$V_2^{(4)} = \frac{C_2^{(4)}}{3} k_{3z}^{\perp q} \Big[k_{3z}^{\perp q} \Big(56 Y_4^0 N_4^0 + 110 Y_2^0 N_2^0 + 34 Y_0^0 N_0^0 \Big)$$
$$- k_{3x}^{\perp q} \Big(21\, \mathrm{Re}\, Y_4^1 N_4^1 - 30\, \mathrm{Re}\, Y_2^1 N_2^1 \Big) \Big]. \tag{7.104}$$

In a general form, the expression can be written as:

$$V_1^{(J)} = \sum_{n=0}^{J} C_3^{(J)} k_{3z}^{\perp q} \left[k_{3z}^{\perp q} Y_n^0 R_n^0 (P_2 P_J'') + 3 k_{3x}^{\perp q} \mathrm{Re}\ Y_n^1 R_n^1 (z P_J'') \right],$$
$$V_2^{(J)} = \sum_{n=0}^{J} C_2^{(J)} k_{3z}^{\perp q} \Big[\frac{2}{3} k_{3z}^{\perp q} Y_n^0 R_n^0 (z (P_J' + 2z P_J'')) +$$
$$+ k_{3x}^{\perp q} \mathrm{Re}\ Y_n^1 R_n^1 (P_J' + 2z P_J'') \Big], \tag{7.105}$$

where

$$R_n^0(f) = \int \frac{d\Omega}{4\pi} f(z) Y_n^0(z, \Theta),$$
$$R_n^1(f) = 2 \int \frac{d\Omega}{4\pi} f(z) \cos\varphi \sin\Theta \mathrm{Re}\, Y_n^1(z, \Theta). \tag{7.106}$$

The P-vector amplitudes for the reggeized π_2-exchanges read:

$$A_{-2}^{(J)}(s) = \hat{P}_{-2}^{(J)} (I - i\hat{\rho}^J(s) \hat{K}^J)^{-1},$$
$$A_0^{(J)}(s) = \hat{P}_0^{(J)} (I - i\hat{\rho}^J(s) \hat{K}^J)^{-1}. \tag{7.107}$$

The P-vector components are parametrized in the form:

$$\left(P_{(-2)}^{(J)}\right)_n = \sum_\alpha \frac{1}{B_{J-2}(-p_\perp^2, r_\alpha)} \left(\frac{G_{-2}^{(J)\alpha} g_n^{\alpha(J)}}{M_\alpha^2 - s} \right) \frac{1}{B_J(-k_{n\perp}^2, r_\alpha)}$$
$$+ \frac{F_{(-2)n}^{(J)}}{B_{J-2}(-p_\perp^2, r_0) B_J(-k_{n\perp}^2, r_0)},$$
$$\left(P_{(0)}^{(J)}\right)_n = \sum_\alpha \frac{1}{B_J(-p_\perp^2, r_\alpha)} \left(\frac{G_0^{(J)\alpha} g_n^{\alpha(J)}}{M_\alpha^2 - s} \right) \frac{1}{B_J(-k_{n\perp}^2, r_\alpha)}$$
$$+ \frac{F_{(0)n}^{(J)}}{B_J(-p_\perp^2, r_0) B_J(-k_{n\perp}^2, r_0)} \tag{7.108}$$

The total amplitude of the π_2-exchange can be rewritten as an expansion over spherical functions:

$$A^{(\pi_2)}_{\pi p\to\pi\pi n} = \sum_{n=0}^{N}\left(Y^0_n A^{0(n)}_{\rm tot}(s) + Y^1_n A^{1(n)}_{\rm tot}\right) R_{\pi_2}(s_{\pi N}, q^2)\left(\varphi^+_n(\boldsymbol{\sigma}\mathbf{p}_\perp)\varphi_p\right) g^{(\pi_2)}_{pn}, \tag{7.109}$$

where

$$\begin{aligned}
A^{0(n)}_{\rm tot}(s) &= \frac{1}{s^2_{\pi N}}(k^{\perp q}_{3z})^2\sum_J\Big[R^0_n(P_2P''_J)C^{(J)}_3 A^{(J)}_{-2}(s) \\
&\quad - \frac{2}{3}R^0_n(z(P'_J + 2zP''_J))C^{(J)}_2 A^{(J)}_0(s)\Big], \\
A^{1(n)}_{\rm tot}(s) &= \frac{1}{s^2_{\pi N}}k^{\perp q}_{3z}k_{3x}\sum_J\Big[3R^1_n(P_2P''_J)C^{(J)}_3 A^{(J)}_{-2}(s) \\
&\quad - R^1_n(z(P'_J + 2zP''_J))C^{(J)}_2 A^{(J)}_0(s)\Big].
\end{aligned} \tag{7.110}$$

Then the final expression is:

$$\begin{aligned}
N(M,t)\langle Y^0_J\rangle &= \frac{\rho(s)\sqrt{s}}{\pi|\mathbf{p}_2|^2 s_{\pi N}} R_{\pi_2}(s_{\pi N},q^2)R^*_{\pi_2}(s_{\pi N},q^2)(-t)(g^{(\pi_2)}_{pn})^2 \\
&\quad\times\sum_{n,m}\Big[d^{000}_{n,m,J}A^{0(n)}_{\rm tot}(s)A^{0(m)*}_{\rm tot}(s) + d^{110}_{n,m,J}A^{1(n)}_{\rm tot}(s)A^{1(m)*}_{\rm tot}(s)\Big], \\
N(M,t)\langle Y^1_J\rangle &= \frac{\rho(s)\sqrt{s}}{\pi|\mathbf{p}_2|^2 s_{\pi N}} R_{\pi_2}(s_{\pi N},q^2)R^*_{\pi_2}(s_{\pi N},q^2)(-t)(g^{(\pi_2)}_{pn})^2 \\
&\quad\times\sum_{n,m}\Big[d^{101}_{n,m,J}A^{1(n)}_{\rm tot}(s)A^{0(m)*}_{\rm tot}(s) + d^{011}_{n,m,J}A^{0(n)}_{\rm tot}(s)A^{1(m)*}_{\rm tot}(s)\Big].
\end{aligned} \tag{7.111}$$

7.7 Appendix C. Cross sections for the reactions $\pi N \to \pi\pi N$, KKN, $\eta\eta N$

We discuss in detail the reactions at incident pion momenta 20–50 GeV/c, such as measured in [23, 24, 25, 26, 27, 17]: (*i*) $\pi^-p\to\pi^+\pi^-+n$, (*ii*) $\pi^-p\to\pi^0\pi^0+n$, (*iii*) $\pi^-p\to K_SK_S+n$, (*iv*) $\pi^-p\to\eta\eta+n$. At these energies, the mesons in the states $J^{PC}=0^{++}$, 1^{--}, 2^{++}, 3^{--}, 4^{++} are produced via t-channel exchange by reggeized mesons belonging to the leading and daughter π, a_1 and ρ trajectories.

But, first, let us present the notations used below.

We consider the process of the Fig. 7.1 type, that is, πN interaction at large momenta of the incoming pion with the production of a two-meson system with a large momentum in the beam direction. This is a peripheral production of two mesons.

The cross section is defined as follows:

$$d\sigma = \frac{(2\pi)^4|A|^2}{8\sqrt{s_{\pi N}}|\mathbf{p}_2|_{\mathrm{cm}(\pi p)}}\, d\phi(p_1+p_2;k_1,k_2,k_3),$$
$$d\phi(p_1+p_2;k_1,k_2,k_3) = (2\pi)^3 d\Phi(P;k_1,k_2)\, d\Phi(p_1+p_2;P,k_3)\, ds, \quad (7.112)$$

where $|\mathbf{p}_2|_{\mathrm{cm}(\pi p)}$ is the pion momentum in the c.m. frame of the incoming hadrons. Taking into account that invariant variables s and t are inherent in the meson peripheral amplitude, we rewrite the phase space in a more convenient form:

$$d\Phi(p_1+p_2;P,k_3) = \frac{1}{(2\pi)^5}\frac{dt}{8|\mathbf{p}_2|_{\mathrm{cm}(\pi p)}\sqrt{s_{\pi N}}}, \qquad t=(k_3-p_2)^2,$$
$$d\Phi(P;k_1,k_2) = \frac{1}{(2\pi)^5}\rho(s)d\Omega\,, \qquad \rho(s) = \frac{1}{16\pi}\frac{2|\mathbf{k}_1|_{\mathrm{cm}(12)}}{\sqrt{s}}. \quad (7.113)$$

The momentum $|\mathbf{k}_1|_{\mathrm{cm}(12)}$ is calculated in the c.m. frame of the outgoing mesons: in this system one has $P = (M,0,0,0,) \equiv (\sqrt{s},0,0,0)$ and $g^{\perp P}_{\mu\nu}k_{1\nu} = -g^{\perp P}_{\mu\nu}k_{2\nu} = (0, k\,\sin\Theta\sin\varphi,\, k\,\cos\Theta\sin\varphi,\, k\,\cos\Theta\, k)$ while $d\Omega = d(\cos\Theta)d\varphi$. We have

$$d\sigma = \frac{(2\pi)^4|A|^2(2\pi)^3}{8|\mathbf{p}_2|_{\mathrm{cm}(\pi p)}\sqrt{s_{\pi N}}}\,\frac{1}{(2\pi)^5}\,\frac{dt\,dM^2\,d\Phi(P,k_1,k_2)}{8|\mathbf{p}_2|_{\mathrm{cm}(\pi p)}\sqrt{s_{\pi N}}} = \frac{|A|^2\rho(M^2)\,MdM\,dt\,d\Omega}{32(2\pi)^3|\mathbf{p}_2|^2_{\mathrm{cm}(\pi p)}\,s_{\pi N}}. \quad (7.114)$$

The cross section can be expressed in terms of the spherical functions:

$$\frac{d^4\sigma}{dtd\Omega dM} = N(M,t)\sum_l\left(\langle Y_l^0\rangle Y_l^0(\Theta,\varphi) + 2\sum_{m=1}^{l}\langle Y_l^m\rangle \mathrm{Re}\, Y_l^m(\Theta,\varphi)\right). \quad (7.115)$$

The coefficients $N(M,t)$, $\langle Y_l^0\rangle$, $\langle Y_l^m\rangle$ are subjects of study in the determination of meson resonances.

Before describing the results of analysis based on the reggeon exchange technique, let us comment methods used in other approaches.

7.7.1 *The CERN-Munich approach*

The CERN-Munich model [26] was developed for the analysis of the data on the $\pi^- p \to \pi^+\pi^- n$ reaction. It is based partly on the absorption model but mainly on phenomenological observations. The amplitude squared is written as

$$|A|^2 = \left| \sum_{J=0} A_J^0 Y_J^0(\Theta,\varphi) + \sum_{J=1} A_J^{(-)} ReY_J^1(\Theta,\varphi) \right|^2 + \left| \sum_{J=1} A_J^{(+)} ReY_J^1(\Theta,\varphi) \right|^2, \tag{7.116}$$

and additional assumptions are made:
1) The helicity-1 amplitudes are equal for natural and unnatural exchanges $A_J^{(-)} = A_J^{(+)}$;
2) The ratio of the $A_J^{(-)}$ and the A_J^0 amplitudes is a polynomial over the mass of the two-pion system which does not depend on J up to the total normalization, $A_J^{(-)} = A_J^0 \left(C_J \sum\limits_{n=0}^{3} b_n M^n \right)^{-1}$.
Then, in [26], the amplitude squared was rewritten using density matrices $\rho_{00}^{nm} = A_n^0 A_m^{0*}$, $\rho_{01}^{nm} = A_n^0 A_m^{(-)*}$, $\rho_{11}^{nm} = 2A_n^{(-)} A_m^{(-)}$ as follows:

$$|A|^2 = \sum_{J=0} Y_J^0(\Theta,\varphi) \left(\sum_{n,m} d_{n,m,J}^{0,0,0} \rho_{00}^{nm} + d_{n,m,J}^{1,1,0} \rho_{11}^{nm} \right)$$
$$+ \sum_{J=0} \mathrm{Re}\, Y_J^1(\Theta,\varphi) \left(\sum_{n,m} d_{n,m,J}^{1,0,1} \rho_{10}^{nm} + d_{n,m,J}^{0,1,1} \rho_{11}^{mn} \right),$$
$$d_{n,m,J}^{i,k,l} = \frac{\int d\Omega\, \mathrm{Re}\, Y_n^i(\Theta,\varphi)\, \mathrm{Re}\, Y_m^k(\Theta,\varphi)\, \mathrm{Re}\, Y_J^l(\Theta,\varphi)}{\int d\Omega\, \mathrm{Re}\, Y_J^l(\Theta,\varphi)\, \mathrm{Re}\, Y_J^l(\Theta,\varphi)}. \tag{7.117}$$

Using this amplitude for the cross section, the fitting to the moments $< Y_J^m >$ has been carried out.

The CERN–Munich approach cannot be applied to large t, it does not work for many other final states either.

7.7.2 *GAMS, VES, and BNL approaches*

In GAMS [23, 24], VES [27], and BNL [17] approaches, the πN data are described by a sum of amplitudes squared with an angular dependence defined by spherical functions:

$$|A^2| = \left| \sum_{J=0} A_J^0 Y_J^0(\Theta,\varphi) + \sum_{J=1} A_J^{(-)} \sqrt{2}\, \mathrm{Re}\ Y_J^1(\Theta,\varphi) \right|^2$$
$$+ \left| \sum_{J=1} A_J^{(+)} \sqrt{2}\, \mathrm{Im}\ Y_J^1(\Theta,\varphi) \right|^2 \tag{7.118}$$

The A_J^0 functions are denoted as $S_0, P_0, D_0, F_0 \ldots$, the $A_J^{(-)}$ functions are defined as $P_-, D_-, F_-, \ldots$ and the $A_J^{(+)}$ functions as $P_+, D_+, F_+, \ldots$ The equality of the helicity-1 amplitudes with natural and unnatural exchanges is not assumed in these approaches.

However, the discussed approaches are not free from other assumptions like the coherence of some amplitudes or the dominance of the one-pion exchange. In reality the interference of the amplitudes being determined by t-channel exchanges of different particles leads to a more complicated picture than that given by (7.118), this latter may lead (especially at large t) to a misidentification of quantum numbers for the produced resonances.

For example, in [17] the S wave appears in an unnatural set of amplitudes only. Natural exchanges have moments with $m = 1, 2, 3\ldots$ However, the a_1 exchange is a natural one, therefore it contributes into the S wave and does not interfere with unnatural exchanges – in this point the moment expansion [17] does not coincide with the formula with reggeon exchanges.

7.8 Appendix D. Status of trajectories on (J, M^2) plane

The π, η, a_1, a_0, a_2, a_3, ρ, and P' (or f_2) trajectories on (J, M^2) planes are shown in Fig. 7.12.

Leading π, ρ, a_1 and η trajectories are unambiguously determined together with their daughter trajectories, while for a_2, a_0, and P' only the leading trajectories can be given in a definite way.

In the construction of (J, M^2)-trajectories it is essential that the leading meson trajectories (π, ρ, a_1, a_2 and P') are well known from the analysis of the diffraction scattering of hadrons at $p_{\rm lab} \sim$ 5–50 GeV/c (for example, see [2] and references therein).

The pion and η trajectories are linear with a good accuracy, (see Fig. 7.12). Other leading trajectories (ρ, a_1, a_2, P') can be also considered as linear:

$$\alpha_X(M^2) \simeq \alpha_X(0) + \alpha'_X(0)M^2. \tag{7.119}$$

The parameters of linear trajectories determined by the masses of the $q\bar{q}$ states, are

$$\begin{aligned}
\alpha_\pi(0) \simeq -0.015,\ &\alpha'_\pi(0) \simeq 0.83\ \text{GeV}^{-2};\\
\alpha_\rho(0) \simeq 0.50,\ &\alpha'_\rho(0) \simeq 0.87\ \text{GeV}^{-2};\\
\alpha_\eta(0) \simeq -0.25,\ &\alpha'_\eta(0) \simeq 0.80\ \text{GeV}^{-2};\\
\alpha_{a_1}(0) \simeq -0.10,\ &\alpha'_{a_1}(0) \simeq 0.72\ \text{GeV}^{-2};\\
\alpha_{a_2}(0) \simeq 0.45,\ &\alpha'_{a_2}(0) \simeq 0.93\ \text{GeV}^{-2};\\
\alpha_{P'}(0) \simeq 0.50,\ &\alpha'_{P'}(0) \simeq 0.93\ \text{GeV}^{-2}.
\end{aligned} \tag{7.120}$$

The slopes $\alpha'_X(0)$ of the trajectories are approximately equal. The inverse slope, $1/\alpha'_X(0) \simeq 1.25 \pm 0.15$ GeV^2, roughly equals the parameter μ^2 for trajectories on the (n, M^2) planes:

$$\frac{1}{\alpha'_X(0)} \simeq \mu^2 . \tag{7.121}$$

In the subsequent sections, considering the scattering processes, we use for the Regge trajectories the momentum transfer squared $M^2 \to t$.

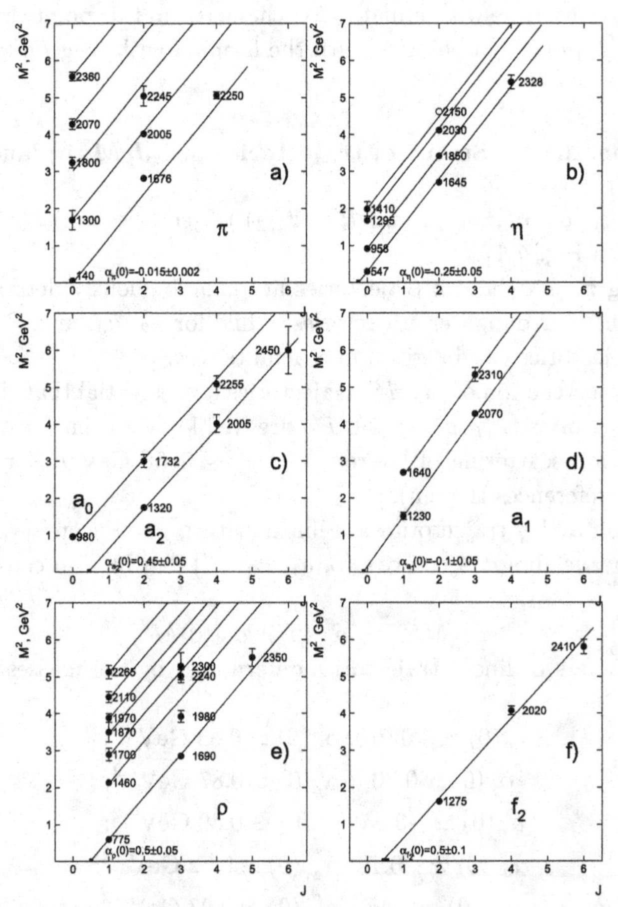

Fig. 7.12 Trajectories in the (J, M^2) plane: a) leading and daughter π-trajectories, b) leading and daughter η-trajectories, c) a_2-trajectories, d) leading and daughter a_1-trajectories, e) ρ-trajectories, f) P'-trajectories.

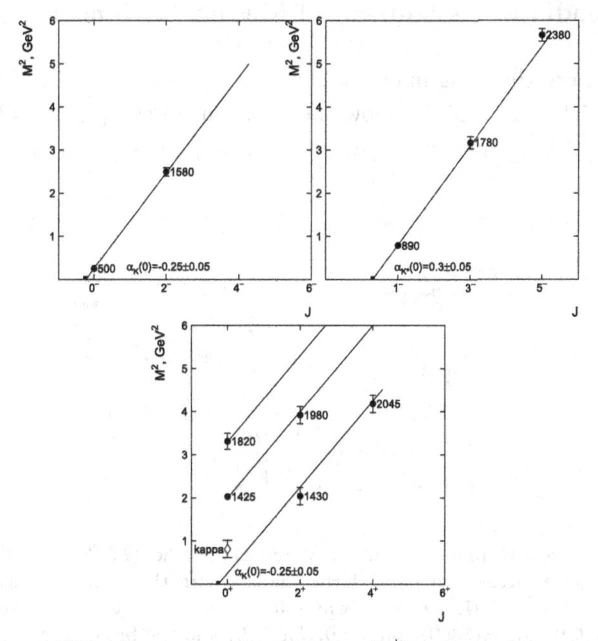

Fig. 7.13 K and K^* leading trajectories as well as K_2^+ leading and daughter trajectories on the (J^P, M^2)-plane. Beyond the linear trajectory, kappa is one of the hypothetical scalar mesons.

7.8.1 *Kaon trajectories on (J, M^2) plane*

As was said above, experimental data in the kaon sector are scarce, so in Fig. 2 we show the leading K-meson trajectory only (the states with $J^P = 0^-$, 2^-), the K^* trajectory ($J^P = 1^-, 3^-, 5^-$) and the leading and daughter trajectories for $J^P = 0^+$, 2^+, 4^+. The parameters of the leading kaon trajectories are as follows:

$$\begin{aligned} \alpha_K(0) \simeq -0.25, \quad & \alpha'_K(0) \simeq 0.90\ \text{GeV}^{-2}; \\ \alpha_{K^*}(0) \simeq 0.30, \quad & \alpha'_{K^*}(0) \simeq 0.85\ \text{GeV}^{-2}; \\ \alpha_{K_{2^+}}(0) \simeq -0.25, \quad & \alpha'_{K_{2^+}}(0) \simeq 1.0\ \text{GeV}^{-2}. \end{aligned} \tag{7.122}$$

The trajectories with $J^P = 1^+$, 3^+, 5^+ cannot be defined unambiguously.

The presented technique may be especially convenient for the study of low-mass singularities in multiparticle production amplitudes. The long-lasting discussions on the sigma-meson observation indicate that this is a problem of current interest.

7.9 Appendix E. Assignment of Mesons to Nonets

We present here the assignment of the mesons to $q\bar{q}$ nonets following [2].

In Figs. 7.14 and 7.15 we show the positions of the $(IJ^{PC} = 00^{++})$ and $(IJ^{PC} = 02^{++})$ resonances in the complex-M planes found in the previous analyses, see [2] for details.

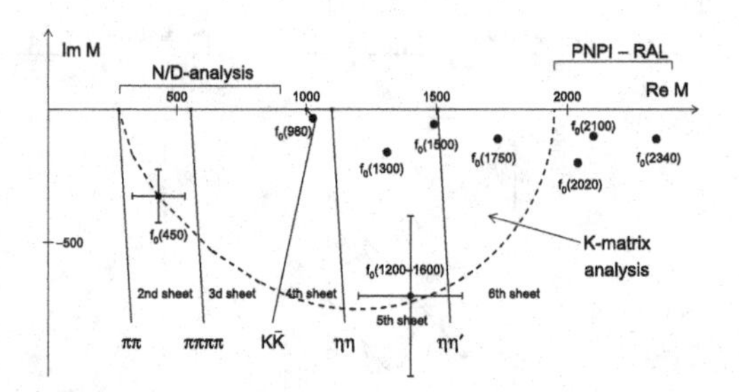

Fig. 7.14 Complex-M plane (M in MeV units) for the $(IJ^{PC} = 00^{++})$ mesons. The dashed line encircles the part of the plane where the K-matrix analyses [6, 8, 10] reconstructs the analytical K-matrix amplitude: in this area the poles corresponding to resonances $f_0(980)$, $f_0(1300)$, $f_0(1500)$, $f_0(1750)$ and the broad state $f_0(1200\text{–}1600)$ are located. Beyond this area, in the low-mass region, the pole of the light σ-meson is located (shown by the point the position of pole, $M = (430 - i320)$ MeV, corresponds to the result of N/D analysis; the crossed bars stand for σ-meson pole found in [12, 28]). In the high-mass region one has resonances $f_0(2030), f_0(2100), f_0(2340)$ [29]. Solid lines stand for the cuts related to the thresholds $\pi\pi, \pi\pi\pi\pi, K\bar{K}, \eta\eta, \eta\eta'$.

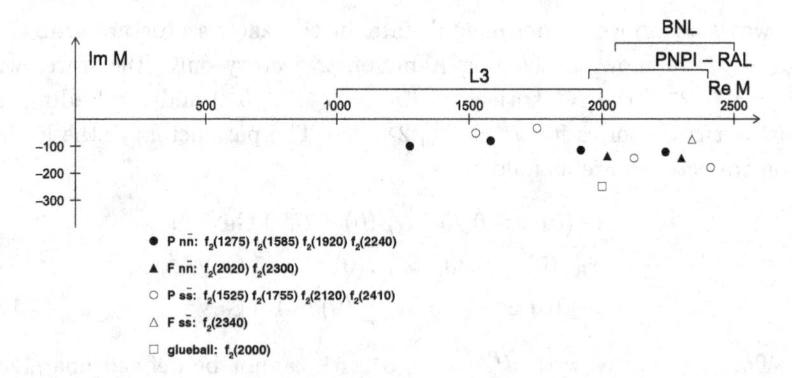

Fig. 7.15 Complex-M plane (M in MeV units) for isoscalar mesons $J^{PC} = 2^{++}$. Position of the f_2-poles on the complex-M plane: states with dominant $^3P_2 n\bar{n}$-component (full circle), $^3F_2 n\bar{n}$-component (full triangle), $^3P_2 s\bar{s}$-component (open circle), $^3F_2 s\bar{s}$-component (open triangle); the position of the tensor glueball is shown by the open square. Mass regions studied by the groups L3 [30], PNPI-RAL [29, 31, 32, 33] and BNL [25, 34] are shown.

In Eq. (7.123), we collected all considered meson $q\bar{q}$ states in nonets according to their $SU(3)_{\text{flavour}}$ attribution (the $^{2S+1}L_J$ states of $q\bar{q}$ mesons with n = 1, 2, 3, and 4). The singlet and octet states, with the same values of the total angular momentum, are mixed.

$q\bar{q}$-mesons	n=1				n=2			
	I=1	I=0	I=0	I=$\frac{1}{2}$	I=1	I=0	I=0	I=$\frac{1}{2}$
$^1S_0(0^{-+})$	$\pi(140)$	$\eta(547)$	$\eta'(958)$	$K(500)$	$\pi(1300)$	$\eta(1295)$	$\eta(1410)$	$K(1460)$
$^3S_1(1^{--})$	$\rho(775)$	$\omega(782)$	$\phi(1020)$	$K^*(890)$	$\rho(1460)$	$\omega(1430)$	$\phi(1650)$	
$^1P_1(1^{+-})$	$b_1(1229)$	$h_1(1170)$	$h_1(1440)$	$K_1(1270)$	$b_1(1620)$	$h_1(1595)$	$\mathbf{h_1(1790)}$	$K_1(1650)$
$^3P_0(0^{++}$	$a_0(980)$	$f_0(980)$	$f_0(1300)$	$K_0(1425)$	$a_0(1474)$	$f_0(1500)$	$f_0(1750)$	$K_0(1820)$
$^3P_1(1^{++})$	$a_1(1230)$	$f_1(1282)$	$f_1(1426)$	$K_1(1400)$	$a_1(1640)$	$f_1(1518)$	$\mathbf{f_1(1780)}$	
$^3P_2(2^{++})$	$a_2(1320)$	$f_2(1275)$	$f_2(1525)$	$K_2(1430)$	$a_2(1675)$	$f_2(1565)$	$f_2(1755)$	$K_2(1980)$
$^1D_2(2^{-+})$	$\pi_2(1676)$	$\eta_2(1645)$	$\eta_2(1850)$	$K_2(1800)$	$\pi_2(2005)$	$\eta_2(2030)$	$\mathbf{\eta_2(2150)}$	
$^3D_1(1^{--})$	$\rho(1700)$	$\omega(1670)$		$K_1(1680)$	$\rho(1970)$	$\omega(1960)$		
$^3D_2(2^{--})$	$\rho_2(1940)$	$\omega_2(1975)$		$K_2(1580)$	$\rho_2(2240)$	$\omega_2(2195)$		$K_2(1773)$
$^3D_3(3^{--})$	$\rho_3(1690)$	$\omega_3(1667)$	$\phi_3(1854)$	$K_3(1780)$	$\rho_3(1980)$	$\omega_3(1945)$	$\mathbf{\phi_3(2140)}$	
$^1F_3(3^{+-})$	$b_3(2032)$	$h_3(2025)$			$b_3(2245)$	$h_3(2275)$		
$^3F_2(2^{++})$	$a_2(2030)$	$f_2(2020)$	$f_2(2340)$		$a_2(2255)$	$f_2(2300)$	$\mathbf{f_2(2570)}$	
$^3F_3(3^{++})$	$a_3(2030)$	$f_3(2050)$		$K_3(2320)$	$a_3(2275)$	$f_3(2303)$		
$^3F_4(4^{++})$	$a_4(2005)$	$f_4(2025)$	$\mathbf{f_4(2100)}$	$K_4(2045)$	$a_4(2255)$	$\mathbf{f_4(2150)}$	$f_4(2300)$	
$^1G_4(4^{-+})$	$\pi_4(2250)$	$\eta_4(2328)$		$K_4(2500)$				
$^3G_3(3^{--})$	$\rho_3(2240)$				$\mathbf{\rho_3(2510)}$			
$^3G_4(4^{--})$								
$^3G_5(5^{--})$	$\rho_5(2300)$			$K_5(2380)$	$\mathbf{\rho_5(2570)}$			
$^3H_6(6^{++})$	$a_6(2450)$	$f_6(2420)$						
$q\bar{q}$-mesons	n=3				n=4			
	I=1	I=0	I=0	I=$\frac{1}{2}$	I=1	I=0	I=0	I=$\frac{1}{2}$
$^1S_0(0^{-+})$	$\pi(1800)$	$\eta(1760)$	$\mathbf{\eta(1880)}$	$K(1830)$	$\pi(2070)$	$\eta(2010)$	$\eta(2190)$	
$^3S_1(1^{--})$	$\rho(1870)$	$\mathbf{\omega(1830)}$	$\mathbf{\phi(1970)}$		$\rho(2110)$	$\omega(2205)$	$\mathbf{\phi(2300)}$	
$^1P_1(1^{+-})$	$b_1(1960)$	$h_1(1965)$	$\mathbf{h_1(2090)}$		$b_1(2240)$	$h_1(2215)$		
$^3P_0(0^{++})$	$\mathbf{a_0(1780)}$	$f_0(2040)$	$f_0(2105)$		$a_0(2025)$	$f_0(2210)$	$f_0(2340)$	
$^3P_1(1^{++})$	$a_1(1930)$	$f_1(1970)$	$\mathbf{f_1(2060)}$		$a_1(2270)$	$\mathbf{f_1(2214)}$	$f_1(2310)$	
$^3P_2(2^{++})$	$a_2(1950)$	$f_2(1920)$	$f_2(2120)$		$a_2(2175)$	$f_2(2240)$	$f_2(2410)$	
$^1D_2(2^{-+})$	$\pi_2(2245)$	$\eta_2(2248)$	$\mathbf{\eta_2(2380)}$			$\mathbf{\eta_2(2520)}$		
$^3D_1(1^{--})$	$\rho(2265)$	$\omega(2330)$						
$^3D_2(2^{--})$				$K_2(2250)$				
$^3D_3(3^{--})$	$\rho_3(2300)$	$\omega_3(2285)$	$\mathbf{\phi_3(2400)}$					

(7.123)

In the lightest nonets we can determine mixing angles more or less reliably, but for the higher excitations the estimates of the mixing angles are very ambiguous. In addition, isoscalar states can contain significant glueball components. For these reasons, we give only the nonet ($9 = 1 \oplus 8$) classification of mesons. States that are predicted but not yet reliably established are shown in boldface.

References

[1] V. V. Anisovich and A. V. Sarantsev, Int. J. Mod. Phys. **A 24**, 2481 (2009); Jad. Fiz. **72**, 1950 (2009), [PAN, **72**, 1889 (2009)]; Jad. Fiz. **72**, 1981 (2009), [PAN, **72**, 1920 (2009)].

[2] A. V. Anisovich, V. V. Anisovich, J. Nyiri, V. A. Nikonov, M. A. Matveev and A. V. Sarantsev, *"Mesons and Baryons. Systematization and Methods of Analysis"* (World Scientific, Singapore, 2008).

[3] B. Hyams *et al.*, Nucl. Phys. B **64** (1973) 134 [AIP Conf. Proc. **13** 206 (1973)].

[4] A. Etkin, *et al.*, Phys. Rev. D **25**, 1786 (1982).

[5] S. U. Chung, Phys. Rev. D **56** (1997) 7299.

[6] V. V. Anisovich and A. V. Sarantsev, Phys. Lett. B **382**, 429 (1996).

[7] V. V. Anisovich, D. V. Bugg and A. V. Sarantsev, Phys. Atom. Nucl. **62**, 1247 (1999) [Yad. Fiz. **62**, 1322 (1999)].

[8] V. V. Anisovich, A. A. Kondashov, Yu. D. Prokoshkin, S. A. Sadovsky, and A. V. Sarantsev, Phys. Atom. Nucl. **63**, 1410 (2000) [Yad. Fiz. **63**, 1489 (2000)]; hep-ph/9711319.

[9] V. V. Anisovich and V. M. Shekhter, Sov. J. Nucl. Phys. **13**, 370 (1971) [Yad. Fiz. **13**, 651 (1971)].

[10] V. V. Anisovich and A. V. Sarantsev, Eur. Phys. J. A **16**, 229 (2003).

[11] V. V. Anisovich, M. N. Kobrinsky, J. Nyiri, and Yu. M. Shabelski, *Quark Model and High Energy Collisions*, 2nd edition (World Scientific, 2004).

[12] V. V. Anisovich, UFN **174**, 49 (2004) [Physics-Uspekhi **47**, 45 (2004)].

[13] A. V. Anisovich, V. V. Anisovich, and A. V. Sarantsev, Phys. Rev. D **62**;051502(R) (2000).

[14] A. B. Kaidalov and B. M. Karnakov, Sov. J. Nucl. Phys. **11**, 121 (1970) [Yad. Fiz. **11**, 216 (1970)].

[15] A. V. Anisovich, V. V. Anisovich, V. N. Markov, M. A. Matveev, and A. V. Sarantsev, J. Phys. G: Nucl. Part. Phys. **28**, 15 (2002).

[16] V. N. Gribov, L. N. Lipatov, and G. V. Frolov, Sov. J. Nucl. Phys. **12**, 543 (1970) [Yad. Fiz. **12**, 994 (1970)].

[17] J. Gunter *et al.* (E582 Collab.) Phys. Rev. D **64**, 072003 (2001).

[18] R. Kaminski, L. Lesniak and K. Rybicki, Eur. Phys. J. C **4**, 4 (2002) [hep-ph/0109268].

[19] J. Z. Bai *et al.* (BES Collab.) Phys. Lett. B **476**, 25 (2000) [hep-ex/0002007].

[20] D. Barberis *et al.* (WA102 Collab.) Phys. Lett. B **479**, 59 (2000) [hep-ex/0003033].

[21] C. A. Baker *et al.*, Phys. Lett. B **467**, 147 (1999).

[22] V. V. Anisovich and A. V. Sarantsev *The analysis of reactions $\pi N \to$ two mesons + N within reggeon exchanges. 2.Basic formulas for fit*, arXiv:0806.1620 [hep-ph].

[23] D. Alde *et al.*, Z. Phys. C **66**, 375 (1995); A. A. Kondashov *et al.*, in *Proceeding 27th Intern. Conf. on High Energy Physics*, Glasgow, 1994, p. 1407;

Yu. D. Prokoshkin *et al.*, Physics-Doklady **342**, 473 (1995);
A. A. Kondashov *et al.*, Preprint IHEP 95-137 (Protvino, 1995).
[24] F. Binon, *et al.*, Nuovo Cimento A **78**, 313 (1983); Nuovo Cimento A **80**, 363 (1984).
[25] S. J. Lindenbaum and R. S. Longacre, Phys. Lett. B **274**, 492 (1992).
[26] G. Grayer, *et al.*, Nucl. Phys. B **75**, 189 (1974);
W. Ochs, PhD Thesis, Münich University, (1974).
[27] D. V. Amelin *et al.*, Phys. Atom. Nucl. **67**, 1408 (2004).
[28] V. V. Anisovich and V. A. Nikonov, Eur. Phys. J. A **8**, 401 (2000).
[29] A. V. Anisovich, C. A. Baker, C. J. Batty *et al.*, Phys. Lett. B **449**, 114 (1999); Phys. Lett. B **452**, 173 (1999); Phys. Lett. B **452**, 180 (1999); Phys. Lett. B **452**, 187 (1999); Phys. Lett. B **472**, 168 (2000); Phys. Lett. B **476**, 15 (2000); Phys. Lett. B **477**, 19 (2000); Phys. Lett. B **491**, 40 (2000); Phys. Lett. B **491**, 47 (2000); Phys. Lett. B **496**, 145 (2000); Phys. Lett. B **507**, 23 (2001); Phys. Lett. B **508**, 6 (2001); Phys. Lett. B **513**, 281 (2001); Phys. Lett. B **517**, 261 (2001); Phys. Lett. B **517**, 273 (2001);
Nucl. Phys. A **651**, 253 (1999); A **662**, 319 (2000); A **662**, 344 (2000).
[30] V. A. Schegelsky, A. V. Sarantsev, V. A. Nikonov, L3 Note 3001, October 27, 2004.
[31] V. V. Anisovich, Pis'ma v ZhETF **80**, 845 (2004) [JETP Letters **80**, 715 (2004)].
[32] V. V. Anisovich and A. V. Sarantsev, Pis'ma v ZhETF **81**, 531 (2005) [JETP Letters **81**, 417 (2005)].
[33] V. V. Anisovich, M. A. Matveev, J. Nyiri, and A. V. Sarantsev, Int. J. Mod. Phys. A **20**, 6327 (2005); Yad. Fiz. **69**, 542 (2000) [Phys. Atom. Nuclei **69**, 520 (2006)].
[34] R. S. Longacre and S. J. Lindenbaum, Report BNL-72371-2004.

Chapter 8

Searching for the Quark–Diquark Systematics of Baryons Composed by Light Quarks $q = u, d$

Assuming that baryons are of quark–diquark structure, we look for the description of baryons composed of light quarks ($q = u, d$). Following the studies of [1], we systematize baryons using the notion of two diquarks: (1) axial–vector state, D_1^1, with spin $S_D = 1$ and isospin $I_D = 1$ and (2) scalar state, D_0^0, with spin $S_D = 0$ and isospin $I_D = 0$. We consider several qD schemes in searching for one with a minimal number of the composed baryon states. The scheme with overlapping qD_0^0 and qD_1^1 states, and obeying the relativistically generalized $SU(6)$-constraint, is describing the data.

In the high-mass region the quark-diquark model predicts several baryon resonances at $M \sim 2.0 - 2.9$ GeV with a double pole structure (*i.e.* two poles with the same ReM but different ImM). We see also that for the description of low-lying baryons with $L = 0$, the $SU(6)$ constraint (or its relativistic generalization) is needed.

In Appendix A we give a relativistic generalization of the quark-diquark wave functions. Correspondingly, relativistic spectral integral equations for the qD_0^0 and qD_1^1 states are presented taking into account the quark-diquark confinement interaction. In Appendix B we discuss the construction of total sets of eigenfunctions for the non-relativistic three-body problem. The presented procedure is based on the invariance of the Laplacian under $O(6)$.

8.1 Diquarks and reduction of baryon states

The experiment gives us a much smaller number of highly excited baryons than the models with three constituent quarks predict. One of the plausible explanation is that the excited baryons do not prefer to be formed as three-body systems of spatially separated colored constituent quarks. Instead, similarly to mesons, they are two-body systems; they consist of a quark and a diquark: $q_\alpha D^\alpha = q_\alpha\left[\varepsilon^{\alpha\beta\gamma} q_\beta q_\gamma\right]$. Here $\varepsilon^{\alpha\beta\gamma}$ is the three-dimensional totally antisymmetrical tensor which acts in the color space. One can omit color indices, imposing the symmetry ansatz for the spin–flavor–coordinate variables of wave functions.

It is an old idea that a qq-system inside the baryon can be regarded as a specific object – a diquark. Thus, interactions with a baryon may be considered as interactions with a quark, q, and a two-quark system, (qq); such a hypothesis was used in [2] for the description of high-energy hadron–hadron collisions. In [3, 4, 5], baryons were described as quark–diquark systems. In hard processes of nucleons (or nuclei), the coherent qq state (composite diquark) can be responsible for interactions in the region of large Bjorken-x values, at $x \sim 2/3$: deep inelastic scatterings were considered in the framework of such an approach in [6, 7, 8, 9, 10]. More detailed considerations of the diquark and the applications to different processes may be found in [11, 12, 13].

Here we suppose that excited baryons are quark–diquark systems. This means that in the space of three colors ($\mathbf{c_3}$) the excited baryons, similarly to excited mesons, are $\left(\bar{\mathbf{c}}_3(D_0^0)\mathbf{c_3}(q)\right)$ or $\left(\bar{\mathbf{c}}_3(D_1^1)\mathbf{c_3}(q)\right)$ systems.

A two-particle system has considerably less degrees of freedom than a three-particle one and, consequently, much less excited states. At the same time, the comparison of experimental data with model calculations [14, 15, 16] demonstrates that the number of predicted three-quark states is much larger than the number of observed ones. The qD-models point to mechanisms which may reduce the number of predicted excited states. Generally, this is the main motivation for developing quark–diquark models, see discussion in [17, 18].

Let us now see what type of states appears in qD_0^0 and qD_1^1 systems. The qD_0^0 systems with total spin $S = 1/2$ and isospin $I = 1/2$ contain the following baryon states J^P at different orbital momenta L (we restrict ourselves to $L \le 6$):

$$
\begin{array}{llllllll}
\mathrm{qD_0^0}: & & & & & & & \\
L=0: & \frac{1}{2}^+ & & & & & & \\
L=2: & & \frac{3}{2}^+ & \frac{5}{2}^+ & & & & \\
L=4: & & & & \frac{7}{2}^+ & \frac{9}{2}^+ & & \\
L=6: & & & & & & \frac{11}{2}^+ & \frac{13}{2}^+ \\
\hline
L=1: & \frac{1}{2}^- & \frac{3}{2}^- & & & & & \\
L=3: & & & \frac{5}{2}^- & \frac{7}{2}^- & & & \\
L=5: & & & & & \frac{9}{2}^- & \frac{11}{2}^- &
\end{array}
\qquad (8.1)
$$

The qD_1^1 systems have quark–diquark total spins $S = 1/2,\ 3/2$ and isospins $I = 1/2$ (nucleons) and $I = 3/2$ (Δ isobars), thus creating the following baryon states J^P at orbital momenta $L \le 6$:

$$
\begin{array}{lcccccccc}
\mathrm{qD}_1^1: & & & & & & & & \\
L=0,\ S=\frac{1}{2}: & \frac{1}{2}^+ & & & & & & & \\
L=0,\ S=\frac{3}{2}: & & \frac{3}{2}^+ & & & & & & \\
L=2,\ S=\frac{1}{2}: & & \frac{3}{2}^+ & \frac{5}{2}^+ & & & & & \\
L=2,\ S=\frac{3}{2}: & \frac{1}{2}^+ & \frac{3}{2}^+ & \frac{5}{2}^+ & \frac{7}{2}^+ & & & & \\
L=4,\ S=\frac{1}{2}: & & & & \frac{7}{2}^+ & \frac{9}{2}^+ & & & \\
L=4,\ S=\frac{3}{2}: & & & \frac{5}{2}^+ & \frac{7}{2}^+ & \frac{9}{2}^+ & \frac{11}{2}^+ & & \\
L=6,\ S=\frac{1}{2}: & & & & & & \frac{11}{2}^+ & \frac{13}{2}^+ & \\
L=6\ S=\frac{3}{2}: & & & & & \frac{9}{2}^+ & \frac{11}{2}^+ & \frac{13}{2}^+ & \frac{15}{2}^+ \\
\hline
L=1,\ S=\frac{1}{2}: & \frac{1}{2}^- & \frac{3}{2}^- & & & & & & \\
L=1,\ S=\frac{3}{2}: & \frac{1}{2}^- & \frac{3}{2}^- & \frac{5}{2}^- & & & & & \\
L=3,\ S=\frac{1}{2}: & & & \frac{5}{2}^- & \frac{7}{2}^- & & & & \\
L=3,\ S=\frac{3}{2}: & & \frac{3}{2}^- & \frac{5}{2}^- & \frac{7}{2}^- & \frac{9}{2}^- & & & \\
L=5,\ S=\frac{1}{2}: & & & & & \frac{9}{2}^- & \frac{11}{2}^- & & \\
L=5,\ S=\frac{3}{2}: & & & & \frac{7}{2}^- & \frac{9}{2}^- & \frac{11}{2}^- & \frac{13}{2}^- &
\end{array}
\tag{8.2}
$$

Symmetry properties, such as those of the $SU(6)$, lead to certain constraints in the realization of these states.

In Eqs. (8.1) and (8.2) only the basic states are included. Actually, every state in (8.1) and (8.2) is characterized also by its radial quantum number $n = 1, 2, 3, \ldots$. So, in (8.1) and (8.2) every state labeled by J^P represents a set of baryons:

$$
J^P \to (n, J^P), \quad n = 1, 2, 3, \ldots \tag{8.3}
$$

The states with different L and S but with the same (n, J^P) can mix with each other. However, the meson systematics tell us that L may be considered as a good quantum number for $q\bar{q}$ systems. Below, investigating the quark–diquark models, we use the same hypothesis and characterize qD_0^0 and qD_1^1 systems by the orbital momentum L. We also consider the total spin S in qD_1^1 systems as another conserved quantum number, though we understand that it should be regarded as a rough approximation only.

Let us now present in more detail the arguments in favor of a possible realization of the quark–diquark structure of highly excited baryons. We use as a guide the spectral integral equation for understanding quark–diquark systems considered here – the equation is shown schematically in

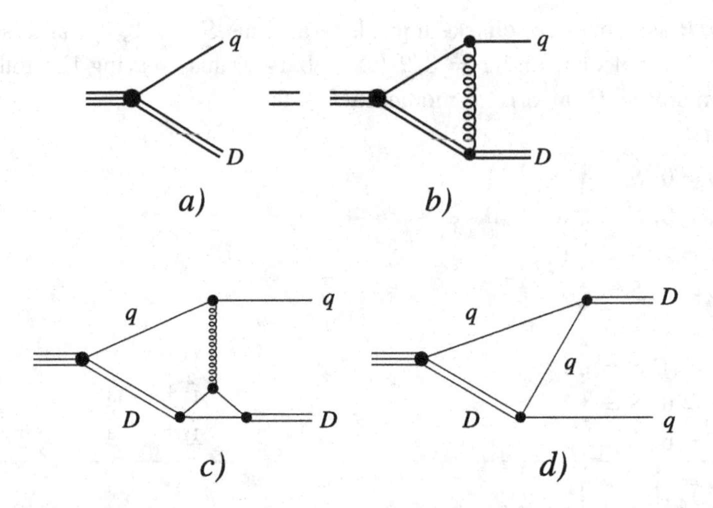

Fig. 8.1 (a,b) Equation for the quark–diquark system (the flavor-neutral interaction is denoted by a helix-type line). (c,d) Processes are considered as negligibly small in the quark–diquark model for highly excited states.

Figs. 8.1a,b. If the interaction (a helix-type line in Fig. 8.1b) is flavor-neutral (gluonic or confinement singularity exchange), diquarks retain their quantum numbers, $qD^0_0 \to qD^0_0$ and $qD^1_1 \to qD^1_1$, and the states qD^1_1 and qD^0_0 do not mix. In the equation shown in Figs. 8.1a,b, it was supposed that three-quark intermediate states are absent. This means that, first, the diquarks should be effectively point-like (diquark form factors lead to a qqq state, see Fig. 8.1c) and, second, the quark-exchange processes, Fig. 8.1d, are suppressed (these processes include three-quark states). Both requirements can be fulfilled, if the diquark size is much less than the baryon one, $R_{diquark} \ll R_{baryon}$, which may happen for highly excited states. Regretfully, we do not know for which states $R_{diquark} \ll R_{baryon}$, thus we consider here several versions.

The chapter is organized as follows.

In Section 8.2, we consider wave functions for quark–diquark systems in the non-relativistic approximation (the relativization of vertices $B \to qqq$ is not difficult, it can be found, for example, in [18, 19]). In this section, we also demonstrate the way to transform the quark–diquark wave function into a three-quark $SU(6)$-symmetrical one.

In Sections 8.3, 8.4 and 8.5 we consider different versions of the classification of baryon states.

First, on the basis of the absence of $N^{L=0}_{\frac{3}{2}^+}(\sim 1200)$ and $\Delta^{L=0}_{\frac{3}{2}^+}(\sim 950)$, predicted by the quark–diquark model in its general form (Section 8.4), we justify the $(L = 0)$ states to obey the $SU(6)$ symmetry rules (Section 8.3).

In Section 8.4 the model with qD^0_0 and qD^1_1 systems is considered in a general form and the overall predictions are given. With an exception for $(L = 0)$ states, we suggest in Section 8.4 the setting of quark–diquark baryons which are in a qualitative agreement with data. Still, some uncertainties exist in the $\Delta_{\frac{5}{2}^-}$ sector owing to certain contradictions in the data.

To be illustrative, we present in this section the (J, M^2) and (n, M^2) plots.

In Section 8.5 we consider the model with qD^0_0 resonances overlapping with those of qD^1_1, having the same $J, L, S = 1/2$. It reduces, though not substantially, the number of easily visible bumps.

We see that even in the quark–diquark model the number of resonances is noticeably larger than presently observed in the experiment. Also, the model predicts a set of overlapping resonances, resulting in a hide of some of them in visible bumps. It is a common prediction inherent in all considered schemes, being therefore a challenge for the experiment.

In the Conclusion, we summarize the problems which appear in the consideration of the quark–diquark scheme.

8.2 Baryons as quark–diquark systems

Here, to be illustrative, we consider wave functions of quark-diquark systems, qD^0_0 and qD^1_1, in the non-relativistic limit. The relativistic generalization of the $B \to 3q$ vertices may be found, for example, in Chapter 7 of [18].

8.2.1 *S-wave diquarks and baryons*

Recall that we have two S-wave diquarks with color numbers $\bar{\mathbf{c}} = 3$: scalar diquark D^0_0 and axial–vector one, $D^{1I_z}_{1S_z}$. The diquark spin–flavor wave functions with $I_D = 1$, $S_D = 1$ and $I_D = 0$, $S_D = 0$ read as follows:

$$D^{11}_{11}(ij) = u^{\uparrow}(i)u^{\uparrow}(j),$$
$$D^{11}_{10}(ij) = \frac{1}{\sqrt{2}}\Big(u^{\uparrow}(i)u^{\downarrow}(j) + u^{\downarrow}(i)u^{\uparrow}(i)\Big),$$

$$D^{10}_{11}(ij) = \frac{1}{\sqrt{2}}\Big(u^{\uparrow}(i)d^{\uparrow}(j) + d^{\uparrow}(i)u^{\uparrow}(j)\Big),$$
$$D^{10}_{10}(ij) = \frac{1}{2}\Big(u^{\uparrow}(i)d^{\downarrow}(j) + u^{\downarrow}(i)d^{\uparrow}(j) + d^{\uparrow}(i)u^{\downarrow}(j) + d^{\downarrow}(i)u^{\uparrow}(j)\Big),$$
$$D^{0}_{0}(ij) = \frac{1}{2}\Big(u^{\uparrow}(i)d^{\downarrow}(j) - u^{\downarrow}(i)d^{\uparrow}(j) - d^{\uparrow}(i)u^{\downarrow}(j) + d^{\downarrow}(i)u^{\uparrow}(j)\Big). \quad (8.4)$$

8.2.2 *Wave functions of quark–diquark systems with $L = 0$*

In the general case, we have the following sets of baryon states:

$$(qD^1_1)_{J^{\pm},L,S(=1/2,\,3/2),n}\,,$$
$$(qD^0_0)_{J^{\pm},L,S(=1/2),n}. \quad (8.5)$$

Recall that positive and negative parities $P = \pm$ are determined by the orbital momentum L between quark and diquark: $(+)$ and $(-)$ for even and odd L. The total spin of the quark–diquark states runs $S = 1/2,\ 3/2$. The states with the same (I, J^P) may have different radial excitation numbers n.

Here we consider the wave functions of quark–diquark systems with $L = 0$, namely, $\Delta^{++}_{3/2}$, $\Delta^{+}_{3/2}$ and p, as well as $\Delta^{++}_{1/2}$, $N^{+}_{3/2}$ and corresponding radial excitations. These examples give us a guide for writing other wave functions of the quark–diquark states composed by light quarks (u, d).

8.2.2.1 *The Δ isobar: quark–diquark wave function for arbitrary n and its transformation into the $SU(6)$ wave function*

The wave function, up to the normalizing coefficient, for $\Psi(\Delta^{++}_{\uparrow\uparrow\uparrow})$ with arbitrary n reads

$$u^{\uparrow}(1)D^{11}_{11}(23)\Phi_1(1;23) + u^{\uparrow}(2)D^{11}_{11}(13)\Phi_1(2;13) + u^{\uparrow}(3)D^{11}_{11}(12)\Phi_1(3;12)$$
$$= u^{\uparrow}(1)u^{\uparrow}(2)u^{\uparrow}(3)\Phi_1(1;23) + u^{\uparrow}(2)u^{\uparrow}(1)u^{\uparrow}(3)\Phi_1(2;13)$$
$$+u^{\uparrow}(3)u^{\uparrow}(1)u^{\uparrow}(2)\Phi_1(3;12) \equiv \Psi(\Delta^{++}_{\uparrow\uparrow\uparrow}). \quad (8.6)$$

Here the indices (1,2,3) label the momenta (or coordinates) of quarks. The wave functions of quarks are symmetrical in the spin, coordinate and flavor spaces. The momentum/coordinate component of the wave function is normalized in a standard way: $\int |\Psi(\Delta^{++}_{\uparrow\uparrow\uparrow})|^2 d\Phi_3 = 1$, where $d\Phi_3$ is the three-particle phase space.

Let us consider $n = 1$. If the momentum/coordinate wave function of the basic state is symmetrical,

$$\Phi_1(1;23) = \Phi_1(2;13) = \Phi_1(3;12) \equiv \varphi_1^{(sym)}(1,2,3), \quad (8.7)$$

we have the $SU(6)$ symmetry for $\Delta^{+}_{\uparrow\uparrow\uparrow}$:

$$\Psi_{SU(6)}(\Delta^{++}_{\uparrow\uparrow\uparrow}) =$$

$$= \Big[u^{\uparrow}(1)u^{\uparrow}(2)u^{\uparrow}(3) + u^{\uparrow}(2)u^{\uparrow}(1)u^{\uparrow}(3) + u^{\uparrow}(3)u^{\uparrow}(1)u^{\uparrow}(2)\Big]\varphi_1(1,2,3)$$

$$= u^{\uparrow}(1)u^{\uparrow}(2)u^{\uparrow}(3)\; 3\varphi_1^{(sym)}(1,2,3) \equiv \{u^{\uparrow}u^{\uparrow}u^{\uparrow}\}\varphi_1(1,2,3). \qquad (8.8)$$

Here and below we omit the index $^{(sym)}$, *i.e.* $\varphi_1^{(sym)}(1,2,3) \to \varphi_1(1,2,3)$.

The $\Delta^{++}_{\uparrow}$ wave function with arbitrary n reads

$$\Psi(\Delta^{++}_{\uparrow}) = C^{\frac{3}{2}\frac{3}{2}}_{1\,1\;\frac{1}{2}\frac{1}{2}}\, C^{\frac{3}{2}\frac{1}{2}}_{1\,0\;\frac{1}{2}\frac{1}{2}} \Big[u^{\uparrow}(1)D^{11}_{10}(23)\Phi_1(1;23)$$

$$+ u^{\uparrow}(2)D^{11}_{10}(31)\Phi_1(2;31) + u^{\uparrow}(3)D^{11}_{10}(12)\Phi_1(3;12)\Big]$$

$$+ C^{\frac{3}{2}\frac{3}{2}}_{1\,1\;\frac{1}{2}\frac{1}{2}}\, C^{\frac{3}{2}\frac{1}{2}}_{1\,1\;\frac{1}{2}-\frac{1}{2}} \Big[u^{\downarrow}(1)D^{11}_{11}(23)\Phi_1(1;23)$$

$$+ u^{\downarrow}(2)D^{11}_{11}(31)\Phi_1(2;31) + u^{\downarrow}(3)D^{11}_{11}(12)\Phi_1(3;12)\Big]\,. \qquad (8.9)$$

If at $n = 1$, as previously, the momentum/coordinate wave function is symmetrical, see Eq. (8.7), we deal with the $SU(6)$ symmetry for basic $\Delta^{+}_{\uparrow}$:

$$\Psi_{SU(6)}(\Delta^{++}_{\uparrow}) = \frac{1}{\sqrt{3}}\Big[u^{\downarrow}(1)u^{\uparrow}(2)u^{\uparrow}(3) + u^{\uparrow}(1)u^{\uparrow}(2)u^{\downarrow}(3) + u^{\uparrow}(1)u^{\downarrow}(2)u^{\uparrow}(3)\Big]$$

$$\times\, \varphi_1(1,2,3). \qquad (8.10)$$

In a more compact form, it reads

$$\Psi_{SU(6)}(\Delta^{++}_{\uparrow}) = \frac{1}{\sqrt{3}}\Big[u^{\downarrow}u^{\uparrow}u^{\uparrow} + u^{\uparrow}u^{\uparrow}u^{\downarrow} + u^{\uparrow}u^{\downarrow}u^{\uparrow}\Big]\varphi_1(1,2,3)$$

$$\equiv \{u^{\downarrow}u^{\uparrow}u^{\uparrow}\}\varphi_1(1,2,3). \qquad (8.11)$$

The $\Delta^{+}_{\uparrow}$ wave function, $\Psi(\Delta^{+}_{\uparrow})$, is proportional to

$$C^{\frac{3}{2}\frac{1}{2}}_{1\,0\;\frac{1}{2}\frac{1}{2}}\, C^{\frac{3}{2}\frac{1}{2}}_{1\,0\;\frac{1}{2}\frac{1}{2}} \Big[u^{\uparrow}(1)D^{10}_{10}(23)\Phi_1(1;23) + u^{\uparrow}(2)D^{10}_{10}(31)\Phi_1(2;31)$$

$$+ u^{\uparrow}(3)D^{10}_{10}(12)\Phi_1(3;12)\Big]$$

$$+C^{\frac{3}{2}\frac{1}{2}}_{1\,0\;\frac{1}{2}\frac{1}{2}}\, C^{\frac{3}{2}\frac{1}{2}}_{1\,1\;\frac{1}{2}-\frac{1}{2}} \Big[u^{\downarrow}(1)D^{10}_{11}(23)\Phi_1(1;23) + u^{\downarrow}(2)D^{10}_{11}(31)\Phi_1(2;31)$$

$$+ u^{\downarrow}(3)D^{10}_{11}(12)\Phi_1(3;12)\Big]$$

$$+C^{\frac{3}{2}\frac{1}{2}}_{1\,1\;\frac{1}{2}-\frac{1}{2}}\, C^{\frac{3}{2}\frac{1}{2}}_{1\,0\;\frac{1}{2}\frac{1}{2}} \Big[d^{\uparrow}(1)D^{11}_{10}(23)\Phi_1(1;23) + d^{\uparrow}(2)D^{11}_{10}(31)\Phi_1(2;31)$$

$$+ d^{\uparrow}(3)D^{11}_{10}(12)\Phi_1(3;12)\Big]$$

$$+C^{\frac{3}{2}\frac{1}{2}}_{1\,1\;\frac{1}{2}-\frac{1}{2}}\, C^{\frac{3}{2}\frac{1}{2}}_{1\,1\;\frac{1}{2}-\frac{1}{2}} \Big[d^{\downarrow}(1)D^{11}_{11}(23)\Phi_1(1;23) + d^{\downarrow}(2)D^{11}_{11}(31)\Phi_1(2;31)$$

$$+ d^{\downarrow}(3)D^{11}_{11}(12)\Phi_1(3;12)\Big], \qquad (8.12)$$

so we have:

$$\begin{aligned}
\Psi(\Delta^+_\uparrow) = &\frac{2}{3}\Big(u^\uparrow(1)D^{10}_{10}(23)\Phi_1(1;23) + u^\uparrow(2)D^{10}_{10}(31)\Phi_1(2;31) \\
&+u^\uparrow(3)D^{10}_{10}(12)\Phi_1(3;12)\Big) \\
&+\frac{\sqrt{2}}{3}\Big(u^\downarrow(1)D^{10}_{11}(23)\Phi_1(1;23) + u^\downarrow(2)D^{10}_{11}(31)\Phi_1(2;31) \\
&+u^\downarrow(3)D^{10}_{11}(12)\Phi_1(3;12)\Big) \\
&+\frac{\sqrt{2}}{3}\Big(d^\uparrow(1)D^{11}_{10}(23)\Phi_1(1;23) + d^\uparrow(2)D^{11}_{10}(31)\Phi_1(2;31) \\
&+d^\uparrow(3)D^{11}_{10}(12)\Phi_1(3;12)\Big) \\
&+\frac{1}{3}\Big(d^\downarrow(1)D^{11}_{11}(23)\Phi_1(1;23) + d^\downarrow(2)D^{11}_{11}(31)\Phi_1(2;31) \\
&+d^\downarrow(3)D^{11}_{11}(12)\Phi_1(3;12)\Big).
\end{aligned} \tag{8.13}$$

The $SU(6)$-symmetrical wave function can be written as

$$\Psi_{SU(6)}(\Delta^+_\uparrow) = \Big(\sqrt{\frac{1}{3}}\{u^\uparrow u^\uparrow d^\downarrow\} + \sqrt{\frac{2}{3}}\{u^\uparrow u^\downarrow d^\uparrow\}\Big)\varphi_1(1,2,3), \tag{8.14}$$

where

$$\{q_i q_j q_\ell\} = \frac{1}{\sqrt{6}}\{q_i q_j q_\ell + q_i q_\ell q_j + q_j q_i q_\ell + q_j q_\ell q_i + q_\ell q_i q_j + q_\ell q_j q_i\} \tag{8.15}$$

for $q_j \neq q_i \neq q_\ell$.

Above, to simplify the presentation, we transformed the wave functions to the $SU(6)$-symmetry ones for $n = 1$. One can present certain examples with an easy generalization to $n > 1$. Assuming for $n = 1$ that

$$\varphi_1(1,2,3) \equiv \varphi_1^{(n=1)}(1,2,3) = A_1^{(1)}\exp(-b^{(1)}s), \tag{8.16}$$

where s is the total energy squared $s = (k_1 + k_2 + k_3)^2$, one may have for $n = 2$:

$$\varphi_1^{(n=2)}(1,2,3) = A_2^{(1)}\exp(-b^{(1)}s)(s - B_2^{(1)}). \tag{8.17}$$

Here $B_2^{(1)}$ is chosen to introduce a node into the $(n = 2)$ wave function. Likewise, we can write wave functions for higher n. It is not difficult to construct models with wave functions of the type (8.16), (8.17) – the versions of corresponding models are discussed below.

8.2.2.2 *The state with $I = 3/2$, $J = 1/2$ at $L = 0$ and $S = 1/2$, $\Delta_{J^P=\frac{1}{2}^+}(L=0, S=1/2)$*

Let us present the wave function for $\Delta^{\uparrow}_{J^P=\frac{1}{2}^+}(L=0, S=1/2)$. It reads for $I_Z = \frac{3}{2}$ and $S_Z = \frac{1}{2}$ as follows:

$$
\begin{aligned}
& C^{\frac{3}{2}\frac{3}{2}}_{11\ \frac{1}{2}\frac{1}{2}} C^{\frac{1}{2}\frac{1}{2}}_{10\ \frac{1}{2}\frac{1}{2}} \Big[u^{\uparrow}(1)D^{11}_{10}(23)\Phi_1(1;23) + u^{\uparrow}(2)D^{11}_{10}(31)\Phi_1(2;31) \\
& \qquad + u^{\uparrow}(3)D^{11}_{10}(12)\Phi_1(3;12)\Big] \\
& + C^{\frac{3}{2}\frac{3}{2}}_{11\ \frac{1}{2}\frac{1}{2}} C^{\frac{1}{2}\frac{1}{2}}_{11\ \frac{1}{2}-\frac{1}{2}} \Big[u^{\downarrow}(1)D^{11}_{11}(23)\Phi_1(1;23) + u^{\downarrow}(2)D^{11}_{11}(31)\Phi_1(2;31) \\
& \qquad + u^{\downarrow}(3)D^{11}_{11}(12)\Phi_1(3;12)\Big] = \\
& = -\frac{1}{\sqrt{3}}\Big[u^{\uparrow}(1)\frac{u^{\uparrow}(2)u^{\downarrow}(3) + u^{\downarrow}(2)u^{\uparrow}(3)}{\sqrt{2}}\Phi_1(1;23) + (1 \rightleftharpoons 2) + (1 \rightleftharpoons 3)\Big] \\
& + \sqrt{\frac{2}{3}}\Big[u^{\downarrow}(1)u^{\uparrow}(2)u^{\uparrow}(3)\Phi_1(1;23) + (1 \rightleftharpoons 2) + (1 \rightleftharpoons 3)\Big]. \qquad (8.18)
\end{aligned}
$$

In the $SU(6)$ limit, the wave functions for $\Delta^{\uparrow}_{J^P=\frac{1}{2}^+}(L=0, S=3/2)$, which depend on s only, are equal to zero:

$$
\begin{aligned}
& -\frac{1}{\sqrt{3}}\Big[u^{\uparrow}(1)\frac{u^{\uparrow}(2)u^{\downarrow}(3) + u^{\downarrow}(2)u^{\uparrow}(3)}{\sqrt{2}} + (1 \rightleftharpoons 2) + (1 \rightleftharpoons 3)\Big]\varphi_1(s) \\
& + \sqrt{\frac{2}{3}}\Big[u^{\downarrow}(1)u^{\uparrow}(2)u^{\uparrow}(3) + (1 \rightleftharpoons 2) + (1 \rightleftharpoons 3)\Big]\varphi_1(s) = 0. \qquad (8.19)
\end{aligned}
$$

Radial excitation wave functions in the $SU(6)$ limit, if they depend on s only (for example, see Eq. (8.17)), are also equal to zero. So, in the $SU(6)$ limit we have for $L = 0$ only the state $\Delta_{J^P=\frac{3}{2}^+}$.

8.2.2.3 *The nucleon $N_{1/2}$: quark–diquark wave function for arbitrary n and its transformation into the $SU(6)$ one*

The S-wave functions for $N^{+\uparrow}_{1/2}(qD^0_0)$ state with arbitrary n reads

$$
\begin{aligned}
\Psi^{+\uparrow}_{J=1/2}(qD^0_0) = & \Big[u^{\uparrow}(1)D^0_0(23)\Phi_0(1;23) + u^{\uparrow}(2)D^0_0(31)\Phi_0(2;31) \\
& + u^{\uparrow}(3)D^0_0(12)\Phi_0(3;12)\Big]. \qquad (8.20)
\end{aligned}
$$

For the symmetrical momentum/coordinate wave function,

$$
\Phi_0(1;23) = \Phi_0(2;31) = \Phi_0(3;12) \equiv \varphi_0(1,2,3), \qquad (8.21)
$$

we have:

$$\Psi^{+\uparrow}_{SU(6)}(qD^0_0) = \left(\sqrt{\frac{2}{3}}\{u^\uparrow u^\uparrow d^\downarrow\} - \sqrt{\frac{1}{3}}\{u^\uparrow u^\downarrow d^\uparrow\}\right)\varphi_0(1,2,3). \quad (8.22)$$

Likewise, we can construct a nucleon as a qD^1_1 system – the wave function of $N^{+\uparrow}_{J=1/2}(qD^1_1)$ is written at arbitrary n as

$$\Psi^{+\uparrow}_{J=1/2}(qD^1_1) = \frac{1}{3}\Big(u^\uparrow(1)D^{10}_{10}(23)\Phi_1(1;23) + u^\uparrow(2)D^{10}_{10}(31)\Phi_1(2;31)$$
$$+u^\uparrow(3)D^{10}_{10}(12)\Phi_1(3;12)\Big)$$
$$+(-\frac{\sqrt{2}}{3})\Big(u^\downarrow(1)D^{10}_{11}(23)\Phi_1(1;23) + u^\downarrow(2)D^{10}_{11}(31)\Phi_1(2;31)$$
$$+u^\downarrow(3)D^{10}_{11}(12)\Phi_1(3;12)\Big)$$
$$+(-\frac{\sqrt{2}}{3})\Big(d^\uparrow(1)D^{11}_{10}(23)\Phi_1(1;23) + d^\uparrow(2)D^{11}_{10}(31)\Phi_1(2;31)$$
$$+d^\uparrow(3)D^{11}_{10}(12)\Phi_1(3;12)\Big)$$
$$+\frac{2}{3}\Big(d^\downarrow(1)D^{11}_{11}(23)\Phi_1(1;23) + d^\downarrow(2)D^{11}_{11}(31)\Phi_1(2;31)$$
$$+d^\downarrow(3)D^{11}_{11}(21)\Phi_1(3;12)\Big). \quad (8.23)$$

In the limit of Eq. (8.7), which means the $SU(6)$ symmetry for qD^1_1 states, we have

$$\Psi^{+\uparrow}_{SU(6)}(qD^1_1) = \left(\sqrt{\frac{2}{3}}\{u^\uparrow u^\uparrow d^\downarrow\} - \sqrt{\frac{1}{3}}\{u^\uparrow u^\downarrow d^\uparrow\}\right)\varphi_1(1,2,3). \quad (8.24)$$

One can see that, if

$$\varphi_0(1,2,3) \neq \varphi_1(1,2,3), \quad (8.25)$$

we have two different nucleon states corresponding to two different diquarks, D^0_0 and D^1_1.

If we require

$$\varphi_0(1,2,3) = \varphi_1(1,2,3), \quad (8.26)$$

it makes possible to have only one level, not two, which means the $SU(6)$ symmetry.

Recall that in the $SU(6)$ limit the nucleon can be presented as a mixture of both diquarks: to be illustrative, we rewrite the spin–flavour part of the proton wave function as follows:

$$\sqrt{\frac{2}{3}}\{u^\uparrow u^\uparrow d^\downarrow\} - \sqrt{\frac{1}{3}}\{u^\uparrow u^\downarrow d^\uparrow\} = \frac{1}{\sqrt{2}}u^\uparrow(1)D^0_0(23) + \frac{1}{3\sqrt{2}}u^\uparrow(1)D^{10}_{10}(23)$$
$$-\frac{1}{3}d^\uparrow(1)D^{11}_{10}(23) - \frac{1}{3}u^\downarrow(1)D^{10}_{11}(23) + \frac{\sqrt{2}}{3}d^\downarrow(1)D^{11}_{11}(23)\,. \quad (8.27)$$

So, the nucleon in the $SU(6)$ limit is a mixture of qD_0^0 and qD_1^1 states in equal proportion.

8.2.2.4 *The state with $I = 1/2$, $J = 3/2$, $L = 0$ and $S = 3/2$, $N_{J^P=\frac{3}{2}^+}(L=0, S=3/2)$*

Let us write down the wave function for $N^{\uparrow}_{J^P=\frac{3}{2}^+}(L = 0, S = 3/2)$ with $I_Z = \frac{1}{2}$ and $S_Z = \frac{3}{2}$:

$$
\begin{aligned}
&C^{\frac{1}{2}\frac{1}{2}}_{10\;\frac{1}{2}\frac{1}{2}}\,C^{\frac{3}{2}\frac{3}{2}}_{11\;\frac{1}{2}\frac{1}{2}}\Big[u^{\uparrow}(1)D^{10}_{11}(23)\Phi_1(1;23) + u^{\uparrow}(2)D^{10}_{11}(31)\Phi_1(2;31)\\
&\qquad\qquad + u^{\uparrow}(3)D^{10}_{11}(12)\Phi_1(3;12)\Big]\\
&+ C^{\frac{1}{2}\frac{1}{2}}_{11\;\frac{1}{2}-\frac{1}{2}}\,C^{\frac{3}{2}\frac{3}{2}}_{11\;\frac{1}{2}\frac{1}{2}}\Big[d^{\uparrow}(1)D^{11}_{11}(23)\Phi_1(1;23) + d^{\uparrow}(2)D^{11}_{11}(31)\Phi_1(2;31)\\
&\qquad\qquad + d^{\uparrow}(3)D^{11}_{11}(12)\Phi_1(3;12)\Big] =\\
&= -\frac{1}{\sqrt{3}}\Big[u^{\uparrow}(1)\frac{u^{\uparrow}(2)d^{\uparrow}(3) + d^{\uparrow}(2)u^{\uparrow}(3)}{\sqrt{2}}\Phi_1(1;23) + (1 \rightleftharpoons 2) + (1 \rightleftharpoons 3)\Big]\\
&+ \sqrt{\frac{2}{3}}\Big[d^{\uparrow}(1)u^{\uparrow}(2)u^{\uparrow}(3)\Phi_1(1;23) + (1 \rightleftharpoons 2) + (1 \rightleftharpoons 3)\Big]. \qquad (8.28)
\end{aligned}
$$

In the $SU(6)$ limit, under the constraint of Eq. (8.7), the wave function for the $\Delta^{\uparrow}_{J^P=\frac{1}{2}^+}(L = 0, S = 3/2)$ is equal to zero:

$$
\begin{aligned}
&-\frac{1}{\sqrt{3}}\Big[u^{\uparrow}(1)\frac{u^{\uparrow}(2)d^{\uparrow}(3) + d^{\uparrow}(2)u^{\uparrow}(3)}{\sqrt{2}} + (1 \rightleftharpoons 2) + (1 \rightleftharpoons 3)\Big]\varphi_1(1,2,3)\\
&+ \sqrt{\frac{2}{3}}\Big[d^{\uparrow}(1)u^{\uparrow}(2)u^{\uparrow}(3) + (1 \rightleftharpoons 2) + (1 \rightleftharpoons 3)\Big]\varphi_1(1,2,3) = 0. \qquad (8.29)
\end{aligned}
$$

Radial excitation wave functions in the $SU(6)$ limit, if they depend on s only (see Eq. (8.17) for example), are equal also to zero.

So, in the $SU(6)$ limit the nucleon state with $L = 0$ and $J^P = \frac{3}{2}^+$ does not exist.

8.2.3 Wave functions of quark–diquark systems with $L \neq 0$

Let us consider, first, the Δ isobar at $I_Z = 3/2$ with fixed J, J_Z, total spin S and orbital momentum L. The wave function for this state at arbitrary

n reads

$$\sum_{S_Z,m_z} C^{J\,J_Z}_{L\,J_Z-S_Z\;S\,S_Z} C^{S\,S_Z}_{1\,S_Z-m_z\;\frac{1}{2}\,m_z} C^{\frac{3}{2}\frac{3}{2}}_{1\,1\;\frac{1}{2}\frac{1}{2}} \Big(u^{m_z}(1) D^{1\,1}_{1\,S_Z-m_z}(23)$$
$$\times |\vec{k}_{1\,cm}|^L Y^{J_Z-S_Z}_L(\theta_1,\phi_1)\Phi^{(L)}_1(1;23) + (1\rightleftharpoons 2) + (1\rightleftharpoons 3)\Big) . \qquad (8.30)$$

Here $|\vec{k}_{1\,cm}|$ and (θ_1,ϕ_1) are the momenta and momentum angles of the first quark in the c.m. system.

For other I_Z values, one should include into the wave function the summation over isotopic states, *i.e.* the following substitution in (8.30):

$$C^{\frac{3}{2}\frac{3}{2}}_{1\,1\;\frac{1}{2}\frac{1}{2}} u^{m_z}(1) D^{1\,1}_{1\,S_Z-m_z}(23) \to \sum_{j_z} C^{\frac{3}{2}\,I_Z}_{1\,I_Z-j_z\;\frac{1}{2}\,j_z} q^{m_z}_{j_z}(1) D^{1\,I_Z-j_z}_{1\,S_Z-m_z}(23) \,. \qquad (8.31)$$

One can see that wave functions of neither(8.30) nor (8.31) give us zeros, when $\Phi^{(L)}_1(1;23)$ depends on s only. Indeed, in this limit we have

$$\sum_{S_Z,m_z} C^{J\,J_Z}_{L\,J_Z-S_Z\;S\,S_Z} C^{S\,S_Z}_{1\,S_Z-m_z\;\frac{1}{2}\,m_z} \sum_{j_z} C^{\frac{3}{2}\,I_Z}_{1\,I_Z-j_z\;\frac{1}{2}\,j_z} \Big(q^{m_z}_{j_z}(1) D^{1\,I_Z-j_z}_{1\,S_Z-m_z}(23)$$
$$\times |\vec{k}_{1\,cm}|^L Y^{J_Z-S_Z}_L(\theta_1,\phi_1) + (1\rightleftharpoons 2) + (1\rightleftharpoons 3)\Big)\phi^{(L)}_1(s) \,. \qquad (8.32)$$

The factor $|\vec{k}_{1\,cm}|^L Y^{J_Z-S_Z}_L(\theta_1,\phi_1)$ and analogous ones in $(1\rightleftharpoons 2)$ and $(1\rightleftharpoons 3)$ prevent the cancelation of different terms in big parentheses of Eq. (8.32), which are present in case of $L=0$, see Eq. (8.19).

For nucleon states ($I=1/2$) we write:

$$\sum_{S_Z,m_z} C^{J\,J_Z}_{L\,J_Z-S_Z\;S\,S_Z} C^{S\,S_Z}_{1\,S_Z-m_z\;\frac{1}{2}\,m_z} \sum_{j_z} C^{\frac{1}{2}\,I_Z}_{1\,I_Z-j_z\;\frac{1}{2}\,j_z} \Big(q^{m_z}_{j_z}(1) D^{1\,I_Z-j_z}_{1\,S_Z-m_z}(23)$$
$$\times |\vec{k}_{1\,cm}|^L Y^{J_Z-S_Z}_L(\theta_1,\phi_1)\Phi^{(L)}_1(1;23) + (1\rightleftharpoons 2) + (1\rightleftharpoons 3)\Big). \qquad (8.33)$$

The $SU(6)$ limit, as previously, is realized at $\Phi^{(L)}_1(i;j\ell) \to \varphi^{(L)}_1(s)$. Then, instead of (8.33), one has:

$$\sum_{S_Z,m_z} C^{J\,J_Z}_{L\,J_Z-S_Z\;S\,S_Z} C^{S\,S_Z}_{1\,S_Z-m_z\;\frac{1}{2}\,m_z} \sum_{j_z} C^{\frac{1}{2}\,I_Z}_{1\,I_Z-j_z\;\frac{1}{2}\,j_z} \Big(q^{m_z}_{j_z}(1) D^{1\,I_Z-j_z}_{1\,S_Z-m_z}(23)$$
$$\times |\vec{k}_{1\,cm}|^L Y^{J_Z-S_Z}_L(\theta_1,\phi_1) + (1\rightleftharpoons 2) + (1\rightleftharpoons 3)\Big)\varphi^{(L)}_1(s). \qquad (8.34)$$

For qD^0_0 states the wave function in the general case reads

$$\sum_{m_z} C^{J\,J_Z}_{L\,J_Z-m_z\;\frac{1}{2}\,m_z}\Big(q^{m_z}_{I_z}(1)D^0_0(23)|\vec{k}_{1\,cm}|^L Y^{J_Z-m_z}_L(\theta_1,\phi_1)\Phi^{(L)}_0(1;23)$$

$$+(1\rightleftharpoons 2)+(1\rightleftharpoons 3)\Big)\,. \qquad (8.35)$$

In the $SU(6)$ limit we have:

$$\sum_{m_z} C^{J\,J_Z}_{L\,J_Z-m_z\;\frac{1}{2}\,m_z}\Big(q^{m_z}_{I_z}(1)D^0_0(23)|\vec{k}_{1\,cm}|^L Y^{J_Z-m_z}_L(\theta_1,\phi_1)$$

$$+(1\rightleftharpoons 2)+(1\rightleftharpoons 3)\Big)\varphi^{(L)}_0(s)\,. \qquad (8.36)$$

Baryons are characterized by I and J^P – the states with different S and L can mix. To select independent states, one may orthogonalize wave functions with the same isospin and J^P. The orthogonalization depends on the structure of momentum/coordinate parts $\Phi^{(L)}_1(i;j\ell)$. But in case of the $SU(6)$ limit the momentum/coordinate wave functions transform into a common factor $\Phi^{(L)}_1(i;j\ell)$, $\Phi^{(L)}_0(i;j\ell)\to\varphi^{(L)}_{SU(6)}(s)$, and one can orthogonalize the spin factors. Namely, we can present the $SU(6)$ wave function as follows:

$$\Psi^{(A)}_{J^P}=Q^{(A)}_{J^P}\varphi^{(A)}_{SU(6)}(s), \qquad (8.37)$$

where $Q^{(A)}_{J^P}$ is the spin operator and $A=I,II,III,...$ refer to different (S,L). The orthogonal set of operators $Q^{(A)}_{J^P}$ is constructed in a standard way:

$$Q^{(\perp I)}_{J^P}\equiv Q^{(\perp A)}_{J^P}\,,$$

$$Q^{(\perp II)}_{J^P}=Q^{(II)}_{J^P}-Q^{(\perp I)}_{J^P}\frac{\left(Q^{(\perp I)+}_{J^P}Q^{(II)}_{J^P}\right)}{\left(Q^{(\perp I)+}_{J^P}Q^{(\perp I)}_{J^P}\right)}\,,$$

$$Q^{(\perp III)}_{J^P}=Q^{(III)}_{J^P}-Q^{(\perp I)}_{J^P}\frac{\left(Q^{(\perp I)+}_{J^P}Q^{(III)}_{J^P}\right)}{\left(Q^{(\perp I)+}_{J^P}Q^{(\perp I)}_{J^P}\right)}-Q^{(\perp II)}_{J^P}\frac{\left(Q^{(\perp II)+}_{J^P}Q^{(III)}_{J^P}\right)}{\left(Q^{(\perp II)+}_{J^P}Q^{(\perp II)}_{J^P}\right)}\,, \qquad (8.38)$$

etc. The convolution of operator $\left(Q^{(A)+}_{J^P}Q^{(B)}_{J^P}\right)$ includes both the summation over quark spins and the integration over quark momenta.

8.2.4 *Quarks and diquarks in baryons*

Exploring the notion of the constituent quark and the composite diquark, we propose several schemes for the structure of low-lying baryons. To be illustrative, let us turn to Fig. 8.2.

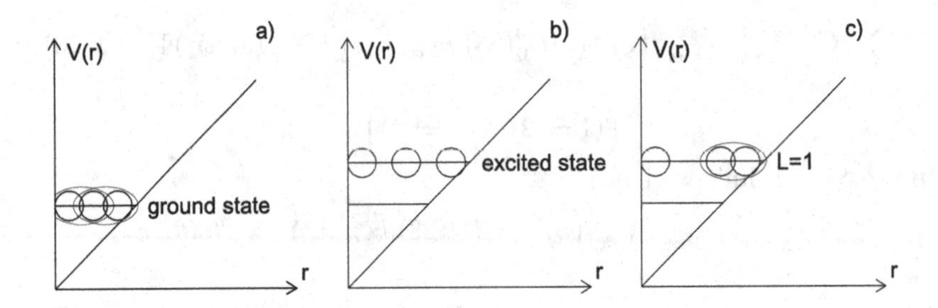

Fig. 8.2 Illustration of the quark–diquark structure of baryon levels. a) Ground state with a complete mixing of constituent quarks: the bound quarks and diquarks, being compressed states, provide us the three-quark $SU(6)$-symmetry structure. b) Conventional picture for an excited state in the standard three-quark model with three spatially separated quarks. c) Example of the excited state in the quark–diquark model: the quark–diquark state with $L = 1$, quark and diquark being spatially separated.

Using the potential picture, the standard scheme of the three-quark baryon is shown in Figs. 8.2a and 8.2b. On the lowest level, there are three S-wave quarks — it is a compact system (the radii of constituent quarks are of the order of ~ 0.2 fm, while the nucleon size is ~ 0.8 fm [17, 20]). So, for the ground state, which is a system with overlapping different quark pairs (Fig. 8.2a), the hypothesis about the $SU(6)$ classification seems to be rather reliable.

Concerning the excited states, the quarks of the standard quark model (see the example in Fig. 8.2b), are in the average located at comparatively large distances from each other. Such a three-quark composite system is characterized by pair excitations – the number of pair excitations may be large, thus resulting in a fast increase of the number of excited baryons.

The quark–diquark structure of levels, supposed in our consideration for $L > 0$, is demonstrated in Fig. 8.2c. For excited states we assume the following quark–diquark picture: two quarks are at comparatively small distances, being a diquark state, and the third quark is separated from this diquark. The number of quark–diquark excitations is noticeably less than the number of excitations in the three-quark system.

8.2.4.1 *Diquark composite systems and mass distributions of diquarks*

Constituent quarks and diquarks are effective particles. We assume that propagators of the diquark composite systems can be well described using the Källen–Lehman representation [21]. For scalar and axial–vector diquarks, the propagators can be written as

$$\Pi^{(D_0^0)}(p) = \int\limits_{m^2_{min}}^{\infty} dm_D^2 \frac{\rho_{D_0^0}(m_D^2)}{m_D^2 - p^2 - i0},$$

$$\Pi^{(D_1^1)}_{\mu\nu}(p) = -g_{\mu\nu}^{\perp p} \int\limits_{m^2_{min}}^{\infty} dm_D^2 \frac{\rho_{D_1^1}(m_D^2)}{m_D^2 - p^2 - i0}, \tag{8.39}$$

where the mass distributions $\rho_{D_0^0}(m_D^2)$ and $\rho_{D_1^1}(m_D^2)$ are characterized by the compactness of the scalar and axial–vector diquarks. The use of the mass propagator (8.39) is definitely needed in the calculation of the subtle effects in baryonic reactions. However, in a rough approximation one may treat diquarks, similarly to constituent quarks, as effective particles:

$$\rho_{D_0^0}(m_D^2) \to \delta(m_D^2 - M^2_{D_0^0}),$$

$$\rho_{D_1^1}(m_D^2) \to \delta(m_D^2 - M^2_{D_1^1}). \tag{8.40}$$

We expect the diquark mass to be in the region of the 600-900 MeV [22].

Mass distributions for three-quark systems in the approximation of (8.40) at fixed $s = s_{12} + s_{13} + s_{23} - 3m^2$ can be shown on the Dalitz-plot. In Fig. 8.3, we show the Dalitz-plots for the approach of short-range diquarks – due to the convention, below we use (8.40). In Figs. 8.3a and 8.3b the cases $M^2_{D_0^0} = M^2_{D_1^1}$ and $M^2_{D_0^0} \neq M^2_{D_1^1}$ are demonstrated, correspondingly.

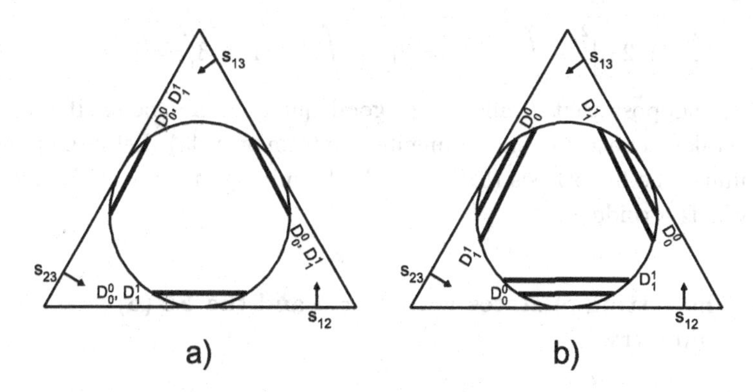

Fig. 8.3 Dalitz plots for three-quark systems at a) $M^2_{D_0^0} = M^2_{D_1^1}$ and b) $M^2_{D_0^0} \neq M^2_{D_1^1}$

8.2.4.2 *Normalization condition for wave functions of quark–diquark systems*

The model with a spatially separated quark and diquark results in a specific orthogonality/normalization condition. The matter is that interference terms with different diquarks provide a small contribution. For example,

$$\int d\Phi_3 |\vec{k}_{1\,cm}|^L Y_L^{J_Z-S_Z}(\theta_1,\phi_1)\Phi_1^{(L)}(1;23) \times \left(|\vec{k}_{2\,cm}|^L Y_L^{J_Z-S_Z}(\theta_2,\phi_2)\Phi_1^{(L)}(2;13)\right)^+ \simeq 0\,. \tag{8.41}$$

Below we neglect such interference terms.

Therefore, the normalization condition,

$$\int d\Phi_3 \left|\Psi_{J,J_Z}^{(L,S)}(1,2,3)\right|^2 = 1\,, \tag{8.42}$$

So, for the qD_1^1 systems we re-write (8.42) in the following form:

$$\int d\Phi_3 \Big| \sum_{S_Z,m_z} C^{J\,J_Z}_{L\,J_Z-S_Z\;\;S\,S_Z} C^{S\,S_Z}_{1\,S_Z-m_z\;\;\frac12\,m_z} \sum_{j_z} C^{I,I_Z}_{1\,I_Z-j_z\;\;\frac12\,j_z} \times q^{m_z}_{j_z}(1) D^{1\,I_Z-j_z}_{1\,S_Z-m_z}(23)|\vec{k}_{1\,cm}|^L Y_L^{J_Z-S_Z}(\theta_1,\phi_1)\Phi_1^{(L)}(1;23)\Big|^2 + \int d\Phi_3 \Big|1\rightleftharpoons 2\Big|^2 + \int d\Phi_3 \Big|1\rightleftharpoons 3\Big|^2 = 1, \tag{8.43}$$

while for qD_0^0 we have

$$\int d\Phi_3 \Big|\sum_{m_z} C^{J\,J_Z}_{L\,J_Z-m_z\;\;1/2\,m_z} q^{m_z}_{j_z}(1) D_0^0(23)|\vec{k}_{1\,cm}|^L Y_L^{J_Z-m_z}(\theta_1,\phi_1) \times \Phi_1^{(L)}(1;23)\Big|^2 + \int d\Phi_3 \Big|1\rightleftharpoons 2\Big|^2 + \int d\Phi_3 \Big|1\rightleftharpoons 3\Big|^2 = 1. \tag{8.44}$$

Here we suppose that L and S are good quantum numbers. If not, one should take into account the mixing in each term of (8.43). Let us emphasize that under the hypothesis (8.43), (8.44) the mixing of terms with different diquarks is forbidden.

8.3 The setting of states with $L = 0$ and the $SU(6)$ symmetry

This section is devoted to the basic $L = 0$ states and their radial excitations. But, first, let us look over the situation of the observed baryons – here some comments are needed.

8.3.1 Baryon spectra for the excited states

The masses of the well-established states (3 or 4 stars in the Particle Data Group classification [23]) are taken as a mean value over the interval given by PDG, with errors covering this interval. But the states established not so definitely require some special discussion.

We have introduced two S_{11} states in the region of 1900 and 2200 MeV, which are classified by PDG as $S_{11}(2090)$. Indeed, the observation of a state with mass 2180 ± 80 MeV by Cutkosky [24] can be hardly compatible with observations [25, 26, 27] of an S_{11} state with the mass around 1900 MeV.

The same procedure has been applied to the states D_{13} around 2000 MeV. Here the first state is located in the region of 1880 MeV and was observed in the analyses [24, 25, 28]. This state is also well compatible with the analysis of photoproduction reactions [29]. The second state is located in the region 2040 MeV: its mass has been obtained as an average value over the results of [24, 26, 30].

The $P_{11}(1880)$ state has been observed by Manley [25] as well as in the analyses of the photoproduction data with open strangeness [29, 31]: we consider this state as well established. Thus, for the state $P_{11}(2100)$ we have taken the mass as an average value over all the measurements quoted by PDG: [24, 26, 27, 30, 32, 33].

We also consider $D_{15}(2070)$ as an established state. It has been observed in the η photoproduction data [34], although we understand that a confirmation of this state by other data is needed. Furthermore, we have taken for $D_{15}(2200)$ the average value, using [24, 30, 33] analyses which give compatible results.

As to the Δ sector, we see that the $\frac{5}{2}^-$ state observed in the analysis of the GWU group [35] with the mass 2233 MeV and quoted as $D_{35}(1930)$ can be hardly compatible with other observations, which give the results in the region of 1930 MeV. Moreover, the GWU result is compatible with the analysis of Manley [25], which is quoted by PDG as $D_{35}(2350)$, though it gives the mass 2171 ± 18 MeV. Thus, we introduce the $\Delta(\frac{5}{2}^-)$ state with the mass 2210 MeV and the error which covers both these results. Then, the mass of the $D_{35}(2350)$ state is taken as an average value over the results of [24, 26, 30].

We also consider the $D_{33}(1940)$ state, which has one star by the PDG classification, as an established one. It is seen very clearly in the analysis of the $\gamma p \to \pi^0 \eta p$ data [36, 37].

One of the most interesting observations concerns $\Delta(\frac{5}{2}^+)$ states. The analyses of Manley [25] and Vrana [26] give a state in the region 1740 MeV with compatible widths. However, this state was confirmed neither by πN elastic nor by photoproduction data. This state is listed by PDG as $F_{35}(2000)$ together with the observation [24] of a state in the elastic πN scattering at 2200 MeV. Here we consider these results as a possible indication to two states: one at 1740 MeV and another at 2200 MeV.

8.3.2 The setting of (L=0) states

We consider $N_{J^P=\frac{1}{2}^+}$ and $\Delta_{J^P=\frac{3}{2}^+}$ states in two versions:
(1) $M^2_{D^0_0} = M^2_{D^1_1}$ (see Fig. 8.3a), the $SU(6)$ symmetry being imposed, and
(2) $M^2_{D^0_0} \neq M^2_{D^1_1}$ (see Fig. 8.3b) with the broken $SU(6)$ symmetry constraints.

8.3.2.1 The $SU(6)$ symmetry for the nucleon $N_{\frac{1}{2}^+}(940)$, isobar $\Delta_{\frac{3}{2}^+}(1238)$ and their radial excitations

We assume $M^2_{D^0_0} = M^2_{D^1_1}$ (see Fig. 8.3a) and suppose the $SU(6)$ symmetry for the lowest baryons with $L = 0$. It gives us two ground states, the nucleon $N_{\frac{1}{2}^+}(940)$ and the isobar $\Delta_{\frac{3}{2}^+}(1238)$ as well as their radial excitations, see section 2:

$L=0$	$S=\frac{1}{2}$, $N(\frac{1}{2}^+)$	$S=\frac{3}{2}$, $\Delta(\frac{3}{2}^+)$
$n=1$	938 ± 30	1232 ± 4
$n=2$	1440 ± 40	1635 ± 75
$n=3$	1710 ± 30	~ 1920
$n=4$	1900 ± 100	~ 2190

(8.45)

Note that the mass-squared splitting of the nucleon radial excitation states, $\delta_n M^2(N_{\frac{1}{2}^+})$, is of the order of 1.05 ± 0.15 GeV^2. This value is close to that observed in the meson sector [18, 38]:

$$M^2[N_{\frac{1}{2}^+}(1440)] - M^2[N_{\frac{1}{2}^+}(940)]$$
$$\simeq M^2[N_{\frac{1}{2}^+}(1710)] - M^2[N_{\frac{1}{2}^+}(1440)] \equiv \delta_n M^2(N_{\frac{1}{2}^+}) \simeq 1.0\,\mathrm{GeV}^2. \quad (8.46)$$

The state with $n = 4$ cannot be unambiguously determined. Namely, in the region of 1880 MeV a resonance structure is seen, which may be either a nucleon radial excitation ($n = 4$) or a ($S = 3/2, L = 2, J^P = 1/2^+$) state. It is also possible that in the region ~ 1900 there are two poles, not one. This means that one pole dives into the complex-M plane and is not observed yet.

One can see that the mass-squared splitting of the $\Delta_{\frac{3}{2}^+}$ isobars, $\delta_n M^2(\Delta_{\frac{3}{2}^+})$, coincides with that of a nucleon, $\delta_n M^2(N_{\frac{1}{2}^+})$, with a good accuracy:

$$\delta_n M^2(\Delta_{\frac{3}{2}^+}) = 1.07 \pm 0.05 \ . \qquad (8.47)$$

Let us emphasize that two states, $\Delta_{\frac{3}{2}^+}(1600)$ and $\Delta_{\frac{3}{2}^+}(1920)$, are considered here as radial excitations of $\Delta_{\frac{3}{2}^+}(1232)$ with $n = 2$ and $n = 3$. However, the resonances $\Delta_{\frac{3}{2}^+}(1600)$ and $\Delta_{\frac{3}{2}^+}(1920)$ can be reliably classified as $S = 1/2, L = 2$ and $S = 3/2, L = 2$ states, with $n = 1$ (see Section 4). Actually, it means that around $\sim$ 1600 MeV one may expect the double-pole structure, while the three-pole structure may be at $\sim$ 1920 MeV.

8.3.2.2 *The setting of (L=0) states with broken SU(6) symmetry, $M_{D_0^0} \neq M_{D_1^1}$*

Here we consider an alternative scheme supposing the diquarks D_0^0 and D_1^1 to have different masses, thus being different effective particles – arguments in favor of different effective masses of D_0^0 and D_1^1 may be found in [6, 22].

In the scheme with two different diquarks, D_0^0 and D_1^1, we have two basic nucleons with corresponding sets of radial excitations.

The first nucleonic set corresponds to the qD_0^0 states, the second one describes the qD_1^1 states:

$L=0$	$S=\frac{1}{2}$, $N(\frac{1}{2}^+)$	$S=\frac{3}{2}$, $N(\frac{1}{2}^+)$	$S=\frac{3}{2}$, $\Delta(\frac{3}{2}^+)$
$n=1$	938 ± 30	1440 ± 40	1232 ± 4
$n=2$	1440 ± 40	1710 ± 30	1635 ± 75
$n=3$	1710 ± 30	2100 ± 100	~ 1920
$n=4$	2100 ± 100		~ 2190

(8.48)

This scheme requires overlapping states (double-pole structure of the partial amplitude) in the regions of M ~1400 MeV, 1700 MeV, 1900 MeV. The double-pole structure may be considered as a signature of the model with two different diquarks, D_0^0 and D_1^1.

The $(L = 0)$ set of isobar states coincides with that defined in the $(M^2_{D_0^0} = M^2_{D_1^1})$ scheme.

8.4 The setting of baryons with $L > 0$ as (qD_1^1, qD_0^0) states

Considering excited states, we analyze several versions, assuming $M_{D_0^0} \neq M_{D_1^1}$ for the $L \geq 1$ states and the $SU(6)$ constraints for $L = 0$ ones.

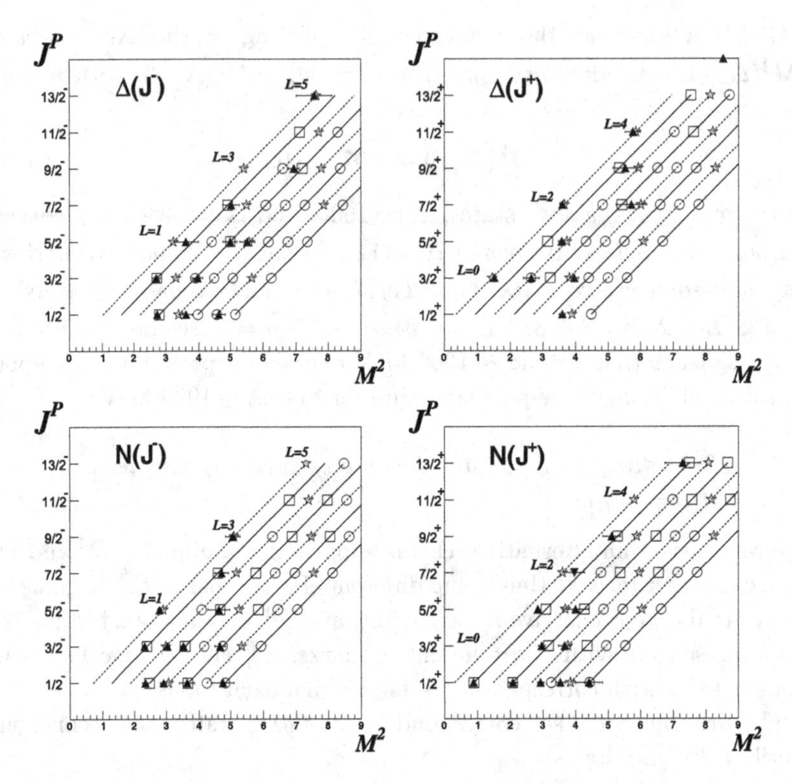

Fig. 8.4 (J^P, M^2) planes for baryons at $M_{D^0_0} \neq M_{D^1_1}$ with the $SU(6)$ constraints for $L = 0$ states. Solid and dashed lines are the trajectories for the states with $S = 1/2$ and $S = 3/2$. Squares: ground states ($n = 1$) with $S = 1/2$; stars and rhombuses: ground states ($n = 1$) with $S = 3/2$; circles: radially exited states ($n > 1$) with $S = 1/2$, 3/2

In Fig. 8.4 we demonstrate the setting of baryons on the (J^P, M^2) planes. We see a reasonably good description of data, although the scheme requires some additional states as well as double pole structures in many cases.

The deciphering of baryon setting shown in Fig. 8.4 is given in (8.49), (8.50), (8.51), (8.52) – the mass values (in MeV units) are taken from [23, 39, 40, 41].

Let us comment the trajectories in Fig. 8.4. The states belonging to the same J^P trajectories have $\delta J^{\pm} = 2^{\pm}$ and $M^2_{(J+2)^{\pm}} - M^2_{J^{\pm}} \simeq 2$ GeV^2. Clear examples are given by $\Delta(\frac{3}{2}^+)$ trajectory (the states $\Delta_{\frac{3}{2}^+}(1231)$, $\Delta_{\frac{7}{2}^+}(1895)$, $\Delta_{\frac{11}{2}^+}(2400)$, $\Delta_{\frac{13}{2}^+}(2920)$) and the $N(\frac{5}{2}^-)$ trajectory (the

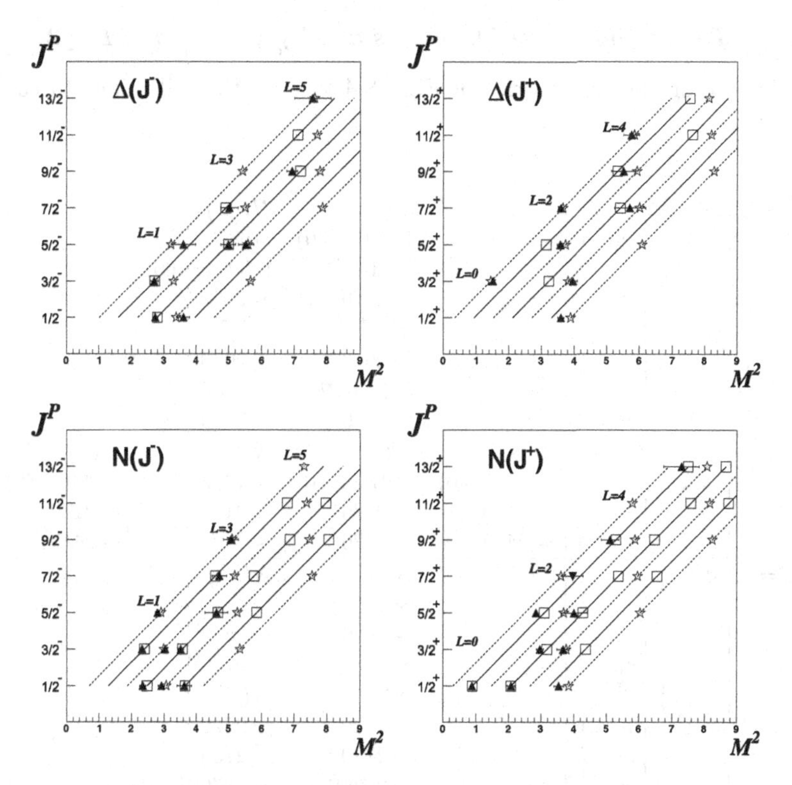

Fig. 8.5 (J, M^2) planes for baryons at $M_{D_0^0} \neq M_{D_1^1}$ within the $SU(6)$ constraints for wave functions of the $L = 0$ states – the basic states are shown only (notations are as in Fig. 8.4).

states $N_{\frac{5}{2}^-}(1670)$, $N_{\frac{9}{2}^-}(2250)$, $N_{\frac{13}{2}^-}(2270)$). At the same time, in Fig. 8.4 we see the lines with $\delta J^\pm = 1^\pm$ and $M^2_{(J+2)^\pm} - M^2_{J^\pm} \simeq 1$ GeV^2: actually, such a line represents two overlapping trajectories.

For a better presentation of the model, let us re-draw the (J, M^2) planes keeping the basic ($n = 1$) states only – they are shown in Fig. 8.5.

For $L = 0$ we see two basic states: $N(1/2^+)$ and $\Delta(3/2^+)$. In the $I = 3/2$ sector we have five states for $L = 1$ and six states for each L at $L > 1$. For ($L = 1, I = 1/2$) states we have five basic states with $J^P = \frac{1}{2}^-, \frac{3}{2}^-$, $J^P = \frac{1}{2}^-, \frac{3}{2}^-, \frac{5}{2}^-$, $J^P = \frac{1}{2}^-, \frac{3}{2}^-$, while for the states with $L > 1$ we have $J^P = (L + \frac{1}{2})^P, (L - \frac{1}{2})^P$, $J^P = (L + \frac{3}{2})^P, (L + \frac{1}{2})^P, (L - \frac{1}{2})^P, (L - \frac{3}{2})^P$, $J^P = (L + \frac{1}{2})^P, (L - \frac{1}{2})^P$.

8.4.1 *The setting of $N(J^+)$ states at $M_{D_0^0} \neq M_{D_1^1}$ and $L \geq 2$*

In this sector the trajectories of Fig. 8.4 give us the following states at $L \geq 2$:

L	S	n				
$L=2$	$S=\frac{1}{2}$			$N(\frac{3}{2}^+)$	$N(\frac{5}{2}^+)$	
		$n=1$		(1720 ± 30)	(1685 ± 5)	
		$n=2$		$\sim 2040^{(*)}$	(2000 ± 100)	
		$n=3$		$\sim 2300^{(*)}$	$\sim 2300^{(*)}$	
	$S=\frac{1}{2}$			$N(\frac{3}{2}^+)$	$N(\frac{5}{2}^+)$	
		$n=1$		$\sim 2040^{(*)}$	~ 2040	
		$n=2$		$\sim 2300^{(*)}$	$\sim 2300^{(*)}$	
		$n=3$		$\sim 2530^{(*)}$	$\sim 2530^{(*)}$	
	$S=\frac{3}{2}$		$N(\frac{1}{2}^+)$	$N(\frac{3}{2}^+)$	$N(\frac{5}{2}^+)$	$N(\frac{7}{2}^+)$
		$n=1$	(1880 ± 40)	(1915 ± 50)	~ 1900	(1990 ± 80)
		$n=2$	~ 2180	~ 2180	~ 2170	~ 2170
		$n=3$	~ 2420	~ 2420	$\sim 2420^{(*)}$	$\sim 2410^{(*)}$
$L=4$	$S=\frac{1}{2}$			$N(\frac{7}{2}^+)$	$N(\frac{9}{2}^+)$	
		$n=1$		~ 2290	(2260 ± 60)	
		$n=2$		~ 2520	$\sim 2520^{(*)}$	
		$n=3$		$\sim 2740^{(*)}$	$\sim 2730^{(*)}$	
	$S=\frac{1}{2}$			$N(\frac{7}{2}^+)$	$N(\frac{9}{2}^+)$	
		$n=1$		~ 2520	$\sim 2520^{(*)}$	
		$n=2$		$\sim 2740^{(*)}$	$\sim 2730^{(*)}$	
		$n=3$		$\sim 2930^{(*)}$	$\sim 2930^{(*)}$	
	$S=\frac{3}{2}$		$N(\frac{5}{2}^+)$	$N(\frac{7}{2}^+)$	$N(\frac{9}{2}^+)$	$N(\frac{11}{2}^+)$
		$n=1$	$\sim 2420^{(*)}$	$\sim 2410^{(*)}$	~ 2410	~ 2400
		$n=2$	$\sim 2640^{(*)}$	$\sim 2630^{(*)}$	~ 2630	~ 2630
		$n=3$	$\sim 2840^{(*)}$	$\sim 2840^{(*)}$	$\sim 2830^{(*)}$	$\sim 2830^{(*)}$
$L=6$	$S=\frac{1}{2}$			$N(\frac{11}{2}^+)$	$N(\frac{13}{2}^+)$	
		$n=1$		~ 2730	(2700 ± 100)	
		$n=2$		~ 2930	~ 2920	
	$S=\frac{1}{2}$			$N(\frac{11}{2}^+)$	$N(\frac{13}{2}^+)$	
		$n=1$		~ 2930	~ 2920	
		$n=2$		~ 3110	~ 3110	
	$S=\frac{3}{2}$		$N(\frac{9}{2}^+)$	$N(\frac{11}{2}^+)$	$N(\frac{13}{2}^+)$	$N(\frac{15}{2}^+)$
		$n=1$	$\sim 2830^{(*)}$	$\sim 2830^{(*)}$	~ 2830	~ 2820
		$n=2$	$\sim 3020^{(*)}$	$\sim 3020^{(*)}$	~ 3020	~ 3010

(8.49)

The symbol $^{(*)}$ means that in this mass region we should have two poles.

There is a reasonable agreement of our predictions with the data.

8.4.2 *The setting of the $N(J^-)$ states at $M_{D_0^0} \neq M_{D_1^1}$*

In the $N(J^-)$ sector the lightest states have $L = 1$, and we see that these states are in agreement with model predictions. But let us stress that the scheme requires a series of radial excitation states at $J^P = \frac{1}{2}^+, \frac{3}{2}^+, \frac{5}{2}^+$ at $M \simeq 2010$ MeV.

$L=1$	$S=\frac{1}{2}$			$N(\frac{1}{2}^-)$	$N(\frac{3}{2}^-)$	
		$n=1$		(1530 ± 30)	(1524 ± 5)	
		$n=2$		(1905 ± 60)	(1875 ± 25)	
		$n=3$		$(2180 \pm 80)^{(*)}$	$(2160 \pm 40)^{(*)}$	
		$n=4$		$\sim 2390^{(*)}$	$\sim 2390^{(*)}$	
	$S=\frac{1}{2}$			$N(\frac{1}{2}^-)$	$N(\frac{3}{2}^-)$	
		$n=1$		~ 1870	~ 1860	
		$n=2$		$\sim 2140^{(*)}$	$\sim 2140^{(*)}$	
		$n=3$		$\sim 2390^{(*)}$	$\sim 2390^{(*)}$	
		$n=4$		$\sim 2610^{(*)}$	$\sim 2610^{(*)}$	
	$S=\frac{3}{2}$			$N(\frac{1}{2}^-)$	$N(\frac{3}{2}^-)$	$N(\frac{5}{2}^-)$
		$n=1$		(1705 ± 30)	(1740 ± 20)	(1670 ± 20)
		$n=2$		~ 2010	~ 2000	~ 2000
		$n=3$		~ 2270	$\sim 2270^{(*)}$	$\sim 2260^{(*)}$
		$n=4$		~ 2500	$\sim 2500^{(*)}$	$\sim 2500^{(*)}$
$L=3$	$S=\frac{1}{2}$			$N(\frac{5}{2}^-)$	$N(\frac{7}{2}^-)$	
		$n=1$		(2150 ± 80)	(2170 ± 50)	
		$n=2$		$\sim 2380^{(*)}$	$\sim 2380^{(*)}$	
		$n=3$		$\sim 2610^{(*)}$	$\sim 2600^{(*)}$	
		$n=4$		$\sim 2810^{(*)}$	$\sim 2810^{(*)}$	
	$S=\frac{1}{2}$			$N(\frac{5}{2}^-)$	$N(\frac{7}{2}^-)$	
		$n=1$		$\sim 2380^{(*)}$	$\sim 2380^{(*)}$	
		$n=2$		$\sim 2610^{(*)}$	$\sim 2600^{(*)}$	
		$n=3$		$\sim 2810^{(*)}$	$\sim 2810^{(*)}$	
	$S=\frac{3}{2}$		$N(\frac{3}{2}^-)$	$N(\frac{5}{2}^-)$	$N(\frac{7}{2}^-)$	$N(\frac{9}{2}^-)$
		$n=1$	$\sim 2270^{(*)}$	$\sim 2260^{(*)}$	~ 2260	(2250 ± 50)
		$n=2$	$\sim 2500^{(*)}$	$\sim 2500^{(*)}$	~ 2490	~ 2490
		$n=3$	$\sim 2720^{(*)}$	$\sim 2710^{(*)}$	$\sim 2710^{(*)}$	$\sim 2700^{(*)}$
$L=5$	$S=\frac{1}{2}$			$N(\frac{9}{2}^-)$	$N(\frac{11}{2}^-)$	
		$n=1$		~ 2600	~ 2600	
		$n=2$		$\sim 2810^{(*)}$	$\sim 2800^{(*)}$	
		$n=3$		$\sim 3000^{(*)}$	$\sim 3000^{(*)}$	
	$S=\frac{1}{2}$			$N(\frac{9}{2}^-)$	$N(\frac{11}{2}^-)$	
		$n=1$		$\sim 2810^{(*)}$	$\sim 2800^{(*)}$	
		$n=2$		$\sim 3000^{(*)}$	$\sim 3000^{(*)}$	
	$S=\frac{3}{2}$		$N(\frac{7}{2}^-)$	$N(\frac{9}{2}^-)$	$N(\frac{11}{2}^-)$	$N(\frac{13}{2}^-)$
		$n=1$	$\sim 2710^{(*)}$	$\sim 2700^{(*)}$	~ 2700	~ 2700
		$n=2$	$\sim 2910^{(*)}$	$\sim 2900^{(*)}$	~ 2900	~ 2900

(8.50)

8.4.3 *The setting of $\Delta(J^+)$ states at $M_{D^0_0} \neq M_{D^1_1}$ and $L \geq 2$*

Considering the equation (8.51), we should remember that the ($L = 0$) states are excluded from the suggested classification.

$L=2$	$S=\frac{1}{2}$			$\Delta(\frac{3}{2}^+)$	$\Delta(\frac{5}{2}^+)$	
		$n=1$		~ 1760	~ 1760	
		$n=2$		~ 2060	~ 2050	
		$n=3$		~ 2310	~ 2310	
		$n=4$		~ 2540	~ 2540	
	$S=\frac{3}{2}$		$\Delta(\frac{1}{2}^+)$	$\Delta(\frac{3}{2}^+)$	$\Delta(\frac{5}{2}^+)$	$\Delta(\frac{7}{2}^+)$
		$n=1$	(1895 ± 25)	$(1990 \pm 35)^{(*)}$	(1890 ± 25)	(1895 ± 20)
		$n=2$	~ 2190	$\sim 2190^{(*)}$	~ 2180	~ 2180
		$n=3$	~ 2430	$\sim 2430^{(*)}$	$\sim 2430^{(*)}$	$\sim 2420^{(*)}$
		$n=4$	~ 2650	$\sim 2650^{(*)}$	$\sim 2650^{(*)}$	$\sim 2640^{(*)}$
$L=4$	$S=\frac{1}{2}$			$\Delta(\frac{7}{2}^+)$	$\Delta(\frac{9}{2}^+)$	
		$n=1$		~ 2300	~ 2300	
		$n=2$		~ 2530	~ 2530	
		$n=3$		~ 2750	~ 2740	
		$n=4$		~ 2940	~ 2940	
	$S=\frac{3}{2}$		$\Delta(\frac{5}{2}^+)$	$\Delta(\frac{7}{2}^+)$	$\Delta(\frac{9}{2}^+)$	$\Delta(\frac{11}{2}^+)$
		$n=1$	$\sim 2430^{(*)}$	$(2390 \pm 100)^{(*)}$	(2400 ± 50)	(2420 ± 100)
		$n=2$	$\sim 2650^{(*)}$	$\sim 2640^{(*)}$	~ 2640	~ 2640
		$n=3$	$\sim 2850^{(*)}$	$\sim 2850^{(*)}$	$\sim 2840^{(*)}$	$\sim 2840^{(*)}$
		$n=4$	$\sim 3040^{(*)}$	$\sim 3040^{(*)}$	$\sim 3030^{(*)}$	$\sim 3030^{(*)}$
$L=6$	$S=\frac{1}{2}$			$\Delta(\frac{11}{2}^+)$	$\Delta(\frac{13}{2}^+)$	
		$n=1$		~ 2740	~ 2740	
		$n=2$		~ 2940	~ 2930	
		$n=3$		~ 3120	~ 3120	
		$n=4$		~ 3300	~ 3290	
	$S=\frac{3}{2}$		$\Delta(\frac{9}{2}^+)$	$\Delta(\frac{11}{2}^+)$	$\Delta(\frac{13}{2}^+)$	$\Delta(\frac{15}{2}^+)$
		$n=1$	$\sim 2840^{(*)}$	$\sim 2840^{(*)}$	~ 2840	(2920 ± 100)
		$n=2$	$\sim 3030^{(*)}$	$\sim 3030^{(*)}$	~ 3030	~ 3020
		$n=3$	$\sim 3210^{(*)}$	$\sim 3210^{(*)}$	~ 3210	~ 3200
		$n=4$	$\sim 3380^{(*)}$	$\sim 3380^{(*)}$	~ 3380	~ 3370

(8.51)

8.4.4 *The setting of $\Delta(J^-)$ states at $M_{D^0_0} \neq M_{D^1_1}$*

In the $\Delta(J^-)$ sector we face a problem with the $\frac{5}{2}^-$ state observed in the analysis [35] with mass 2233 MeV and quoted as $D_{35}(1930)$ [23]. However, it can be hardly compatible with other observations, which are in the region of 1930 MeV. In addition, the result of [35] is compatible with the analysis of Manley [25], quoted by PDG as $D_{35}(2350)$. Thus, we introduce the $\Delta(\frac{5}{2}^-)$ state with the mass 2210 MeV and with the error which covers the results of [25, 35]. Then, the mass of the $D_{35}(2350)$ state is taken as an average value over the results of [24, 26, 30].

$L=1$	$S=\frac{1}{2}$			$\Delta(\frac{1}{2}^-)$	$\Delta(\frac{3}{2}^-)$	
		$n=1$		(1650 ± 25)	(1640 ± 40)	
		$n=2$		(1900 ± 50)	(1990 ± 40)	
		$n=3$		(2150 ± 50)	~ 2210	
		$n=4$		~ 2460	~ 2450	
	$S=\frac{3}{2}$			$\Delta(\frac{1}{2}^-)$	$\Delta(\frac{3}{2}^-)$	$\Delta(\frac{5}{2}^-)$
		$n=1$		~ 1800	~ 1800	(1910 ± 80)
		$n=2$		~ 2090	~ 2080	~ 2080
		$n=3$		~ 2340	$\sim 2340^{(*)}$	$\sim 2330^{(*)}$
		$n=4$		~ 2570	$\sim 2560^{(*)}$	$\sim 2560^{(*)}$
$L=3$	$S=\frac{1}{2}$			$\Delta(\frac{5}{2}^-)$	$\Delta(\frac{7}{2}^-)$	
		$n=1$		~ 2210	(2240 ± 60)	
		$n=2$		~ 2450	~ 2440	
		$n=3$		~ 2670	~ 2660	
	$S=\frac{3}{2}$		$\Delta(\frac{3}{2}^-)$	$\Delta(\frac{5}{2}^-)$	$\Delta(\frac{7}{2}^-)$	$\Delta(\frac{9}{2}^-)$
		$n=1$	$\sim 2340^{(*)}$	$(2350 \pm 50)^{(*)}$	~ 2330	~ 2320
		$n=2$	$\sim 2560^{(*)}$	$\sim 2560^{(*)}$	~ 2560	~ 2550
		$n=3$	$\sim 2770^{(*)}$	$\sim 2770^{(*)}$	$\sim 2770^{(*)}$	$\sim 2760^{(*)}$
$L=5$	$S=\frac{1}{2}$			$\Delta(\frac{9}{2}^-)$	$\Delta(\frac{11}{2}^-)$	
		$n=1$		(2633 ± 30)	~ 2660	
		$n=2$		~ 2860	~ 2860	
		$n=3$		~ 3050	~ 3050	
	$S=\frac{3}{2}$		$\Delta(\frac{7}{2}^-)$	$\Delta(\frac{9}{2}^-)$	$\Delta(\frac{11}{2}^-)$	$\Delta(\frac{13}{2}^-)$
		$n=1$	$\sim 2770^{(*)}$	$\sim 2760^{(*)}$	~ 2760	(2750 ± 100)
		$n=2$	$\sim 2960^{(*)}$	$\sim 2960^{(*)}$	~ 2950	~ 2950
		$n=3$	$\sim 3140^{(*)}$	$\sim 3140^{(*)}$	~ 3140	~ 3140

(8.52)

8.4.5 *Overlapping of baryon resonances*

In equations (8.49), (8.50), (8.51), (8.52), the overlapping resonances are denoted by the symbol (*) – the model predicts a lot of such states. Decay processes lead to a mixing of the overlapping states (owing to the transition *baryon*(1) $\to$ *hadrons* $\to$ *baryon*(2)). It results in a specific phenomenon, that is, when several resonances overlap, one of them accumulates the widths of neighbouring resonances and transforms into the broad state, see [18, 42] and references therein.

This phenomenon had been observed in [43, 44] for meson scalar–isoscalar states and applied to the interpretation of the broad state $f_0(1200-1600)$, being a descendant of a pure glueball [45, 46].

In meson physics this phenomenon can play an important role, in particular, for exotic states which are beyond the $q\bar{q}$ systematics. Indeed, being among $q\bar{q}$ resonances, the exotic state creates a group of overlapping resonances. The exotic state, after accumulating the "excess" of widths, turns into a broad one. This broad resonance should be accompanied by narrow states which are the descendants of states from which the widths have been borrowed. In this way, the existence of a broad resonance accompanied by narrow ones may be a signature of exotics. This possibility, in context of searching for exotic states, was discussed in [47].

In the considered case of quark–diquarks baryons (equations (8.49), (8.50), (8.51), (8.52)), we deal mainly with two overlapping states: it means that we should observe one comparatively narrow resonance together with the second one which is comparatively broad. The experimental observation of the corresponding two-pole singularities in partial amplitudes looks as a rather intricate problem.

8.4.5.1 *The setting of baryons with $M_{D^0_0} \neq M_{D^1_1}$ on (n, M^2) planes*

Equations (8.49), (8.50), (8.51), (8.52) allow us to present the setting of baryons ($M_{D^0_0} \neq M_{D^1_1}$) on (n, M^2) planes – they are shown in Figs. 8.6 and 8.7.

We have three trajectories on the (n, M^2) plot for $N(\frac{1}{2}^+)$ with the starting states shown in Fig. 8.5 in the $L = 0$ group.

In the plot for $N(\frac{3}{2}^+)$ we also have three trajectories, with the starting $L = 2$ states, see Fig. 8.5. Likewise, all other (n, M^2) plots are constructed: the starting states are those shown in Fig. 8.5.

We have a lot of predicted radial excitation states, though not many observed ones – the matter is that in the case of overlapping resonances the broad state is concealed under the narrow one. As it is well known, the mixing states repulse from one another. The mixing of overlapping states, due to the transition into real hadrons $baryon(1) \to real\ hadrons \to baryon(2)$ also leads to the repulsing of resonance poles in the complex-M plane: one is moving to the real M axis (*i.e.* reducing the width), another is diving deeper into the complex-M plane (*i.e.* increasing the width) – for more detail see [18], Sections 3.4.2 and 3.4.3. To separate two overlapping poles, one needs to carry out the measurements of decays into different channels – different resonances have, as a rule, different partial widths, so the total width of the "two-pole resonance" should depend on the reaction type.

Radially excited states are seen in Figs. 8.6 and 8.7, namely,
$N(\frac{1}{2}^+)$ sector (four states on the lowest trajectory),
$N(\frac{5}{2}^+)$ sector (two states on the lowest trajectory),
$N(\frac{1}{2}^-)$ sector (three states on the lowest trajectory),
$N(\frac{3}{2}^-)$ sector (three states on the lowest trajectory),
$\Delta(\frac{3}{2}^+)$ sector (three states on the lowest trajectory),
$\Delta(\frac{1}{2}^-)$ sector (three states on the lowest trajectory),
$\Delta(\frac{3}{2}^-)$ sector (two states on the lowest trajectory),
$\Delta(\frac{5}{2}^-)$ sector (two states on the lowest trajectory).

However, in Figs. 8.6 and 8.7 we do not mark specially the resonances which are "shadowed" by the observed ones. The slopes of all trajectories in Figs. 8.6 and 8.7 coincide with each other and with those of meson trajectories (see [18, 38]).

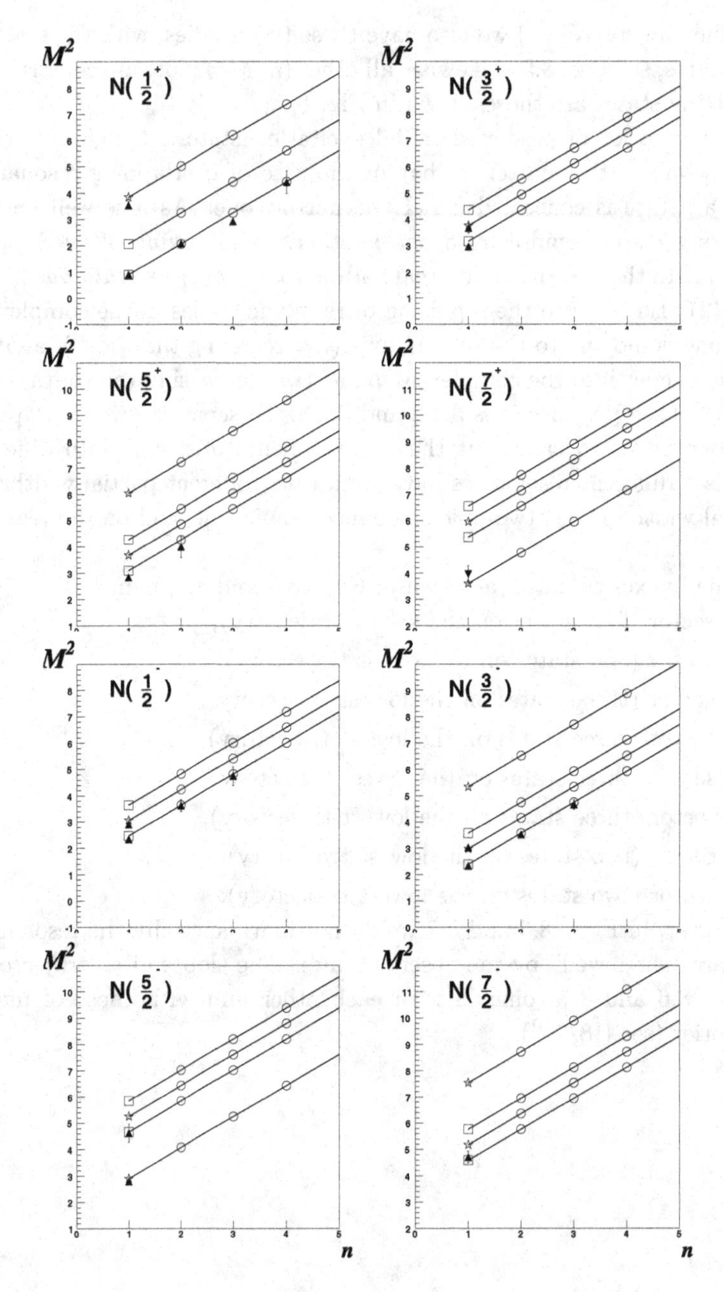

Fig. 8.6 (n, M^2) planes for $N(J^{\pm})$ states, $M_{D_0^0} \neq M_{D_1^1}$.

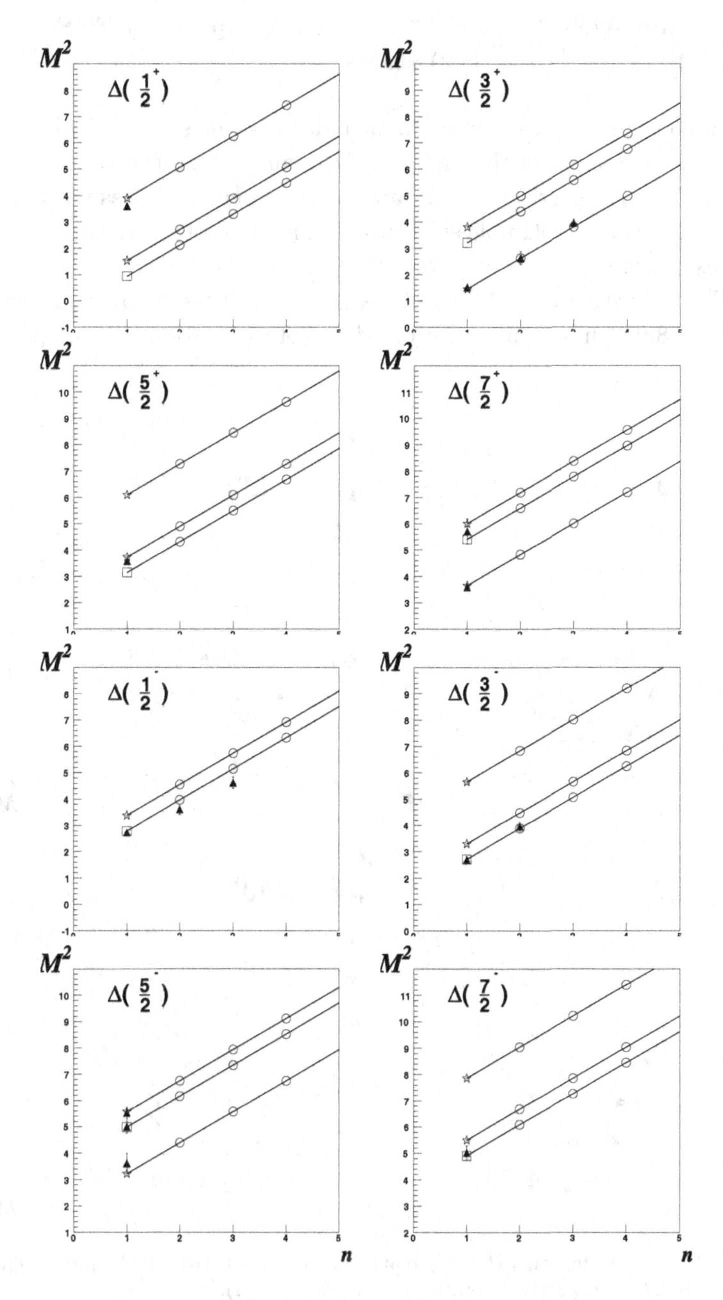

Fig. 8.7 (n, M^2) planes for $\Delta(J^{\pm})$ states, $M_{D_0^0} \neq M_{D_1^1}$.

8.5 Version with $M_{D_0^0} = M_{D_1^1}$ and overlapping $qD_0^0(S = 1/2)$ and $qD_1^1(S = 1/2)$ states

Here we consider the case with further decrease of states which can be easily seen. We suppose that $M_{D_0^0} = M_{D_1^1}$ and the states $qD_0^0(S = 1/2)$ and $qD_1^1(S = 1/2)$ overlap. So, for a naive observer, who does not perform an analysis of the double pole structure, the number of states with $S = 1/2$ decreases twice.

In Fig. 8.8 we show (J, M^2) plots as they look like for "naive observers", while Fig. 8.9 demonstrates the (J, M^2) plots for ground states ($n = 1$) only.

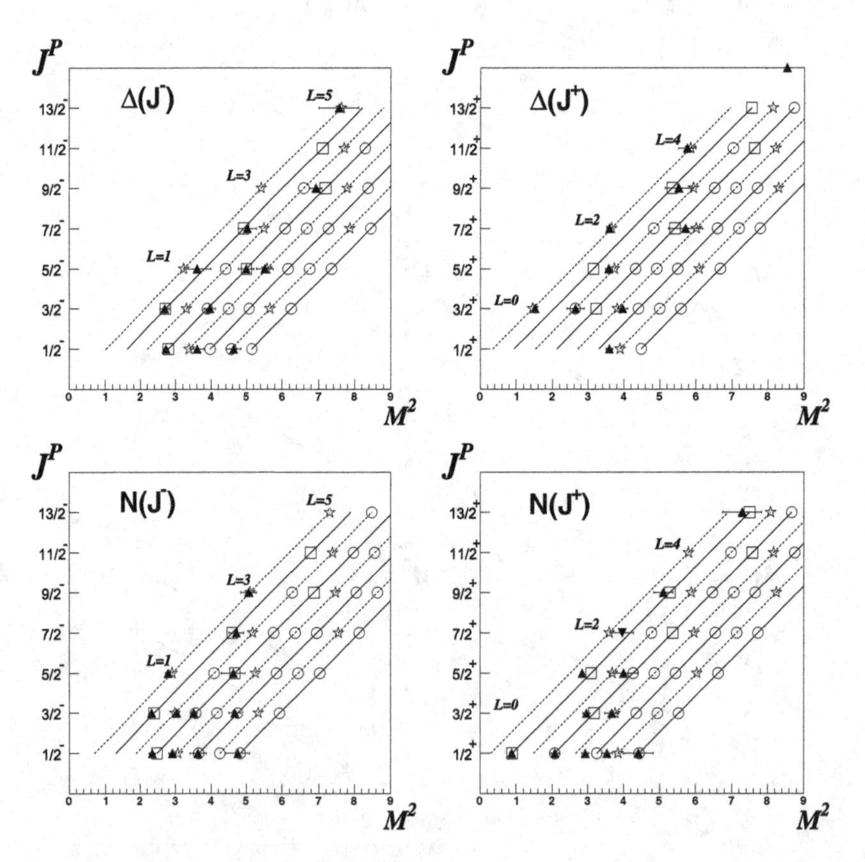

Fig. 8.8 Baryon setting on (J^P, M^2) planes in the model with overlapping $qD_0^0(S = 1/2)$ and $qD_1^1(S = 1/2)$ states (notations are as in Fig. 8.4).

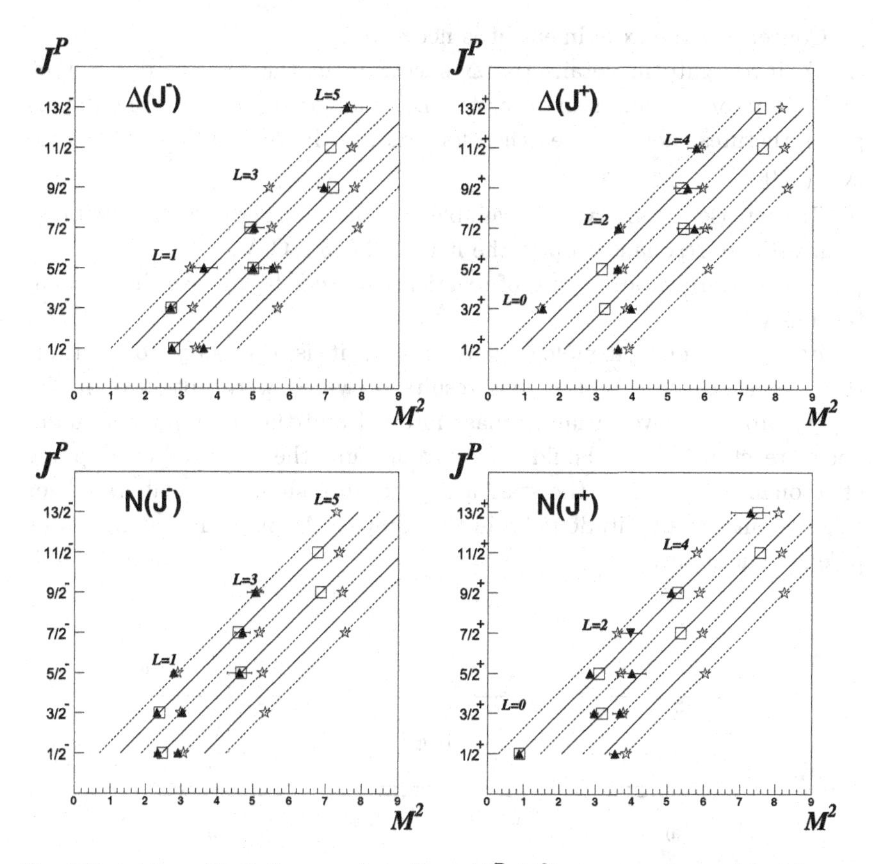

Fig. 8.9 Setting of basic baryons ($n = 1$) on (J^P, M^2) planes in the model with overlapping $qD_0^0(S = 1/2)$ and $qD_1^1(S = 1/2)$ states (notations are as in Fig. 8.4).

Figures 8.8, 8.9 present us a more compact scheme than that given in Figs. 8.4, 8.5.

8.6 Conclusion

We have systematized all baryon states in the framework of the hypothesis of their quark–diquark structure. We cannot say whether such a systematization is unambiguous, so we discuss possible versions of setting baryons upon multiplets. To carry out a more definite systematization, additional efforts are needed in the experiments and the phenomenological comprehension of data.

Concerning the experiment, it is necessary:
(i) To investigate in details the Δ spectrum in the region around 1700 MeV. Here one should search for the D_{15} and/or F_{15} states. The double pole structures should be searched for, first, in the regions $N_{\frac{1}{2}^+}(1400)$ and $\Delta_{\frac{3}{2}^+}(1600)$.
(ii) To increase the interval of available energies in order to get a possibility to investigate resonances up to the masses 3.0–3.5 GeV.
(iii) To measure various types of reactions in order to analyze them simultaneously.

As to the phenomenology and theory, it is necessary to continue the K-matrix analysis, the first results of which were obtained in [36, 48], in order to cover a larger mass interval and the most possible number of reactions. One should take into account the expected overlapping of resonances. Namely, the standard procedure should be elaborated for singling out the amplitude poles in the complex-M plane in case when one pole is under another.

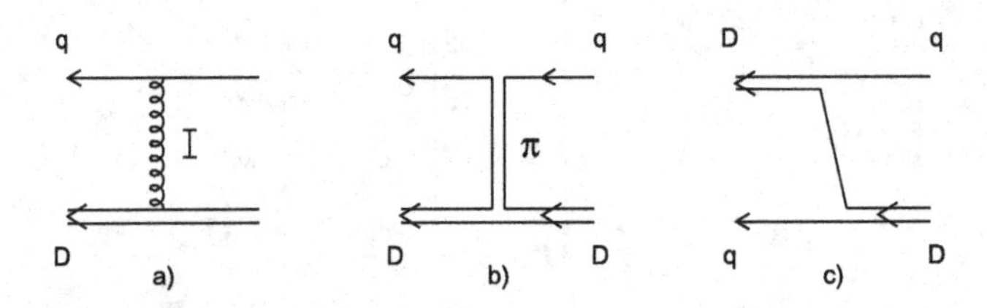

Fig. 8.10 (a) Confinement interaction taken into account in Eqs. (8.66) and (8.77); (b,c) pion-exchange and quark-exchange interactions which are considered small for highly excited qD_0^0 and qD_1^1 systems.

In Appendix A we have derived relativistic spectral integral equations for the simplest case, when the quark–diquark system is characterized by one channel: this is qD_0^0 for nucleon states and qD_1^1 for Δ_J states.

Considering quark–diquark states, we take into account only the confinement interaction (Fig. 8.10a) which is a rather rough approximation. Still, the above-performed classification of baryon states gives us a hint that such an approximation may work on a qualitative level. More precise results need including and investigating other interactions, for example, the pion and u-channel quark exchanges (Figs. 8.10b and 8.10c) as a perturbative admixture.

8.7 Appendix A. Spectral integral equations for pure qD_0^0 and qD_1^1 systems

In the present appendix we give a relativistic description of quark-diquark systems, qD_0^0 and qD_1^1. Here we concentrate our efforts on writing equations for pure qD_0^0 and qD_1^1 systems; the contribution of three-quark states is neglected. Spectral integral equations for qD_0^0 and qD_1^1 systems are shown in Fig. 8.1: the double line means diquark (qD_0^0 or qD_1^1), the flavor-neutral singular interaction is denoted as the helix-type line. The flavor-neutrality of interaction results in the absence of mixing of the qD_0^0 and qD_1^1 states.

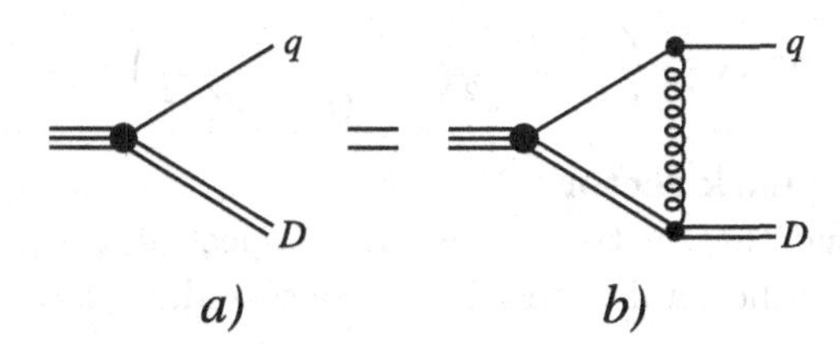

Fig. 8.11 (a,b) Equation for quark–diquark system (the flavor-neutral interaction denoted as the helix-type line).

8.7.1 *Confinement singularities*

Confinement singularities were applied to the calculation of levels in the meson sector, to $q\bar{q}$ systems [49] and heavy quark ones $Q\bar{Q}$ [50, 51]. Here we apply confinement singularities to quark-diquark compound states.

(i) Meson sector

The linearity of the $q\bar{q}$-trajectories in the (n, M^2) planes in the meson sector [38] (experimentally up to large n values, $n \le 7$) provides us the t-channel singularity $V_{conf} \sim 1/q^4$ which creates in the coordinate representation the barrier $V_{conf} \sim r$. The confinement interaction is a two-component one [49]:

$$V_{conf} = (I \otimes I)\, b_S\, r + (\gamma_\mu \otimes \gamma_\mu)\, b_V\, r\ ,$$
$$b_S \simeq -b_V \simeq 0.15\ \text{GeV}^{-2}\ . \tag{8.53}$$

The position of $q\bar{q}$ levels and data on radiative decays tell us that singular t-channel exchanges are necessary both in the scalar $(I \otimes I)$ and vector $(\gamma_\mu \otimes \gamma_\mu)$ channels. The t-channel exchange interactions (8.53) can take place both for white and color states, $\mathbf{c} = \mathbf{1} + \mathbf{8}$, though, of course, the color-octet interaction plays a dominant role.

The spectral integral equation for the meson-$q\bar{q}$ vertex (or for $q\bar{q}$ wave function of the meson) was solved by introducing a cut-off into the interaction (8.53): $r \to re^{-\mu r}$. The cut-off parameter is small: $\mu \sim 1 - 10$ MeV; if μ is changing in this interval, the $q\bar{q}$-levels with $n \le 7$ remain practically the same.

In [49, 50, 51], the spectral integral equations were solved in the momentum representation – this is natural, since we used dispersion integration technics (see the discussion in [18]). In this representation, the interaction is re-written as follows:

$$re^{-\mu r} \to 8\pi \left(\frac{4\mu^2}{(\mu^2 - q_\perp^2)^3} - \frac{1}{(\mu^2 - q_\perp^2)^2} \right). \tag{8.54}$$

(ii) Quark–diquark sector

Bearing in mind that in the framework of spectral integration (as in dispersion technics) the total energy is not conserved, we have to write

$$q_\perp^2 = (k_1^\perp - k_1^{\prime\perp})_\mu (-k_2^\perp + k_2^{\prime\perp})_\mu \tag{8.55}$$

for the momentum transfer, where k_1 and k_2 are the momenta of the initial quark and diquark, while k_1' and k_2' are those after the interaction. The index $\perp$ means that we use components perpendicular to total momentum p for the initial state and to p' for the final state:

$$k_{i\,\mu}^\perp = g_{\mu\nu}^{\perp p} k_{i\,\nu}\,, \quad g_{\mu\nu}^{\perp p} = g_{\mu\nu} - \frac{p_\mu p_\nu}{p^2}\,, \quad p = k_1 + k_2\,, \quad p^2 = s\,,$$

$$k_{i\,\mu}^{\prime\perp} = g_{\mu\nu}^{\perp p'} k_{i\,\nu}'\,, \quad g_{\mu\nu}^{\perp p'} = g_{\mu\nu} - \frac{p_\mu' p_\nu'}{p'^2}\,, \quad p' = k_1' + k_2'\,, \quad p'^2 = s'. \tag{8.56}$$

Generally, we can write for t-channel interaction block:

$$I_N(q_\perp^2) = \frac{4\pi(N+1)!}{(\mu^2 - q_\perp^2)^{N+2}} \sum_{n=0}^{N+1} \left(\mu + \sqrt{q_\perp^2}\right)^{N+1-n} \left(\mu - \sqrt{q_\perp^2}\right)^n. \tag{8.57}$$

8.7.2 *The qD_0^0 systems*

First, let us present vertices for the transitions $N_{J^P} \to qD_0^0$ and the block of the confinement interaction (transition $qD_0^0 \to qD_0^0$). Convoluting them, see Fig. 8.11, we obtain spectral integral equation for the qD_0^0 system. An alternative way is to write the Bethe–Salpeter equation [52] or its modifications [53, 54, 55].

8.7.2.1 *Vertices for the transition $N_{J^P} \to qD_0^0$*

(i) The N_{J^+} vertices with even L

The vertex for $J = L + 1/2$ with even L is equal to:

$$\Phi^+(k_2)\bar{u}(k_1)X^{(L)}_{\mu_1\ldots\mu_L}(k^\perp)G^{(L,J^+)}(k_\perp^2)\psi_{(J^+)\mu_1\ldots\mu_j}(p) \equiv$$
$$\equiv \Phi^+(k_2)\bar{u}(k_1)S^{(L,J^+)}_{\mu_1\ldots\mu_j}(k_1,k_2)G^{(L,J^+)}(k_\perp^2)\psi_{(J^+)\mu_1\ldots\mu_j}(p), \quad j = L, \quad (8.58)$$

while for $J = L - 1/2$ it is

$$\Phi^+(k_2)\bar{u}(k_1)i\gamma_5\gamma_\nu X^{(L)}_{\mu_1\ldots\mu_{L-1}\nu}(k^\perp)G^{(J^+)}(k_\perp^2)\psi_{(J^+)\mu_1\ldots\mu_j}(p) \equiv$$
$$\equiv \Phi^+(k_2)\bar{u}(k_1)S^{(L,J^+)}_{\mu_1\ldots\mu_j}(k_1,k_2)G^{(L,J^+)}(k_\perp^2)\psi_{(J^+)\mu_1\ldots\mu_j}(p),$$
$$j = L - 1. \quad (8.59)$$

(ii) The N_{J^-} vertices with odd L

For odd L, the vertex for $J = L + 1/2$ is equal to:

$$\Phi^+(k_2)\bar{u}(k_1)X^{(L)}_{\mu_1\ldots\mu_L}(k^\perp)G^{(J^-)}(k_\perp^2)\psi_{(J^-)\mu_1\ldots\mu_j}(p) \equiv$$
$$\equiv \Phi^+(k_2)\bar{u}(k_1)S^{(L,J^-)}_{\mu_1\ldots\mu_j}(k_1,k_2)G^{(L,J^-)}(k_\perp^2)\psi_{(J^+)\mu_1\ldots\mu_j}(p), \quad j = L, \quad (8.60)$$

while for $J = L - 1/2$:

$$\Phi^+(k_2)\bar{u}(k_1)i\gamma_5\gamma_\nu X^{(L)}_{\mu_1\ldots\mu_{L-1}\nu}(k^\perp)G^{(J^-)}(k_\perp^2)\psi_{(J^-)\mu_1\ldots\mu_j}(p) \equiv$$
$$\equiv \Phi^+(k_2)\bar{u}(k_1)S^{(L,J^-)}_{\mu_1\ldots\mu_j}(k_1,k_2)G^{(L,J^-)}(k_\perp^2)\psi_{(J^+)\mu_1\ldots\mu_j}(p),$$
$$j = L - 1, \quad (8.61)$$

We denote vertices (8.58)-(8.61) as follows:

$$G^{(L,J^P)}_{qD_0^0}(k_1,k_2;p) = \Phi^+(k_2)\bar{u}(k_1)G^{(L,J^P)}_{(qD_0^0)\mu_1\ldots\mu_j}(k_1,k_2)\psi^{(J^P)}_{\mu_1\ldots\mu_j}(p) \equiv$$
$$\equiv \Phi^+(k_2)\bar{u}(k_1)S^{(L,J^P)}_{(qD_0^0)\mu_1\ldots\mu_j}(k_1,k_2)G^{(L,J^P)}_{qD_0^0}(k_\perp^2)\psi^{(J^P)}_{\mu_1\ldots\mu_j}(p),$$
$$j = J - \frac{1}{2}. \quad (8.62)$$

The wave function of the qD_0^0 system reads:

$$\psi^{(L,J^P)}_{(qD_0^0)\mu_1\ldots\mu_j}(k_1,k_2) = S^{(L,J^P)}_{(qD_0^0)\mu_1\ldots\mu_j}(k_1,k_2)\frac{G^{(L,J^P)}_{qD_0^0}(k_\perp^2)}{s - M_{J^P}^2}$$
$$\equiv S^{(L,J^P)}_{(qD_0^0)\mu_1\ldots\mu_j}(k_1,k_2)\psi^{(L,J^P)}_{qD_0^0}(s), \quad (8.63)$$

where $\psi^{(L,J^P)}_{qD_0^0}(s)$ is its invariant part.

8.7.2.2 *Confinement singularities in the qD_0^0 interaction amplitude*

Introducing the momenta of quarks and diquarks, we must remember that total energies are not conserved in the spectral integrals (like in dispersion relations). Hence, in the general case $s \neq s'$.

The interaction amplitude in qD_0^0 system is written in the following form:

$$\text{S}: \quad \sum_N \alpha_S^{(N)}(q_\perp^2)\Big(\bar{u}(k_1)\, I\, u(k_1')\Big)\, I_N^{(\mu\to 0)}(q_\perp^2)\Big(\Phi^+(k_2)\Phi(k_2')\Big),$$
$$\text{V}: -\sum_N \alpha_V^{(N)}(q_\perp^2)\Big(\bar{u}(k_1)\,\gamma_\nu\, u(k_1')\Big)\, I_N^{(\mu\to 0)}(q_\perp^2)\Big(\Phi^+(k_2)\,(k_2+k_2')_\nu\,\Phi(k_2')\Big). \tag{8.64}$$

Recall tha the singular block, $I_N^{(\mu\to 0)}(q_\perp^2)$, is given in (8.57), and the operator $\Phi(k_2)$ refers to scalar diquarks.

Diquarks should be considered as composite particles, hence one can expect the appearance of form factors in the interaction block, correspondingly, $\alpha_S^{(N)}(q_\perp^2)$ and $\alpha_V^{(N)}(q_\perp^2)$.

The sum of interaction terms in (8.64) can be re-written in a compact form:

$$\Phi^+(k_2)\bar{u}(k_1)\, V_{D_0^0}^q(k_1,k_2;k_1',k_2')\, u(k_1')\Phi(k_2') \tag{8.65}$$

8.7.2.3 *Spectral integral equation for qD_0^0 system*

The spectral integral equation for qD_0^0 system reads (see Fig. 8.11):

$$\Phi^+(k_2)\bar{u}(k_1)G^{(L,J^P)}_{(qD_0^0)\mu_1\ldots\mu_j}(k_1,k_2)\psi^{(J^P)}_{\mu_1\ldots\mu_j}(p) =$$
$$= \Phi^+(k_2)\bar{u}(k_1)\int\limits_{(m+M_{D_0^0})^2}^{\infty}\frac{ds'}{\pi}\frac{d\phi_2(P';k_1',k_2')}{s'-M^2_{L,J^P}}V^q_{D_0^0}(k_1,k_2;k_1',k_2')\frac{\hat{k}_1'+m}{2m}$$
$$\times G^{(L,J^P)}_{(qD_0^0)\mu_1'\ldots\mu_j'}(k_1',k_2')\psi^{(J^P)}_{\mu_1'\ldots\mu_j'}(p)\,. \tag{8.66}$$

Here the interaction block (the right-hand side of (8.66)) is presented using the spectral (dispersion relation) integral over ds', and $d\phi_2(P';k_1',k_2')$ is a standard phase-space integral for the qD_0^0 system in the intermediate state.

It is suitable to work with an equation re-written in the following way: **(i)** the left-hand and right-hand sides of Eq. (8.66) are convoluted with the spin operator of vertex (8.62),

(ii) the convoluted terms are integrated over final-state phase space of the qD^0_0 system:

$$d\phi_2(p;k_1,k_2)=\frac{1}{2}(2\pi)^4\delta^{(4)}(p-k_1-k_2)\prod_{a=1,2}\frac{d^3k_a}{(2\pi)^3 2k_{a0}}\,. \tag{8.67}$$

We obtain:

$$\begin{aligned}&\int d\phi_2(p;k_1,k_2)Sp\Big[S^{(L,J^P)}_{(qD^0_0)\nu_1\ldots\nu_j}(k_1,k_2)\frac{\hat{k}_1+m}{2m}G^{(L,J^P)}_{(qD^0_0)\mu_1\ldots\mu_j}(k_1,k_2)\\ &\qquad\times F^{\nu_1\ldots\nu_j}_{\mu_1\ldots\mu_j}(p)\Big]\\ &=\int d\phi_2(p;k_1,k_2)Sp\Big[S^{(L,J^P)}_{(qD^0_0)\nu_1\ldots\nu_j}(k_1,k_2)\frac{\hat{k}_1+m}{2m}\int\limits_{(m+M_{D^0_0})^2}^{\infty}\frac{ds'}{\pi}\frac{d\phi_2(P';k'_1,k'_2)}{s'-M^2_{L,J^P}}\\ &\times V^q_{D^0_0}(k_1,k_2;k'_1,k'_2)\frac{\hat{k}'_1+m}{2m}G^{(L,J^P)}_{(qD^0_0)\mu'_1\ldots\mu'_j}(k'_1,k'_2)F^{\nu_1\ldots\nu_j}_{\mu'_1\ldots\mu'_j}(p)\Big]\,.\end{aligned} \tag{8.68}$$

We can re-write (8.68) incorporating the wave function $\psi^{(L,J^P)}_{qD^0_0}(s)$, which is determined in Eq. (8.63):

$$\begin{aligned}&\int d\phi_2(p;k_1,k_2)Sp\Big[S^{(L,J^P)}_{(qD^0_0)\nu_1\ldots\nu_j}(k_1,k_2)\frac{\hat{k}_1+m}{2m}S^{(L,J^P)}_{(qD^0_0)\mu_1\ldots\mu_j}(k_1,k_2)\\ &\qquad\times F^{\nu_1\ldots\nu_j}_{\mu_1\ldots\mu_j}(p)\Big](s-M^2_{J^P})\psi^{(L,J^P)}_{(qD^0_0)}(s)\\ &=\int d\phi_2(p;k_1,k_2)Sp\Big[S^{(L,J^P)}_{(qD^0_0)\nu_1\ldots\nu_j}(k_1,k_2)\frac{\hat{k}_1+m}{2m}\int\limits_{(m+M_{D^0_0})^2}^{\infty}\frac{ds'}{\pi}d\phi_2(P';k'_1,k'_2)\\ &\times V^q_{D^0_0}(k_1,k_2;k'_1,k'_2)\frac{\hat{k}'_1+m}{2m}S^{(L,J^P)}_{(qD^0_0)\mu'_1\ldots\mu'_j}(k'_1,k'_2)\psi^{(L,J^P)}_{(qD^0_0)}(s')F^{\nu_1\ldots\nu_j}_{\mu'_1\ldots\mu'_j}(p)\Big].\end{aligned} \tag{8.69}$$

8.7.3 *Spectral integral equations for qD^1_1 systems with $I=3/2$*

Here we present equations for vertices, or wave functions, for the transitions $\Delta_{J^P}\to qD^1_1$ with $P=\pm$. Below, for the sake of simplicity, we consider the $\Delta^{++}_{J^P}$ state: in this case it is necessary to take into account one quark–diquark channel only, namely, $uD^{1,I_3=3/2}_1$. We omit isotopic indices, denoting the quark–diquark state as qD^1_1.

8.7.3.1 Vertices for the transition $\Delta_{J^P} \to qD_1^1$

The qD_1^1 states are characterized by the total spin of the quark and diquark ($S = \frac{1}{2}, \frac{3}{2}$), orbital momentum ($L$) and total angular momentum (J). The parity (P) is determined by L.

The systematization performed in [49] favors, in the first approximation, the consideration of quantum numbers S and L as good ones. Below, we follow this result.

Outgoing quark–diquark states with fixed S read:

$$S = \frac{1}{2} : \Phi_\nu^+(k_2)\bar{u}(k_1) i\gamma_5\gamma_\nu \frac{\hat{P}+\sqrt{s}}{\sqrt{s}} \equiv \Phi_\nu^+(k_2)\bar{u}(k_1) S_\nu^{1/2}(P) \,,$$

$$S = \frac{3}{2} : \Phi_\nu^+(k_2)\bar{u}(k_1)\Big(-\frac{2}{3} g_{\nu\mu}^{\perp P} + \frac{1}{3}\sigma_{\nu\mu}^{\perp P}\Big)\frac{\hat{P}+\sqrt{s}}{\sqrt{s}} \equiv \Phi_\nu^+(k_2)\bar{u}(k_1) S_{\nu\mu}^{3/2}(P). \tag{8.70}$$

Here we use the operator (8.64) for spins 1/2 and 3/2, substituting $M_{J^P}^2 \to s$.

(i) The transition vertices $\Delta_{J^\pm}$ at $S = \frac{1}{2}$

The transition vertices for $J = L \pm 1/2$ read:

$$\begin{aligned}
J = L + \frac{1}{2} &: \Phi_\nu^+(k_2)\bar{u}(k_1) S_\nu^{1/2}(P) X^{(L)}_{\mu_1\ldots\mu_L}(k^\perp) G^{(S,L,J^\pm)}(k_\perp^2)\psi^{(J^\pm)}_{\mu_1\ldots\mu_j}(p) \\
&\equiv \Phi_\nu^+(k_2)\bar{u}(k_1) S^{(1/2,L,J^\pm)}_{\nu\mu_1\ldots\mu_j}(k_1,k_2) G^{(S,L,J^\pm)}(k_\perp^2)\psi^{(J^\pm)}_{\mu_1\ldots\mu_j}(p), \quad j = L\,, \\
J = L - \frac{1}{2} &: \Phi_\nu^+(k_2)\bar{u}(k_1) S_\nu^{1/2}(P)\, i\gamma_5\gamma_{\nu'} X^{(L)}_{\mu_1\ldots\mu_{L-1}\nu'}(k^\perp) \\
&\times G^{(S,L,J^\pm)}(k_\perp^2)\psi^{(J^\pm)}_{\mu_1\ldots\mu_j}(p) \equiv \Phi_\nu^+(k_2)\bar{u}(k_1) S^{(1/2,L,J^\pm)}_{\nu\mu_1\ldots\mu_j}(k_1,k_2) \\
&\times G^{(S,L,J^\pm)}(k_\perp^2)\psi^{(J^\pm)}_{\mu_1\ldots\mu_j}(p), \quad j = L-1\,.
\end{aligned} \tag{8.71}$$

Note that in (8.71), as in (8.59) and (8.61), we use the axial–vector operator $i\gamma_5\gamma_{\nu'}$ for the decreasing rank of the vertex at fixed L; recall also that $P = +$ corresponds to even L and $P = -$ to odd ones.

(ii) The transition vertices $\Delta_{J^\pm}$ at $S = \frac{3}{2}$

The transition vertices for $J = |\mathbf{L} + \frac{\mathbf{3}}{\mathbf{2}}|$ and $P = (-1)^L$ are written as follows:

$$\begin{aligned}
J = L + \frac{3}{2} : \quad & \Phi_\nu^+(k_2)\bar{u}(k_1) S^{3/2}_{\nu\mu_{L+1}}(P) X^{(L)}_{\mu_1\ldots\mu_L}(k^\perp) \\
& \times G^{(3/2,L,J^\pm)}(k_\perp^2)\psi^{(J^\pm)}_{\mu_1\ldots\mu_j}(p) \\
\equiv \Phi_\nu^+(k_2)\bar{u}(k_1) & S^{(3/2,L,J^\pm)}_{\nu\mu_1\ldots\mu_j}(k_1,k_2) G^{(3/2,L,J^\pm)}(k_\perp^2)\psi^{(J^\pm)}_{\mu_1\ldots\mu_j}(p), \quad j = L+1\,,
\end{aligned}$$

$$J = L + \frac{1}{2}: \quad \Phi^+_\nu(k_2)\bar{u}(k_1)S^{3/2}_{\nu\mu_L}(P)i\gamma_5\gamma_{\nu_1}X^{(L)}_{\nu_1\mu_1\ldots\mu_{L-1}}(k^\perp)$$
$$\times G^{(3/2,L,J^\pm)}(k^2_\perp)\psi^{(J^\pm)}_{\mu_1\ldots\mu_j}(p)$$
$$\equiv \Phi^+_\nu(k_2)\bar{u}(k_1)S^{(3/2,L,J^\pm)}_{\nu\mu_1\ldots\mu_j}(k_1,k_2)G^{(3/2,L,J^\pm)}(k^2_\perp)\psi^{(J^\pm)}_{\mu_1\ldots\mu_j}(p), \quad j = L\,,$$
$$J = L - \frac{1}{2}: \quad \Phi^+_\nu(k_2)\bar{u}(k_1)S^{3/2}_{\nu\mu_{L-1}}(P)\prod_{a=1}^{2}(i\gamma_5\gamma_{\nu_a})X^{(L)}_{\nu_1\nu_2\mu_1\ldots\mu_{L-2}}(k^\perp)$$
$$\times G^{(3/2,L,J^\pm)}(k^2_\perp)\psi^{(J^\pm)}_{\mu_1\ldots\mu_j}(p)$$
$$\equiv \Phi^+_\nu(k_2)\bar{u}(k_1)S^{(3/2,L,J^\pm)}_{\nu\mu_1\ldots\mu_j}(k_1,k_2)G^{(3/2,L,J^\pm)}(k^2_\perp)\psi^{(J^\pm)}_{\mu_1\ldots\mu_j}(p), \quad j = L-1,$$
$$J = L - \frac{3}{2}: \quad \Phi^+_\nu(k_2)\bar{u}(k_1)S^{3/2}_{\nu\mu_{L-2}}(P)\prod_{a=1}^{3}(i\gamma_5\gamma_{\nu_a})X^{(L)}_{\nu_1\nu_2\nu_3\mu_1\ldots\mu_{L-3}}(k^\perp)$$
$$\times G^{(3/2,L,J^\pm)}(k^2_\perp)\psi^{(J^\pm)}_{\mu_1\ldots\mu_j}(p) \equiv \Phi^+_\nu(k_2)\bar{u}(k_1)S^{(3/2,L,J^\pm)}_{\nu\mu_1\ldots\mu_j}(k_1,k_2)$$
$$\times G^{(3/2,L,J^\pm)}(k^2_\perp)\psi^{(J^\pm)}_{\mu_1\ldots\mu_j}(p), \quad j = L-2\,. \tag{8.72}$$

(iii) The invariant wave functions of the qD^1_1 systems

The invariant wave functions of the qD^1_1 systems are determined by vertices $G^{(S,L,J^\pm)}(k^2_\perp)$ as follows:

$$\frac{G^{(S,L,J^\pm)}(k^2_\perp)}{s - M^2_{S,L,J^P}} = \psi^{(S,L,J^P)}_{qD^1_1}(s). \tag{8.73}$$

Remind that the relative momentum squared $k^2_\perp$ depends on s only.

8.7.3.2 *Confinement singularities in qD^1_1 interaction block*

Introducing the momenta of quarks and diquarks, we should remember that total energies are not conserved in the spectral integrals (just as in dispersion relations), so $s \neq s'$.

The interaction amplitude for qD^1_1 system is written as follows.
S-exchange:

$$\Phi^+_\nu(k_2)\bar{u}(k_1)\,V^{(S-ex)}_{qD^1_1;\nu\nu'}(k_1,k_2;k'_1,k'_2)\,u(k'_1)\Phi_{\nu'}(k'_2)$$
$$= \sum_N \beta^{(N)}_S(q^2_\perp)\Big(\bar{u}(k_1)\,I\,u(k'_1)\Big)\,I^{(\mu\to 0)}_N(q^2_\perp)\Big(\Phi^+_\nu(k_2)g_{\nu\nu'}\Phi_{\nu'}(k'_2)\Big), \tag{8.74}$$

V-exchange:

$$\Phi^+_\nu(k_2)\bar{u}(k_1)\,V^{(V-ex)}_{qD^1_1;\nu\nu'}(k_1,k_2;k'_1,k'_2)\,u(k'_1)\Phi_{\nu'}(k'_2) = -\sum_N \beta^{(1N)}_V(q^2_\perp)$$
$$\times\Big(\bar{u}(k_1)\gamma_\xi u(k'_1)\Big)I^{(\mu\to 0)}_N(q^2_\perp)\Big(\Phi^+_\nu(k_2)g_{\nu\nu'}\Phi_{\nu'}(k'_2)\Big)(k_2+k'_2)_\xi$$
$$-\sum_N \beta^{(2N)}_V(q^2_\perp)\Big(\bar{u}(k_1)\gamma_\xi u(k'_1)\Big)I^{(\mu\to 0)}_N(q^2_\perp)\Big[\Phi^+_\xi(k_2)\Big(k_{2\nu}g_{\nu\nu'}\Phi_{\nu'}(k'_2)\Big)$$
$$+\Big(\Phi^+_\nu(k_2)g_{\nu\nu'}k'_{2\nu'}\Big)\Phi_\xi(k'_2)\Big]. \qquad (8.75)$$

We can re-write the sum of interaction terms in (8.74) in a compact form:

$$V_{qD^1_1;\nu\nu'}(k_1,k_2;k'_1,k'_2) = V^{(S-ex)}_{qD^1_1;\nu\nu'}(k_1,k_2;k'_1,k'_2) + V^{(V-ex)}_{qD^1_1;\nu\nu'}(k_1,k_2;k'_1,k'_2),$$
$$\Phi^+(k_2)\bar{u}(k_1)\,V_{qD^1_1;\nu\nu'}(k_1,k_2;k'_1,k'_2)\,u(k'_1)\Phi(k'_2). \qquad (8.76)$$

8.7.3.3 *Spectral integral equation for qD^1_1 system*

The spectral integral equation for qD^1_1 system with fixed total spin $S = 1/2, 3/2$, orbital momentum L and total angular momentum J^P reads:

$$\Phi^+_\nu(k_2)\bar{u}(k_1)S^{(S,L,J^P)}_{\nu\mu_1\ldots\mu_j}(k_1,k_2)G^{(S,L,J^P)}(k^2_\perp)\psi_{(J^P)\mu_1\ldots\mu_j}(p)$$
$$= \Phi^+_\nu(k_2)\bar{u}(k_1)\int\limits_{(m+M_{D^1_1})^2}^{\infty}\frac{ds'}{\pi}\frac{d\phi_2(P';k'_1,k'_2)}{s'-M^2_{S,L,J^P}}V_{qD^1_1;\nu\nu'}(k_1,k_2;k'_1,k'_2)\frac{\hat{k}'_1+m}{2m}$$
$$\times S^{(S,L,J^P)}_{\nu'\mu'_1\ldots\mu'_j}(k'_1,k'_2)G^{(S,L,J^P)}(k'^2_\perp)\psi^{(J^P)}_{\mu'_1\ldots\mu'_j}(p). \qquad (8.77)$$

Here, as in case of the qD^0_0 system, the interaction block (the right-hand side of (8.77)) is written with the use of spectral integral over ds', and $d\phi_2(P';k'_1,k'_2)$ is the phase space, see Eq. (8.67), for the qD^1_1 system in the intermediate state.

As in Eq. (8.66), we can transform Eq. (8.77) convoluting the left-hand and right-hand sides with spin operator of vertices, see (8.71) and (8.72), and integrating the convoluted terms over final-state phase space of the

qD_1^1 system, $d\phi_2(p;k_1,k_2)$. We obtain:

$$\int d\phi_2(p;k_1,k_2)Sp\Big[S^{(S,L,J^P)}_{(qD_1^1)\nu\nu_1\dots\nu_j}(k_1,k_2)\frac{\hat{k}_1+m}{2m}S^{(S,L,J^P)}_{(qD_1^1)\nu\mu_1\dots\mu_j}(k_1,k_2)$$
$$\times F^{\nu_1\dots\nu_j}_{\mu_1\dots\mu_j}(p)\Big](s-M^2_{S,L,J^P})\psi^{(L,J^P)}_{(qD_1^1)}(s)$$
$$=\int d\phi_2(p;k_1,k_2)Sp\Big[S^{(S,L,J^P)}_{(qD_1^1)\nu\nu_1\dots\nu_j}(k_1,k_2)\frac{\hat{k}_1+m}{2m}$$
$$\times\int\limits_{(m+M_{D_1^1})^2}^{\infty}\frac{ds'}{\pi}d\phi_2(P';k'_1,k'_2)V_{qD_1^1;\nu\nu'}(k_1,k_2;k'_1,k'_2)\frac{\hat{k}'_1+m}{2m}$$
$$\times S^{(S,L,J^P)}_{(qD_1^1)\nu'\mu'_1\dots\mu'_j}(k'_1,k'_2)\psi^{(L,J^P)}_{(qD_1^1)}(s')F^{\nu_1\dots\nu_j}_{\mu'_1\dots\mu'_j}(p)\Big]. \qquad (8.78)$$

8.8 Appendix B. Group theoretical description. Symmetrical basis in the three-body problem

In the theory of representation of the $O(n)$ group one usually considers the canonical Gelfand–Zeitlin chain $O(n)\supset O(n-1)\supset\dots$. Such a chain is, however, inconvenient, when the subject is the many-body problem where the chain $O(n)\supset O(n-3)\supset\dots$ is more natural. This chain corresponds to the decrease of the number of particles by one (*i.e* that of the degree of freedom by three). In addition, in the many-body problem it is reasonable to follow the schemes in the framework of which the angular momenta are summed up, and make sure that the total angular momentum J and its projection M remain conserved at all rotations. Of course, the number of group parameters (*i.e.* that of the Euler angles) will decrease.

Indeed, there is no need to consider all the rotations. It is sufficient to take into account only those which do not mix up the components of different vectors.

In the three particle problem the two vectors η and ξ form the $O(6)$ group. Here, in order to include the angular momentum J into the number of observables, one has to separate the $O(3)$ group. If so, there remain only 5 angles of the 15 of the $O(6)$ group; they characterize the position of vectors η and ξ in the six-dimensional space. Three angles define the position of the plane of the vectors η, ξ in the three-dimensional space, two angles determine the angle between these two vectors and their lengths (with the condition $\xi^2+\eta^2=\rho^2$).

In several works [56, 57, 58, 59, 60, 61] attempts were made to construct total sets of eigenfunctions for the three-body problem. This was based on the invariance of the Laplacian under $O(6)$. The explicit calculations were carried out differently. One of the possibilities is that the eigenfunctions correspond to the classification of the three-particle states according to the chain

$$O(3) \supset SU(3) \supset O(6)$$

which is characterized by five quantum numbers. Four of them – K, J, M, ν – are the six-dimensional angular momentum, corresponding to the eigenvalue of the 6-Laplacian, the usual three-dimensional momentum and its projection, and a number, characterizing the permutation symmetry. Generally speaking, a number of states of the system belongs to the given set K, J, M and ν. The fifth quantum number Ω which solves this problem is the eigenvalue of the hermitian operator $\hat{\Omega}$, commuting with the $O(3)$ generators. (The operator was introduced by Racah [62]; its explicit form was obtained in Descartes coordinates by Badalyan, in polar coordinates it is given in [57].) In order to find the eigenfunctions $\Phi^{KJM}_{\nu\Omega}$ corresponding to the above choice of quantum numbers, we have to carry out the common solution of the complicated equations $\Delta\Phi = -K(K+4)\Phi$ and $\hat{\Omega}\Phi = \Omega\Phi$.

A different way of constructing the set of functions is based on the graphical tree method. With the help of a simple algorithm a function with quantum numbers K, J, M, j_1, j_2 is built up. If so, $\Phi^{j_1 j_2}_{KJM}$ are, by definition, eigenfunctions of the six-dimensional Laplacian $O(6)$; they do not have, however, definite permutation symmetry properties. Hence, we have to turn from the obtained set of eigenfunctions to functions which change simply when the coordinates of the particles are permutated, *i.e.* to the system with the quantum numbers K, J, M, ν, Ω.

We could not calculate explicitly the transformation coefficients yet. However, making use of the result obtained in [58], one can carry out a transformation which leads to a system with quantum numbers $K, J, M, \nu, (j_1 j_2)$. Generally speaking, this system is not orthonormalized yet. It can be orthonormalized either by a standard method, or by the construction of eigenfunctions of the operator $\hat{\Omega}$ from the function $\Phi^{\nu(j_1 j_2)}_{KJM}$. The solution of the corresponding equations is not a difficult task. However, these equations turn out to be high order equations. Thus the solutions of these equations can hardly be presented explicitly.

The obtained solution provides a natural description for the basis of three particles.

8.8.1 *Group theoretical properties. Parametrization*

The group theoretical features as well as the choice of coordinates and the parametrization were already given in detail [56, 57]. Let us write the formulae once more, in a somewhat different form.

For a system of three particles the radius vectors x_i $(i = 1, 2, 3)$ are fixed by the condition

$$x_1 + x_2 + x_3 = 0. \tag{8.79}$$

In the case of equal masses the Jacobi-coordinates are written as

$$\xi = -\sqrt{\frac{3}{2}}(x_1 + x_2), \quad \eta = \frac{1}{\sqrt{2}}(x_1 - x_2), \tag{8.80}$$

while

$$\xi^2 + \eta^2 = x_1^2 + x_2^2 + x_3^2 = \rho^2, \tag{8.81}$$

where ρ is the radius of a five-dimensional sphere.

We are considering basis functions for which a quantum number ν, connected with the permutation symmetry, exists. The permutations mix up the components of ξ and η, and hence, it is useful to consider the six-dimensional vector $X = \binom{\eta}{\xi}$, for which

$$P_{12}\binom{\eta}{\xi} = \binom{-\eta}{\xi},$$

$$P_{13}\binom{\eta}{\xi} = \begin{pmatrix} \frac{1}{2} & \frac{\sqrt{3}}{2} \\ \frac{\sqrt{3}}{2} & -\frac{1}{2} \end{pmatrix}\binom{\eta}{\xi} = \begin{pmatrix} \cos\frac{2\pi}{3} & \sin\frac{2\pi}{3} \\ -\sin\frac{2\pi}{3} & \cos\frac{2\pi}{3} \end{pmatrix} P_{12}\binom{\eta}{\xi},$$

$$P_{14}\binom{\eta}{\xi} = \begin{pmatrix} \frac{1}{2} & -\frac{\sqrt{3}}{2} \\ -\frac{\sqrt{3}}{2} & -\frac{1}{2} \end{pmatrix}\binom{\eta}{\xi} = \begin{pmatrix} \cos\frac{4\pi}{3} & \sin\frac{4\pi}{3} \\ -\sin\frac{4\pi}{3} & \cos\frac{4\pi}{3} \end{pmatrix} P_{12}\binom{\eta}{\xi}. \tag{8.82}$$

Further, the complex vector

$$z = \xi + i\eta, \tag{8.83}$$

is introduced, for which the permutations can be expressed especially simply:

$$P_{12}z = z^*, \quad P_{13}z = e^{2/3\pi i}z^*, \quad P_{23}z = e^{4/3\pi i}z^*,$$
$$P_{12}z^* = z, \quad P_{13}z^* = e^{4/3\pi i}z, \quad P_{23}z^* = e^{2/3\pi i}z^*. \tag{8.84}$$

The transition from the vectors η and ξ to z and z^* can be written as the product of two transformations:

$$\binom{z}{z^*} = \frac{1}{\sqrt{2}}\begin{pmatrix} \cos\frac{\pi}{4} & \sin\frac{\pi}{4} \\ -\sin\frac{\pi}{4} & \cos\frac{\pi}{4} \end{pmatrix}\begin{pmatrix} i & 0 \\ 0 & 1 \end{pmatrix}\binom{\eta}{\xi} \tag{8.85}$$

or, introducing $\zeta = i\eta$,

$$\begin{pmatrix} z \\ z^* \end{pmatrix} = \frac{1}{\sqrt{2}} \begin{pmatrix} \cos\frac{\pi}{4} & \sin\frac{\pi}{4} \\ -\sin\frac{\pi}{4} & \cos\frac{\pi}{4} \end{pmatrix} \begin{pmatrix} \zeta \\ \xi \end{pmatrix}. \tag{8.86}$$

(The first of the matrices describes, obviously, the turn by $\pi/4$.)

On the base of the vectors η, ξ and z, z^* two groups, $O(6)$ and $SU(3)$, can be constructed. The generators of these groups can be written, accordingly, as

$$a_{ik} = X_i \frac{\partial}{\partial X_k} - X_k \frac{\partial}{\partial X_i} \quad (i,k = 1,2,...,6) \tag{8.87}$$

and

$$A_{ik} = iz_i \frac{\partial}{\partial z_k} - iz_k^* \frac{\partial}{\partial z_i} \quad (i,k = 1,2,3). \tag{8.88}$$

The vector z can be parameterized as

$$z = \rho e^{-i\lambda/2} \left(\cos\frac{a}{2} l_1 + i \sin\frac{a}{2} l_2 \right). \tag{8.89}$$

It is convenient to make use of the fact that the three vectors x_1, x_2 and x_3 form a triangle. Here a and λ characterize the form of this triangle the vertices of which are determined by the position of the three particles. The position of the plane of the triangle in the space is defined by the unit vectors I_1 and I_2 which are on the plane; together with the third vector $I = [I_1 I_2]$ they form a moving coordinate system. The orientation of the vectors I_1, I_2 and I compared to the fixed system of coordinates is determined with the help of the Euler angles ϕ_1, θ, ϕ_2. Let us remark that the five parameters (angles) ϕ_1, θ, ϕ_2 and a, λ determine the rotation on the five-dimensional sphere.

Finally, let us present the $O(3)$ generators

$$J_{ik} = \frac{1}{2}(A_{ik} - A_{ki}) = \frac{1}{2}\left(iz_i \frac{\partial}{\partial z_k} - iz_k \frac{\partial}{\partial z_i} + iz_i^* \frac{\partial}{\partial z_k^*} - iz_k \frac{\partial}{\partial z_i^*} \right), \tag{8.90}$$

the generators of the deformation group of the fixed triangle

$$B_{ik} = \frac{1}{2}(A_{ik} + A_{ki}) = \frac{1}{2}\left(iz_i \frac{\partial}{\partial z_k} + iz_k \frac{\partial}{\partial z_i} - iz_i^* \frac{\partial}{\partial z_k^*} - iz_k^* \frac{\partial}{\partial z_i^*} \right) \tag{8.91}$$

and the scalar operator

$$N = \frac{1}{2i} \text{ Sp } A = \frac{1}{2} \sum_k \left(z_k \frac{\partial}{\partial z_k} - z_k^* \frac{\partial}{\partial z_k} \right) \tag{8.92}$$

the eigenvalue of which is the quantum number ν.

8.8.2 *Basis functions*

Systems of basis functions characterized by different quantum numbers correspond to different parameterizations. What concerns the eigenfunctions symmetrized over permutations, it is convenient to consider them in the z, z^* space, while the functions $\Phi^{j_1 j_2}_{KJM}$ are determined in the space of η and ξ. The latter are well known as the tree functions [63].

$$\Phi^{j_1 j_2}_{KJM}(\eta,\xi)= \sum_{m_1+m_2=M} C^{JM}_{j_1 m_1 j_2 m_2}\Phi^{j_1 j_2 m_1 m_2}_{K}(\eta,\xi)=Y^{j_1 j_2}_{JM}(m,n)\Psi_{K j_1 j_2}(\Phi). \tag{8.93}$$

Here

$$Y^{j_1 j_2}_{JM}(m,n) = \sum_{m_1 m_2} C^{JM}_{j_1 m_1 j_2 m_2} Y_{j_1 m_1}(m) Y_{j_2 m_2}(n)\,, \tag{8.94}$$

with

$$m = \eta/\eta\,, \qquad n = \xi/\xi\,;$$

j_1 and j_2 are momenta, related to the vectors η and ξ. For simplicity, in the following we shall suppose $\rho = 1$ (*i.e.* $\xi^2 + \eta^2 = 1$). Let us introduce the angle Φ:

$$\cos\Phi = \xi, \qquad \sin\Phi = \eta.$$

$$\Psi_{K j_1 j_2}(\phi) = N_{K j_1 j_2}(\sin\Phi)^{j_1}(\cos\Phi)^{j_2} P^{(j_1+1/2,\,j_2+1/2)}_{(K-j_1-j_2)/2}(\cos 2\Phi). \tag{8.95}$$

The normalization factor, known from the theory of the Jacobi polynomials, is

$$N_{K j_1 j_2} = \sqrt{\frac{2(K+2)\Gamma((K-j_1-j_2)/2+1)\Gamma((K+j_1+j_2)/2+2)}{\Gamma((K-j_1+j_2)/2+3/2)\,\Gamma((K+j_1-j_2)/2+3/2)}}\,. \tag{8.96}$$

Instead of the Jacobi polynomial we can introduce the Wigner d- function [1]

[1] Carrying out the transition from the Jacobi polynomial to the Wigner d-function, it is worth mentioning that there exist different notations. Indeed, in [64] the definition

$$P^{\alpha,\beta}_k(\cos 2\theta) = i^{b-1}\left[\frac{(l-b)!(l+b)!}{(l-a)!(l+a)!}\right]^{1/2}(\sin\theta)^{b-a}(\cos\theta)^{-b-a} d^l_{ab}(\cos 2\theta).$$

is given. Here i appears because in [64] the unitary matrices are defined as

$$u(\varphi,\theta,\psi) = \begin{pmatrix} e^{i\varphi/2} & 0 \\ 0 & e^{-i\varphi/2}\end{pmatrix}\begin{pmatrix} \cos\frac{\theta}{2} & i\sin\frac{\theta}{2} \\ i\sin\frac{\theta}{2} & \cos\frac{\theta}{2}\end{pmatrix}\begin{pmatrix} e^{i\varphi/2} & 0 \\ 0 & e^{-i\varphi/2}\end{pmatrix},$$

where the second matrix is a unitary one while in [65] it is orthogonal:

$$u(\varphi,\theta\psi) = \begin{pmatrix} e^{i\varphi/2} & 0 \\ 0 & e^{-i\varphi/2}\end{pmatrix}\begin{pmatrix} \cos\frac{\theta}{2} & \sin\frac{\theta}{2} \\ -\sin\frac{\theta}{2} & \cos\frac{\theta}{2}\end{pmatrix}\begin{pmatrix} e^{i\varphi/2} & 0 \\ 0 & e^{-i\varphi/2}\end{pmatrix}.$$

According to the definition in [65],

$$P_k^{\alpha,\beta}(\cos 2\theta) = (-1)^{b-a}\left[\frac{(l-b)!(l+b)!}{(l-a)!(l+a)!}\right]^{1/2}$$
$$\times(\sin\theta)^{b-a}(\cos\theta)^{-b-a}d^l_{ab}(\cos 2\theta)\,,$$
$$l = k+\frac{\alpha+\beta}{2}\,,\quad a = \frac{\alpha+\beta}{2}\,,\quad b = \frac{\beta-\alpha}{2}\,. \tag{8.97}$$

In terms of the d-function the eigenfunction (8.93) obtains the form

$$\Phi^{j_1j_2}_{KJM}(\eta,\xi) = 2(K+2)^{1/2}(-1)^{-j_1-1/2}\sum_{m_1m_1} C^{JM}_{j_1m_1j_2m_2} \tag{8.98}$$
$$\times Y_{j_1m_1}(m)Y_{j_2m_2}(n)\frac{1}{(\sin 2\Phi)^{1/2}}d^{(K+1)/2}_{(j_1+j_2+1)/2,(j_2-j_1)/2}(\cos 2\Phi)\,.$$

The factor $(\sin 2\Phi)^{-1/2}$ appears due to the different normalizations over the angle Φ in the Jacobi polynomial and in the d-function. The Jacobi polynomial is normalized in the six-dimensional space, the element of the volume is

$$\cos^2\Phi\sin^2\Phi\, d\Phi = \frac{1}{4}\sin^2 2\Phi\, d\Phi\,;$$

while for the d-function, normalized in the usual space, we have $\sin 2\Phi d\sin 2\Phi$.

The transition to the basis function constructed on the vector pair η', ξ' which is related to η, ξ by the transformation

$$\begin{pmatrix}\eta'\\ \xi'\end{pmatrix} = \begin{pmatrix}\cos\varphi & \sin\varphi\\ -\sin\varphi & \cos\varphi\end{pmatrix}\begin{pmatrix}\eta\\ \xi\end{pmatrix}, \tag{8.99}$$

can be carried out with the help of the coefficient $\langle j_1'j_2'|j_1j_2\rangle^{\varphi}_{KJM}$ obtained in [66, 67]:

$$\Phi^{j_1j_2}_{KJM}(\eta',\xi') = \sum_{j_1'j_2'}\langle j_1'j_2'|j_1j_2\rangle^{\varphi}_{KJM}\,\Phi^{j_1'j_2'}_{KJM}(\eta,\xi)\,. \tag{8.100}$$

Comparing (8.99) and (8.86), we see that for the transition of the basis function from η, ξ to the function of the vector z, x^* the coefficient $\langle j_1'j_2'|j_1j_2\rangle^{\varphi}_{KJM}$ has to be used at the value $\varphi=\pi/4$, after substituting ξ by ζ. In the following it will be shown how the obtained function $\Phi^{j_1j_2}_{KJM}(z,z^*)$ can be transformed into $\Phi^{\nu(j_1j_2)}_{KJM}(z,z^*)$.

But let us first investigate the transformation coefficient (8.100) in detail.

8.8.3 *Transformation coefficients* $\langle j_1' j_2' | j_1 j_2 \rangle^{\varphi}_{KJM}$

In [66] the transformation coefficient is defined as the overlap integral of $\Phi^{j_1' j_2'}_{KJM}(\eta, \xi)$ and $\Phi^{j_1 j_2}_{KJM}(\eta', \xi')$:

$$\langle j_1' j_2' | j_1 j_2 \rangle^{\varphi}_{KLM} = \sum_{\substack{m_1 m_2 \\ m_1' m_2'}} C^{JM}_{j_1 m_1 j_2 m_2} C^{JM}_{j_1' m_1' j_2' m_2'}$$

$$\times \int \left(\Phi_K^{j_1' j_2' m_1' m_2'}(\eta, \xi) \right)^* \Phi_K^{j_1 j_2 m_1 m_2}(\eta', \xi')\, d\eta\, d\xi. \qquad (8.101)$$

As it was already mentioned, the transformation takes place at given J and M values. The explicit form of the coefficient is calculated with the help of the production function for $\Phi_K^{j_1 i_2 m_1 m_2}(\eta \xi)$.

The calculation is carried out in two steps [67]. First we consider the coefficient corresponding to the representation in which the vector η is expanded over the "new" vectors ξ' and η'. After that, we do the same for ξ. As it is presented in [67], the transformation coefficient, corresponding to the transition $\Phi^{j_1 0}_{K_1 j_1 m 1}(\eta', 0)$ to $\Phi^{pq}_{K_1 j_1 m_1}(\eta', \xi')$, has the form

$$\langle pq | j_1 0 \rangle^{\varphi}_{K j_1 m_1} = (-1)^{(K_1 + j_1)/2} 2^{(K_1 - j_1)/2}$$

$$\times \left[\frac{\left(\frac{K_1 - p - q}{2}\right)! \left(\frac{K_1 + p + q}{2}\right)! \left(\frac{K_1 - j_1}{2}\right)! (K_1 + 1)!!}{(K_1 - p + q + 1)!! (K_1 + p - q + 1)!! \left(\frac{1 + j_1}{2} + 1\right)! (K_1 - j_1 + 1)!!} \right]^{1/2}$$

$$\times \begin{pmatrix} p & q & j_1 \\ 0 & 0 & 0 \end{pmatrix} [(2p+1)(2q+1)]^{1/2} (\cos\varphi)^p (\sin\varphi)^q\, P^{(p+1/2, q+1/2)}_{(K_1 - p - q)/2}(-\cos 2\varphi)$$

$$(8.102)$$

and is in fact the tree function in $O(6)$. The expression (8.102) is similar to the formula describing the $O(2)$-rotation, which transforms the Legendre-polynomial with the help of the function $D^J_{0M} = P_{JM}$, being the tree function in $O(2)$. The same procedure has to be carried out for the coefficient of the transformation of $\Phi^{0 j_2}_{K_2 j_2 m_2}(0, \xi)$ into $\Phi^{0r}_{K_2 j_2 m_2}(\eta' \xi')$; after that, collecting the momenta $p + r = j_1'$ and $q + s = j_2'$ (with the total momentum J and a given $K = K_1 + K_2$), we obtain the general expression for the coefficient

$$\langle j_1' j_2' | j_1 j_2 \rangle_{KJM}^{\varphi} = \frac{\pi}{4} (-1)^{J+(K+j_1+j_2)/2}$$
$$\times \left(\frac{K_1 - j_1}{2}\right)! \left(\frac{K_1 + j_1 + 1}{2}\right)! \left(\frac{K_2 - j_2}{2}\right)! \left(\frac{K_2 + j_2 + 1}{2}\right)!$$
$$\times \left[\frac{\left(\frac{K-j_1'-j_2'}{2}\right)! \left(\frac{K+j_1'+j_2'}{2} + 1\right)! (K + j_1' + j_2' + 1)!! (K + j_1' - j_2' + 1)!!}{\left(\frac{K-j_1-j_2}{2}\right)! \left(\frac{K-j_1+j_2}{2} + 1\right)! (K - j_1 + j_2 + 1)!! (K + j_1 - j_2 + 1)!!} \right]^{1/2}$$
$$\times \sum_{pr\,qs} \left\{\!\!\left\{ \begin{matrix} p & r & j_1' \\ q & s & j_2' \\ j_1 & j_2 & J \end{matrix} \right\}\!\!\right\} \left[\Gamma\left(\frac{K_1 - p + q}{2} + \frac{3}{2}\right) \Gamma\left(\frac{K_1 + p - 1}{2} + \frac{3}{2}\right) \right.$$
$$\left. \times \Gamma\left(\frac{K_2 - r + s}{2} + \frac{3}{2}\right) \Gamma\left(\frac{K_2 + r - s}{2} + \frac{3}{2}\right) \right]^{-1} (\cos\varphi)^{p+s} (\sin\varphi)^{q+r}$$
$$\times P_{(K_1-p-q)/2}^{(p+1/2,q+1/2)}(-\cos 2\varphi) P_{(K_2-s-r)/2}^{(s+1/2,r+1/2)}(-\cos 2\varphi). \tag{8.103}$$

Here we introduce the notation

$$\left\{\!\!\left\{ \begin{matrix} p & r & j_1' \\ q & s & j_2 \\ j_1 & j_2 & J \end{matrix} \right\}\!\!\right\} = [(2j_1 + 1)(2j_2 + 1)(2j_1' + 1)(2j_2' + 1)]^{1/2} (2p + 1)(2q + 1)$$
$$\times (2s + 1)(2r + 1) \begin{pmatrix} p & q & j_1 \\ 0 & 0 & 0 \end{pmatrix} \begin{pmatrix} s & r & j_2 \\ 0 & 0 & 0 \end{pmatrix} \begin{pmatrix} p & r & j_1' \\ 0 & 0 & 0 \end{pmatrix} \begin{pmatrix} q & s & j_2' \\ 0 & 0 & 0 \end{pmatrix} \left\{ \begin{matrix} p & r & j_1' \\ q & s & j_2' \\ j_1 & j_2 & J \end{matrix} \right\}, \tag{8.104}$$

which underlines the way how the momenta are transformed.

In [66] and [67] not only the technique of the calculations differs but also the form of the final expressions. Thus it is reasonable to find the explicit connection between the two forms of the transformation coefficients $\langle j_1' j_2' | j_1 j_2 \rangle_{KJM}^{\phi}$ and prove their analogousness. Just this was done in [58], where a simpler way of obtaining this coefficient was presented, namely, the substitution of the overlap integral by the matrix element of an exponential function which in fact equals unity.

Instead of calculating the matrix element from the exponent, we expand it in a series over the orders of $\cos\phi$ and $\sin\phi$ and transform this series into one over the Jacobi polynomials, *i.e.* the eigenfunctions of the Laplacian. It is convenient to carry out this expansion in two steps: first expand the exponent over the spherical Bessel functions and then collect these functions in Jacobi polynomials. Since this way of calculating the transformation coefficient may turn out to be useful also for other coefficients in the theory of angular momenta, we present it below in detail.

Let us consider the six-dimensional vector

$$\begin{pmatrix} P_i \\ i\,Q_i \end{pmatrix}.$$

We calculate the matrix element of the exponential function

$$\exp\Big[-2(P_i^2+Q_i^2)\Big] \tag{8.105}$$

over J and M, with the condition $P_i^2 - Q_i^2 = 0$. Let us transform one of the multiplication factors in P_i^2 and Q_i^2, expressing P_i and Q_i in terms of P_k and Q_k with the help of the transformation (21). Then

$$\langle j_1'j_2'\,|\exp\Big[-2(P_i^2+Q_i^2)\Big]j_1j_2\,\rangle_{KJM} = \langle j_1'j_2'|\exp\Big[-2P_iP_k\cos\varphi$$

$$-2Q_iP_k\cos\varphi + 2iQ_iP_k\sin\varphi + 2iP_iQ_k\sin\varphi\Big]|j_1j_2\rangle_{KJM}. \tag{8.106}$$

The matrix element can be written as

$$\sum_{m_1m_2} C^{JM}_{j_1mj_2m}C^{JM}_{j_1'm_1'j_2'm_2'}\int\exp\Big[-2P_iP_k\cos\varphi - 2Q_iQ_k\cos\varphi + 2iQ_iP_k\sin\varphi$$

$$+2iP_iQ_k\sin\varphi\Big]Y^*_{j_1'm_1'}(\hat{P}_k)Y^*_{j_2'm_2'}(\hat{Q}_i)d\hat{P}_id\hat{Q}_id\hat{P}_kd\hat{Q}_k\,, \tag{8.107}$$

where $\hat{P}_i$, $\hat{Q}_i$, $\hat{P}_k$ and $\hat{Q}_k$ are the usual three-dimensional polar angles. (Let us mention that the expression (8.107) coincides up to the normalization with the formula for the transformation coefficient in [66].) If we now extract in this sum the term with eigenfunctions characterized by the quantum numbers j_1, j_2, j_1', j_2', it turns out to be equal, up to the normalization, to the coefficient we are looking for. Indeed, the overlap integral is, by definition, a matrix element of unity in a mixed representation.

Further, we apply the well-known expansion of the plane wave

$$e^{ipx} = \sum_{\lambda_\mu} i^\lambda j_\lambda(px)\,Y^*_{\lambda_\mu}(p)\,Y_{\lambda_\mu}(x)$$

and, expanding into a series the exponent in the integrand (8.107), we arrive at a rather long, but simple expression

$$\sum_{\substack{m_1m_2\\ m_1'm_2'}} C^{JM}_{j_1m_1j_2m_2}C^{JM}_{j_1'm_1'j_2'm_2'}\int\sum_{\substack{pr\,qs\\ \pi\rho\kappa\sigma}}(-1)^{(p+r+q+s)/2}j_p(2iP_iP_k\cos\varphi)$$

$$\times\, j_r(2Q_iP_k\sin\varphi)j_q(2P_iQ_k\sin\varphi)j_s(2iQ_iQ_k\cos\varphi)Y^*_{p\pi}(\hat{P}_i)Y^*_{q\kappa}(\hat{P}_i)$$

$$\times\, Y_{j_1m_1}(\hat{P}_1)Y^*_{r\rho}(Q_i)Y^*_{s\sigma}(Q_i)Y_{j_2m_2}(Q_i)Y_{p\pi}(\hat{P}_k)Y_{r\rho}(\hat{P}_k)Y^*_{j_1'm_1'}(\hat{P}_k)$$

$$\times Y_{q\kappa}(\hat{Q}_k)Y_{s\sigma}(\hat{Q}_k)Y^*_{j_2'm_2'}(\hat{Q}_k)d\hat{P}_id\hat{Q}_id\hat{P}_kd\hat{Q}_k\,. \tag{8.108}$$

Making use of the well-known features of the spherical functions:

$$
\begin{aligned}
Y^*_{lm}(\vartheta,\varphi) &= (-1)^m Y_{l-m}(\vartheta,\varphi)\\
&\int_0^{2\pi} d\varphi \int_0^{\pi} d\vartheta \sin\vartheta Y_{l_1m_1}(\vartheta,\varphi)Y_{l_2m_2}(\vartheta,\varphi)\\
&= \left[\frac{(2l_1+1)(2l_2+1)}{4\pi(2l_3+1)}\right]^{1/2} C^{l_30}_{l_10l_20}C^{l_3m_3}_{l_1m_1l_2m_2} \qquad (8.109)
\end{aligned}
$$

(see [65]), we re-write (8.109) in the form

$$
\begin{aligned}
&\frac{\pi^2}{16}\sum_{\substack{pr\,qs\\ \pi\rho\kappa\sigma}}\sum_{\substack{m_1m_2\\ m_1'm_2'}} C^{JM}_{j_1m_1j_2m_2}C^{JM}_{j_1'm_1'j_2'm_2'}C^{j_1m_1}_{p\pi\,q\kappa}C^{j_2m_2}_{r\rho s\sigma}C^{j_1'm_1'}_{p\pi r\rho}C^{j_2'm_2'}_{q\kappa s\sigma}C^{j_10}_{p0q0}C^{j_20}_{r0s0}\\
&\times C^{j_1'0}_{p0r0}C^{j_2'0}_{q0s0}\frac{(2p+1)(2r+1)(2q+1)(2s+1)}{[(2j_1+1)(2j_2+1)(2j_1'+1)(2j_2'+1)]^{1/2}}\\
&\times j_p(2iP_iP_k\cos\varphi)j_r(2Q_iP_k\sin\varphi)j_p(2P_iQ_k\sin\varphi)j_s(2iQ_iQ_k\cos\varphi)\\
&\times(-1)^{m_1+m_2-j_1-j_2+1/2(p+r+q+s)}. \qquad (8.110)
\end{aligned}
$$

The summation of the Clebsch–Gordan coefficients leads to the 9j-coefficient

$$
\begin{aligned}
&\sum_{m_sm_{sk}} C^{j_{12}m_{12}}_{j_1m_1j_2m_2}C^{j_{34}m_{34}}_{j_3m_3j_4m_4}C^{jm}_{j_{12}m_{12}j_{34}m_{34}}C^{j_{13}m_{13}}_{j_1m_1j_3m_3}C^{j_{24}m_{24}}_{j_2m_2j_4m_4}C^{j'm'}_{j_{13}m_{13}j_{24}m_{24}}\\
&= \delta_{jj'}\delta_{mm'}\Big[(2j_{12}+1)(2j_{13}+1)(2j_{24}+1)(2j_{34}+1)\Big]^{1/2}\begin{Bmatrix} j_1 & j_2 & j_{12}\\ j_3 & j_4 & j_{34}\\ j_{13} & j_{24} & j\end{Bmatrix}.
\end{aligned}
$$

Hence, in (8.110) we can get rid of the series of sums, including the 9j-coefficient, separating the quantum numbers j_1, j_2, j_1', j_2', J:

$$
\begin{aligned}
&\frac{\pi^2}{16}\sum_{pr\,qs}\begin{Bmatrix} p & r & j_1'\\ q & s & j_2\\ j_1 & j_2 & J\end{Bmatrix} C^{j_1'0}_{p0r0}C^{j_2'0}_{q0s0}C^{j_10}_{p0q0}C^{j_20}_{p0s0}(2p+1)(2r+1)(2q+1)(2s+1)\\
&\times(-1)^{1/2(p+r+q+s)}j_p(2iP_iP_k\cos\varphi)j_r(2Q_iP_k\sin\varphi)\\
&\times j_q(2P_iQ_k\sin\varphi)j_s(2iQ_iQ_k\cos\varphi). \qquad (8.111)
\end{aligned}
$$

In the following, we apply the expression for the expansion of the product of two Bessel functions into the sum of hypergeometric functions (see [68]). Here we substitute in the standard formulae the hypergeometric functions by Jacobi polynomials (or Wigner's d-functions). Substituting also the lengths of the vectors P_i, P_k, Q_i, Q_k by unity, we can write

$$j_p(2i\cos\varphi)\,j_q(2\sin\varphi) = \frac{\pi}{4}\sum_{K_1} P^{(p+1/2,q+1/2)}_{(K_1-p-q)/2}(-\cos 2\varphi)$$
$$\times \frac{(i)^{K_1-q}(\cos\varphi)^p(\sin\varphi)^q}{\Gamma\left(\frac{K_1+p-q}{2}+\frac{3}{2}\left(\Gamma\left(\frac{K_1-p+q}{2}+\frac{3}{2}\right)\right.\right.} j_s(2i\cos\varphi)j_r(2\sin\varphi)$$
$$= \frac{\pi}{4}\sum_{K_2} P^{(s+1/2,r+1/2)}_{(K_2-s-r)/2}(-\cos 2\varphi)\frac{(i)^{K_2-r}(\cos\varphi)^s(\sin\varphi)^r}{\Gamma\left(\frac{K_2+s-r}{2}+\frac{3}{2}\right)\Gamma\left(\frac{K_2-p+q}{2}+\frac{3}{2}\right)}. \tag{8.112}$$

Selecting from these sums only terms with definite $K = K_1 + K_2$, we obtain the expression

$$\frac{1}{(16)^2}\sum_{pr\,qs}\left\{\left\{\begin{matrix} p & r & j_1' \\ q & s & j_2 \\ j_1 & j_2 & J\end{matrix}\right\}\right\}\frac{(-1)^{K/2}}{\Gamma\left(\frac{K_1-p+q}{2}+\frac{3}{2}\right)\Gamma\left(\frac{K_1+p-q}{2}+\frac{3}{2}\right)}$$
$$\times\left[\Gamma\left(\frac{K_2-r+s}{2}+\frac{3}{2}\right)\Gamma\left(\frac{K_2+r-s}{2}+\frac{3}{2}\right)\right](\cos\varphi)^{p+s}$$
$$\times(\sin\varphi)^{q+r}P^{(p+1/2,q+1/2)}_{(K_1-p-q)/2}(-\cos 2\varphi)P^{(s+1/2,r+1/2)}_{(K_2-s-r)/2}(-\cos 2\varphi). \tag{8.113}$$

The formula (8.113) coincides with the general form of the coefficient $\langle j_1'j_2'|j_1j_2\rangle^{\phi}_{KJM}$, given in [67].

Let us, finally, present the orthonormalized transformation coefficient in terms of the d-function which makes the interpretation of different expressions easier with the help of the six-dimensional rotations:

$$\langle j_1'j_2'|j_1j_2\rangle^{\varphi}_{KJM} = \frac{\pi}{2}(-1)^{J+1}\left(\frac{K_1-j_1}{2}\right)!\left(\frac{K_1+j_1+1}{2}\right)!\left(\frac{K_2-j_2}{2}\right)!$$
$$\times\left(\frac{K_2+j_2+1}{2}\right)!$$
$$\times\frac{a_{Kj_1'j_2'}}{a_{Kj_1j_2}}\sum_{pr\,qs}\left[\left(\frac{K_1-p+q+1}{2}\right)!\left(\frac{K_1+p-q+1}{2}\right)!\left(\frac{K_1-p-q}{2}\right)!\right.$$
$$\left.\times\left(\frac{K_1+p+q}{2}+1\right)!\right]^{-\frac{1}{2}}$$
$$\times\left[\left(\frac{K_2-s+r+1}{2}\right)!\left(\frac{K_2+s-r+1}{2}\right)!\left(\frac{K_2-r-s}{2}\right)!\left(\frac{K_2+r+s}{2}\right)!\right]^{1/2}$$
$$\times\left\{\left\{\begin{matrix} p & r & j_1' \\ q & s & j_2' \\ j_1 & j_2 & J\end{matrix}\right\}\right\}\frac{1}{\sin 2\varphi}d^{(K_1+1)/2}_{q+p+1)/2,(p-q)/2}(2\varphi)d^{(K_2+1)/2}_{(r+s+1)/2,(s-r)/2}(2\varphi), \tag{8.114}$$

where

$$a_{Kj_1j_2} = \left[\left(\frac{K-j_1-j_2}{2}\right)!\left(\frac{K+j_1+j_2}{2}+1\right)!\left(\frac{K-j_1+j_2+1}{2}\right)!\right.$$
$$\left.\times\left(\frac{K+j_1-j_2+1}{2}+1\right)!\right]^{\frac{1}{2}}. \tag{8.115}$$

Since the normalizations of the Jacobi polynomial and the d-function are known, we do not have to think about it while carrying out the calculations. (Let us remind once more that the factor $(\sin 2\phi)^{-1}$ is related to the six-dimensional normalization.)

The expression (8.114) is similar to the formula of the "three d-functions" in $SU(2)$; it shows the expansion of one of the $O(6)$ d-functions over the products of two d-functions of a special form. As usual in such expressions, there is a freedom in the choice of K_1 or K_2 ($K_1 + K_2 = K$).

8.8.4 *Applying* $\langle j_1'j_2'|j_1j_2\rangle^{\Phi}_{KLM}$ *to the three-body problem*

Let us use the obtained formulae for the calculation of the coefficients of the transition from $\Phi^{j_1'j_2'}_{KJM}(\eta\xi)$ to $\Phi^{j_1j_2}_{KJM}(z,z^*)$. As it was shown already, this can be done by the application of the transformation coefficient at the $2\phi = \pi'2$ value, first substituting η by $i\eta = \zeta$. The argument of the d-functions in (8.114) is 2ϕ. The turns by $\pi/2$ – the so-called Weyl-coefficients – are included in many formulae of the theory of $O(n)$ representation.) The function of the new arguments

$$\Phi^{j_1j_2}_{KJM}(z,z^*) = \sum_{j_1'j_2'}\langle j_1'j_2'|j_1j_2\rangle^{\pi/4}_{KJM}\Phi^{j_1'j_2'}_{KJM}(\zeta,\xi) \tag{8.116}$$

can be written in the form

$$\Phi^{j_1j_2}_{KJM}(z,z^*) = \frac{\pi}{2}(-1)^{J+1}\left(\frac{K_1-j_1}{2}\right)!\left(\frac{K_1+j_1+1}{2}\right)!\left(\frac{K_2-j_2}{2}\right)!$$
$$\times\left(\frac{K_2+j_2+1}{2}\right)!$$
$$\times\sum_{\substack{j_1'j_2'\\ m_1m_2}} C^{JM}_{j_1'm_1'j_2'm_2'}N_{Kj_1'j_2'}\frac{a_{Kj_1'j_2'}}{a_{Kj_1j_2}}$$
$$\times\sum_{pr\,qs}\left[\left(\frac{K_1-p+q+1}{2}\right)!\left(\frac{K_1+p-q+1}{2}\right)\left(\frac{K_1-p-q}{2}\right)!\left(\frac{K_1+p+q}{2}+1\right)!\right]^{-\frac{1}{2}}$$

$$\times\left[\left(\frac{K_2-s+r+1}{2}\right)!\left(\frac{K_2+s-r+1}{2}\right)\left(\frac{K_2-r-s}{2}\right)!\left(\frac{K_2+r+s}{2}+1\right)!\right]^{-\frac{1}{2}}$$
$$\times\left\{\begin{Bmatrix} p & r & j_1' \\ q & s & j_2 \\ j_1 & j_2 & J\end{Bmatrix}\right\} d^{(K_1+1)/2}_{(q+p+1)/2,(p-q)/2}\left(\frac{\pi}{2}\right)$$
$$\times d^{(K_2+1)/2}_{(r+s+1)/2,(s-r)/2}\left(\frac{\pi}{2}\right)Y_{j_1'm_1'}(z)Y_{j_2'm_2'}(z^*)\frac{1}{z^2-z^{*2}}$$
$$\times d^{(K+1)/2}_{(j_1'+j_2'+1)/2,(j_1'-j_2')/2}(z^2+z^{*2}), \qquad (8.117)$$

where the relations

$$\frac{1}{2}(z^2+z^{*2}) = \xi^2-\eta^2, \quad z^2-z^{*2} = 2i\sin 2\Phi = 4i\xi\eta,$$
$$zz^* = \xi^2+\eta^2. \qquad (8.118)$$

are taken into account.

As it was mentioned in the previous section, it is convenient to use $K_2 = r+s$ in (8.117).

The expression (8.117) presents a series over the degrees of z and Z^*. In each term of this series the degree of z is $p+q$, that of z^* – $r+s$. Considering the parametrization z in (11), we see that each term of z and z^* introduces a factor $exp(-i\lambda/2)$ and $exp(i\lambda/2)$, respectively. Because of this, any term in the series (8.117) will contain a factor $exp(-i\lambda(p+q-r-s)/2)$. The series can be changed into a Fourier series over $exp(-i\nu\lambda)$, if collecting all terms of the series with a given

$$p+q-r-s = 2\nu. \qquad (8.119)$$

In this case each term of the new series will have a definite value of ν and, hence, the Fourier series will be at the same time a series over the eigenfunctions of the operator N. We arrive at functions characterized by the set $K, J, M, \nu, (j_1j_2)$. Their normalization can be easily obtained from that of the d-functions. The only remaining problem is the transition from (j_1j_2) to Ω, which we have already mentioned above. Although a multiplicity of the equations appears practically only at large K, the construction of convenient expressions deserves further efforts.

8.8.5 *The d-function of the O(6) group*

The coefficient $\langle j_1'j_2'|j_1j_2\rangle^{\phi}_{KJM}$ describes the rotations in the six-dimensional space. From these rotations one can, obviously, construct an arbitrary rotation. The $O(n)$ rotations are usually composed of $n(n-1)/2$

rotations on all coordinate planes [69]. With the help of the calculated coefficients rotations can be constructed in any plane, characterized by two arbitrary vectors η and ξ. In other words, these coefficients lead to simultaneous rotations in three two-dimensional planes (η_x, ξ_x), (η_y, ξ_x) and (η_z, ξ_z) and simplify the calculations seriously.

It would be rather interesting to generalize all this to the group $O(n)$ $(n > 6)$, and consider tensors of higher dimensions instead of vectors.

References

[1] A.V. Anisovich, V.V. Anisovich, V.A. Nikonov, M.A. Matveev, A.V. Sarantsev and T.O. Vulfs, Int. J. Mod. Phys. **A25**, 2965 (2010); Phys. Atom. Nucl. **74**, 418 (2011).

[2] V.V. Anisovich, Pis'ma ZhETF **2**, 439 (1965) [JETP Lett. **2**, 272 (1965)].

[3] M. Ida and R. Kobayashi, Progr. Theor. Phys. **36**, 846 (1966).

[4] D.B Lichtenberg and L.J. Tassie, Phys. Rev. **155**, 1601 (1967).

[5] S. Ono, Progr. Theor. Phys. **48** 964 (1972).

[6] V.V. Anisovich, Pis'ma ZhETF **21** 382 (1975) [JETP Lett. **21**, 174 (1975)]; V.V. Anisovich, P.E. Volkovitski, and V.I. Povzun, ZhETF **70**, 1613 (1976) [Sov. Phys. JETP **43**, 841 (1976)].

[7] A. Schmidt and R. Blankenbeckler, Phys. Rev. **D16**, 1318 (1977).

[8] F.E Close and R.G. Roberts, Z. Phys. C **8**, 57 (1981).

[9] T. Kawabe, Phys. Lett. B **114**, 263 (1982).

[10] S. Fredriksson, M. Jandel, and T. Larsen, Z. Phys. C **14**, 35 (1982).

[11] M. Anselmino and E. Predazzi, eds., *Proceedings of the Workshop on Diquarks*, World Scientific, Singapore (1989).

[12] K. Goeke, P.Kroll, and H.R. Petry, eds., *Proceedings of the Workshop on Quark Cluster Dynamics* (1992).

[13] M. Anselmino and E. Predazzi, eds., *Proceedings of the Workshop on Diquarks II*, World Scientific, Singapore (1992).

[14] N. Isgur and G. Karl, Phys. Rev. D**18**, 4187 (1978); D**19**, 2653 (1979); S. Capstick, N. Isgur, Phys. Rev. D**34**, 2809 (1986).

[15] L.Y. Glozman et al., Phys. Rev. D**58**:094030 (1998).

[16] U. Löring, B.C. Metsch, H.R. Petry, Eur.Phys. A**10**, 395 (2001); A**10**, 447 (2001).

[17] V.V. Anisovich, M.N. Kobrinsky, J. Nyiri, Yu.M. Shabelski *Quark Model and High Energy Collisions* , Second Edition, World Scientific, Singapore, 2004.

[18] A.V. Anisovich, V.V. Anisovich, J. Nyiri, V.A. Nikonov, M.A. Matveev and A.V. Sarantsev, *Mesons and Baryons. Systematization and Methods of Analysis*, World Scientific, Singapore, 2008.

[19] V.V. Anisovich, D.I. Melikhov, V.A. Nikonov, Yad. Fiz. **57**, 520 (1994) [Phys. Atom. Nucl. **57**, 490 (1994)]

[20] V.V. Anisovich, E.M. Levin and M.G. Ryskin, Yad. Fiz. **29**, 1311 (1979); [Phys. Atom. Nuclei **29**, 674 (1979)].
[21] G. Källen, Helv. Phys. Acta. **25**, 417 (1952);
H. Lehman, Nuovo Cim. **11**, 342 (1952).
[22] V.V. Anisovich, S.M. Gerasyuta, A.V. Sarantsev, Int. J. Mod. Phys. A**6**, 625 (1991).
[23] J. Beringer *et al.* [Particle Data Group], Phys. Rev. D **86**, 010001 (2012) 1.
[24] R. E. Cutkosky, C. P. Forsyth, J. B. Babcock, R. L. Kelly and R. E. Hendrick,
[25] D. M. Manley and E. M. Saleski, Phys. Rev. D **45**, 4002 (1992).
[26] T. P. Vrana, S. A. Dytman and T. S. H. Lee, Phys. Rept. **328**, 181 (2000) [arXiv:nucl-th/9910012].
[27] R. Plotzke *et al.* [SAPHIR Collaboration], Phys. Lett. B **444** (1998) 555.
[28] K. W. Bell *et al.*, Nucl. Phys. B **222**, 389 (1983).
[29] A. V. Sarantsev, V. A. Nikonov, A. V. Anisovich, E. Klempt and U. Thoma, Eur. Phys. J. A **25**, 441 (2005) [arXiv:hep-ex/0506011].
[30] G. Hohler, In *Bratislava 1973, Proceedings, Triangle Meeting On Hadron Interactions At Low Energies*, Bratislava 1975, 11-75.
[31] A. V. Anisovich, V. Kleber, E. Klempt, V. A. Nikonov, A. V. Sarantsev and U. Thoma, Eur. Phys. J. A **34**, 243 (2007) [arXiv:0707.3596 [hep-ph]].
[32] M. Ablikim *et al.*, Phys. Rev. Lett. **97**, 262001 (2006) [arXiv:hep-ex/0612054].
[33] M. Batinic, I. Slaus, A. Svarc and B. M. K. Nefkens, Phys. Rev. C **51**, 2310 (1995) [Erratum-ibid. C **57**, 1004 (1998)] [arXiv:nucl-th/9501011].
[34] V. Crede *et al.* [CB-ELSA Collaboration], Phys. Rev. Lett. **94**, 012004 (2005) [arXiv:hep-ex/0311045].
[35] R. A. Arndt, W. J. Briscoe, I. I. Strakovsky and R. L. Workman, Phys. Rev. C **74**, 045205 (2006) [arXiv:nucl-th/0605082].
[36] I. Horn *et al.* [CB-ELSA Collaboration], Phys. Rev. Lett. **101**, 202002 (2008) [arXiv:0711.1138 [nucl-ex]].
[37] I. Horn *et al.* [CB-ELSA Collaboration], Eur. Phys. J. A **38**, 173 (2008) [arXiv:0806.4251 [nucl-ex]].
[38] A.V. Anisovich, V.V. Anisovich, and A.V. Sarantsev, Phys. Rev. D **62**:051502(R) (2000).
[39] A.V. Anisovich, E. Klempt, V.A. Nikonov *et al.* arXiv: hep-ph:0911.5277.
[40] A.V. Sarantsev *et al.* Phys. Lett. B **659**:94-100 (2008)
[41] A.V. Anisovich, I.Jaegle, E. Klempt *et al.* Eur. Phys. J. A **41**:13-44 (2009)
[42] V.V. Anisovich, UFN, **174**, 49 (2004) [Physics-Uspekhi, **47**, 45 (2004)].
[43] V.V.Anisovich and A.V.Sarantsev, Int. J. Mod. Phys. A**24**, 2481 (2009).
[44] V.V.Anisovich and A.V.Sarantsev, Phys. Lett. B **382**, 429 (1996).
[45] A.V.Anisovich, V.V.Anisovich, Yu.D.Prokoshkin, and A.V.Sarantsev, Zeit. Phys. A **357**, 123 (1997).
[46] A.V.Anisovich, V.V.Anisovich, and A.V.Sarantsev, Phys. Lett. B **395**, 123 (1997); Zeit. Phys. A **359**, 173 (1997).
[47] V.V.Anisovich, D.V.Bugg, and A.V.Sarantsev, Phys. Rev. D **58**:111503 (1998).

[48] U. Thoma *et al.*, Phys. Lett. B **659** (2008) 87 [arXiv:0707.3592 [hep-ph]].
[49] V.V. Anisovich, L.G. Dakhno, M.A. Matveev, V.A. Nikonov, and A. V. Sarantsev, Yad. Fiz. **70**, 480 (2007) [Phys. Atom. Nucl. **70**, 450 (2007)].
[50] V.V. Anisovich, L.G. Dakhno, M.A. Matveev, V.A. Nikonov, and A. V. Sarantsev, Yad. Fiz. **70**, 68 (2007) [Phys. Atom. Nucl. **70**, 63 (2007)].
[51] V.V. Anisovich, L.G. Dakhno, M.A. Matveev, V.A. Nikonov, and A.V. Sarantsev, Yad. Fiz. **70**, 392 (2007) [Phys. Atom. Nucl. **70**, 364 (2007)].
[52] E. Salpeter and H.A. Bethe, Phys. Rev. **84**, 1232 (1951).
[53] H. Hersbach, Phys. Rev. C **50**, 2562 (1994); Phys. Rev. A **46**, 3657 (1992).
[54] F. Gross and J. Milana, Phys. Rev. D **43**, 2401 (1991).
[55] K.M. Maung, D.E. Kahana, and J.W. Norbury, Phys. Rev. D **47**, 1182 (1993).
[56] J. Nyiri, Ya.A. Smorodinsky, Preprint JINR E4-4043 (1968); Yad. Fiz. **9** 882 (1969) (Sov. J. Nucl. Phys. **9** 515 (1969)); Materials of Symposium on the Nuclear Three Body Problem and Related Topics, Budapest (1971). Also in: Selected Papers of Ya.A. Smorodinsky, Moscow (2001).
[57] J. Nyiri, Ya.A. Smorodinsky, Preprint JINR E2-4809 (1969); Yad. Fiz. **12** 202 (1970) (Sov. J. Nucl. Phys. **12** 109 (1971)). Also in: Selected Papers of Ya.A. Smorodinsky, Moscow (2001).
[58] J. Nyiri, Acta Phys. Slovaca **23** 82 (1973). 1571 (1965).
[59] V.V. Pustovalov, Yu.A. Simonov, JETP **51** 345 (1966).
[60] A.J. Dragt, J. Math. Phys. **6** 533 (1965).
[61] V.V. Pustovalov, Ya.A. Smorodinsky, Yad. Fiz. **10** 1287 (1970).
[62] G. Racah, Rev. Mod. Phys. **21** 494 (1949).
[63] N.Ya. Vilenkin, G.I. Kuznetsov, Ya.A. Smorodinsky, Yad. Fiz. **2** 906 (1965).
[64] N.Ya. Vilenkin, Special Functions and the Theory of Group Representations, Nauka, Moscow (1965).
[65] D.A. Varshalovich, L.N. Moskalev, V.K. Khersonsky, Quantum Theory of the Angular Momentum, Nauka, Moscow (1975).
[66] J. Raynal, J. Revai, Nuovo Cim. **A 68** 612 (1970).
[67] Ya.A. Smorodinsky, V.D. Efros, Yad. Fiz. **17** 210 (1973).
[68] I.S. Gradshtein, I.M. Ryzhik, Table of Integrals, Series and Products, Fizmatgiz, Moscow (1963).
[69] M.K.P. Wong, J. Math. Phys. **19** 713 (1978).

Chapter 9

Conclusion

The dispersion technique allows us to construct analytic and unitary amplitudes which is a necessary condition to study strong interaction physics (*i.e.* the physics of hadrons). Working in the framework of this approach one can be sure that causality is fulfilled in the considered processes, and there exists a complete set of states one can work with.

The dispersion technique, first developed for a description of two-particle states, can be naturally generalized to three-particle and, possibly, to many-particle states. Indeed, if in the beginning it was used for finding resonances in two-particle reactions, now it is more and more applied for the investigation of many-particle systems. A combined analysis of data from two- and many-particle final states is now a usual procedure which helps to resolve many ambiguities and provides a more exact determination of resonance properties. The exact knowledge of such properties as the position of amplitude poles (which is the universal value for all processes) and amplitude residues (the product of universal couplings of the initial and final states at the pole position) is the basis for the quark-gluon systematization of the resonances. Such a systematization is one of the most important steps in understanding the interaction of the effective quarks and the gluons in the soft region.

The ability to handle relativistically both two- and three-particle hadron states is, without doubt, a very serious achievement. However it is not sufficient for a successful analysis of the data, extraction of the resonance parameters and systematization of observed states. First of all one needs a correct understanding of the main problems which such an analysis can face:

(1) The poles of amplitudes with different quantum numbers show a tendency for accumulation around a mass value, forming so called "towers"

[1]. So in many cases a signal comes not from one isolated state but from several states situated nearby.

(2) The overlapping resonances, having the same quantum numbers, mix in a way that one of the resonances accumulates the widths of all others and becomes a rather broad state [2, 3, 4]. It is essential to find broad resonances, especially when searching for exotic states [5].

(3) The scattering (and production) amplitudes have not only pole singularities, which correspond to the resonances, and threshold singularities, which correspond to the two particle scattering, but also singularities defined by the many particle interaction. Typical examples of such singularities are the logarithmic singularity defined by the triangle diagram [6, 7, 8] and the root-singularity defined by box-diagram [9, 10]. If these singularities occur close to the physical region it is important to take them into account [11] or, at least, to study the significance of their contribution.

The systematization of the investigated hadrons may be a very useful tool to resolve the first two problems. Indeed, the meson states – resonances – are, generally speaking, well placed on the linear trajectories in the (n, M^2)- and (J, M^2)-plots [12, 13]. So, one can see which resonances are missing and where there are superfluous states. The case of baryons is a more complicated one: these states have more complicated structure and there are less experimental results in this sector [14, 15]. However, it seems, that these states are also situated on linear trajectories (Chapter 8).

Carrying out the systematization of hadrons one can use two languages: the language of quarks and gluons at small distances, and the language of hadrons at relatively large ones. Interactions at small distances form, in fact, the spectrum of the states. Hadron scattering and transitions at larger distances are responsible for the problems mentioned in topics 2 and 3. Hadronic interactions are, as a rule, many-particle ones, and we are facing inevitably the relativistic many-body problem.

It is rather important to find the components of hadronic molecules, responsible for the formation of the "hadronic coat", when searching for exotic states – the broad resonance $f_0(1200-1600)$ is the best candidate for the scalar glueball [16]. The best candidate for the tensor glueball $f_2(2000)$ [17, 18] also appears as a broad resonance. Considering the doubtlessly important role of the long-range hadron components in the meson states, we can now only guess about the role of the hadron components in the baryon sector – there is no sufficient experimental information here. Still, there is no doubt that the investigation of the baryon states is one of the most actual problems and a large amount of experimental information is expected from the photo-production experiments in the nearest future.

To be more general, we may say that the baryon systematization is knocking on the door. This means not only to place the states on the trajectories, and to carry out the qqq-systematization, but also to solve three-quark equations, to determine effective quark forces and to find the "hadronic coats" (*i.e.* to define the role of the long-distance hadron forces).

References

[1] D.V. Bugg, Phys. Rept. **397**, 257 (2004).
[2] I.S. Shapiro, Nucl. Phys. A **122**, 645 (1968).
[3] I.Yu. Kobzarev, N.N. Nikolaev, and L.B. Okun, Sov. J. Nucl. Phys. **10**, 499 (1970).
[4] L. Stodolsky, Phys. Rev. D **1**, 2683 (1970).
[5] V.V.Anisovich, D.V.Bugg, and A.V.Sarantsev, Phys. Rev. D **58**:111503 (1998).
[6] V.V. Anisovich and L.G. Dakhno, Phys. Lett. **10**, 221 (1964); Nucl. Phys. 76, 657 (1966).
[7] I.J.R. Aitchison, Phys. Rev. **133**, 1257 (1964).
[8] B.N. Valuev, ZhETP **47**, 649 (1964).
[9] V.V. Anisovich, Yad. Fiz. **6**, 146 (1967);
V.V. Anisovich and M.N. Kobrinsky, Yad. Fiz. **13**, 168 (1971), [Sov. J. Nucl. Phys. **13**, 169 (1971)].
[10] P. Collas, R.E. Norton, Phys. Rev. **160**, 1346 (1967).
[11] A.K. Likhoded and M.N. Kobrinsky, Yad. Fiz. **17** , 1310 (1973).
[12] A.V. Anisovich, V.V. Anisovich, M.A. Matveev, V.A. Nikonov, J. Nyiri and A.V. Sarantsev, *Mesons and Baryons. Systematization and Methods of Analysis.*, World Scientific, Singapore (2008).
[13] A.V. Anisovich, V.V. Anisovich, and A.V. Sarantsev, Phys. Rev. D**62**:051502(R) (2000).
[14] J. Beringer *et al.* [Particle Data Group], Phys. Rev. D **86**, 010001 (2012) 1.
[15] E. Klempt and A. Zaitsev, Phys. Rept. **454**, 1 (2007).
[16] V.V. Anisovich, Yu.D. Prokoshkin and A.V. Sarantsev, Phys. Lett. **B 389**, 388 (1996).
[17] R.S. Longacre and S.J. Lindenbaum, Report BNL-72371-2004.
[18] V.V. Anisovich, Pis'ma v ZhETF, **80** 845 (2004), [JETP Letters, **80** 715 (2004)], V.V. Anisovich and A.V. Sarantsev, Pis'ma v ZhETF, **81** 531 (2005), [JETP Letters, **81** 417 (2005)], V.V. Anisovich, M.A. Matveev, J. Nyiri and A.V. Sarantsev, Int. J. Mod. Phys. **A 20** 6327 (2005).